James Hervey Hyslop

The Elements of Logic

Theoretical and Practical

James Hervey Hyslop

The Elements of Logic
Theoretical and Practical

ISBN/EAN: 9783337276812

Printed in Europe, USA, Canada, Australia, Japan

Cover: Foto ©berggeist007 / pixelio.de

More available books at **www.hansebooks.com**

THE ELEMENTS

OF

LOGIC

THEORETICAL AND PRACTICAL

BY

JAMES H. HYSLOP Ph.D.

INSTRUCTOR IN LOGIC PSYCHOLOGY AND ETHICS COLUMBIA COLLEGE
NEW YORK

THIRD EDITION

NEW YORK
CHARLES SCRIBNER'S SONS
1894

PREFACE

ALL who have had anything to do with Logic will recognize, without being told it, the extent to which I am indebted to Jevons for both matter and method in the treatment of this subject. But they will quite as readily perceive the deviations from him, and the additions which I have made with the hope of improving upon his work. Jevons designed his "Elementary Lessons" to direct the student in practical reasoning and correct thinking in professional vocations. I have intended the present work to serve the same end, and, if possible, more completely than Jevons. For this reason I have been determined in the development of the subject by the questions constantly put me by students. As far as possible I have endeavored to answer all questions likely to be raised when framing rules about the logical treatment of conceptions and propositions. It will be apparent, therefore, that the plan has necessitated many judicious omissions both of irrelevant matter in Jevons and others, and of certain theoretical discussions peculiar to the scientific rather than the practical aspect of Logic. With a view to the student's guidance and mental discipline I have added a large number of practical questions and exercises at the close of the book.

It is important to remark, however, that I have aimed to satisfy two wants at the same time. I have tried to produce a work that could be used both for beginners and for advanced students of the subject, but not for those who care to go into it exhaustively. Beginners can be directed to the definitions and illustrations, and the heavier matter omitted, while the special questions developed at length and in a more technical manner can be taken up by the advanced student, or by those who are interested sufficiently to press further inquiries. But only one or two chapters are inserted which do not have reference to the practical advantages of Logic. The full treatment of all subjects is designed to afford students a better guide than Jevons can possibly be. With this in view I have laid considerable stress upon the nature of Conceptions, Propositions, and the Classification of Fallacies. The last subject always gives students a great deal of trouble, because they find no means of distinguishing one fallacy from another as clearly as is desirable. By the manner in which I have tried to treat the subject I hope I have rendered their work easier in this important part of Logic. De Morgan is the only author I know who has approximated what the subject of fallacies deserves; but he leaves very much to be desired. He has not attempted either a classification or a discussion of the way in which several fallacies may coincide at the same time, and differ only in the point of view from which they are regarded.

In the treatment of certain questions it has seemed fit to venture upon some distinctions which may be regarded as innovations. I have distinguished between two kinds of "General Terms," which I have called "Mathematical" and "Logical Generals," for reasons that the text must explain. In pursuance of this distinction and the double signification of the term "Genus," I have also coined the term "Conferentia," as denoting the "logical" as opposed to the "mathematical genus." A term is required to contrast with "differentia," as "genus" contrasts with "species." Whether I am justified in the invention must be a matter between me and my critics when they have examined the work.

If I mistake not, I have somewhat modified the ordinary treatment of Induction. First, I have carefully distinguished between "inductive reasoning" or "inference," and what is ordinarily called "Induction" or "Inductive Method." Then, to complete the thought involved in this distinction, I have discussed "Scientific Method" as somewhat extra-logical, and included therein the process of verification as involving, in some instances at least, more than purely inductive inference.

The importance of this will be apparent when we observe the constant confusion by logicians of formal Induction as a process of reasoning with the methods of discovering and verifying new knowledge. Bacon and Mill have not sufficiently distinguished them, and on that account have done much to misrepresent and to disparage the importance of Deduction. I have endeavored, so far as I am able, to remove this defect from the discussion. But of my success I am probably not a competent judge. Besides, the limits to which I have been compelled to reduce the discussion may appear to make this part of the work unsatisfactory. I am content, however, if I have given a hint in the right direction.

Special acknowledgments are due to Mr. Fowler and Mr. Venn for the valuable help which I have received from their works on Logic. Mr. Venn's recent volume on "Empirical Logic" is unusually sagacious and suggestive in all the most fundamental matters. Jevons receives due acknowledgment in the extent to which I have modelled my work after his.

JAMES H. HYSLOP.

COLUMBIA COLLEGE, April 23, 1892.

PREFACE TO THE SECOND EDITION.

The favorable reception and rapid exhaustion of the first edition offer an opportunity to correct a number of typographical errors and to introduce a few minor changes. The most important of these were made on pages 120 and 170. In the former case the doctrine about negative exclusive propositions has been modified, and in the latter a few changes were made to correct the consequences of an error previously overlooked. Rules for the sorites have been added at the end of Chapter XIV., and a note to Chapter XIX. has also been added on page 274, in order to prevent any possible misunderstanding of what was said about exclusive propositions in the main part of the chapter. No other important changes have been made.

JAMES H. HYSLOP.

Columbia College,
January 11, 1893.

PREFACE TO THE THIRD EDITION.

The third edition of this work offers an opportunity to make some further corrections of statement, and of typographical errors overlooked in the previous editions. Thus page 80 has been wholly rewritten in deference to criticisms in *Mind*, which gained their point wholly from my lack of clear statement. This difficulty, I hope, is now overcome. The note also on page 274 has been recast and lengthened, and a few changes of words or sentences here and there have been made, in order to avoid wrong or ambiguous statement. These are all of the important alterations affecting the book.

JAMES H. HYSLOP.

Columbia College,
October 27, 1894.

CONTENTS

CHAPTER I.
INTRODUCTION, DEFINITION AND DIVISIONS OF LOGIC, . . 1

CHAPTER II.
ELEMENTS OF LOGICAL DOCTRINE, ... 16

CHAPTER III.
TERMS OR CONCEPTS AND THEIR KINDS, ... 31

CHAPTER IV.
THE AMBIGUITY OF TERMS, ... 50

CHAPTER V.
THE INTENSION AND THE EXTENSION OF CONCEPTS, . . . 68

CHAPTER VI.
DEFINITION AND DIVISION, ... 82

CHAPTER VII.
PROPOSITIONS OR JUDGMENTS, ... 105

CHAPTER VIII.
THE RELATION BETWEEN SUBJECT AND PREDICATE, . . 122

CHAPTER IX.
Opposition, ... 141

CHAPTER X.
Immediate Inference, ... 154

CHAPTER XI.
Principles of Mediate Reasoning, ... 171

CHAPTER XII.
Moods and Figures of the Syllogism, ... 181

CHAPTER XIII.
Reduction of Moods and Figures, ... 190

CHAPTER XIV.
Forms of Syllogistic Reasoning, ... 197

CHAPTER XV.
Hypothetical Reasoning, ... 204

CHAPTER XVI.
Disjunctive Syllogisms, ... 212

CHAPTER XVII.
Classification of Fallacies, ... 219

CHAPTER XVIII.
Material Fallacies, ... 228

CHAPTER XIX.
QUANTIFICATION OF THE PREDICATE, ... 262

CHAPTER XX.
MATHEMATICAL AND OTHER REASONING, ... 275

CHAPTER XXI.
THE LAWS OF THOUGHT, ... 290

CHAPTER XXII.
INDUCTIVE REASONING, ... 295

CHAPTER XXIII.
SCIENTIFIC METHOD, ... 336

PRACTICAL QUESTIONS AND PROBLEMS, ... 367

PRACTICAL EXERCISES, ... 383

INDEX, ... 399

ELEMENTS OF LOGIC

CHAPTER I.

INTRODUCTION—DEFINITION AND DIVISIONS OF LOGIC

I. DEFINITION.—Logic has sometimes been defined as a science, sometimes as an art, and sometimes as both a science and an art. Dr. Watts calls it "the Art of Thinking;" Thompson, "the Science of the Laws of Thought;" and Whately, "the Science and also the Art of Reasoning." More elaborate and technical definitions are such as Hamilton's, that "Logic is the Science of the Formal and Necessary Laws of Thought as Thought;" or Ueberweg's, that it is "the Science of the Regulative Laws of Human Knowledge." Most writers define it as a science instead of an art, and in so far as they so regard it their views of it are substantially the same. They begin to diverge from each other only when they speak of it as an art. Yet a careful examination of the various usages of the term "art" will show that the difference between it and a science is less than general disputes would imply. This is apparent if we but reflect that science and art usually have the same subject-matter, although they have different ends in view. A science teaches us *to know;* an art, *to do.* Science endeavors to discover truth or knowledge, art to apply it, or to formulate rules for applying it, to the realization of some other end. But it is evident in such a view that art assumes knowledge as a condition of itself, and hence science and art may go together as complementary of each other. This is true of Logic to the extent that it may be regarded both as a science and as an art, according as it aims at certain truths, or at the application of them in practice. Logic as a science aims to ascertain what are the laws of thought; as an art it aims to apply these laws to the detection of fallacies or for the determination of correct reasoning. As a science Logic will be concerned chiefly, if not wholly, with the general principles of thinking, and only indirectly with truths of any other kind. The laws of thought, the object to which they can be applied, or the kind of phenomena in which they are embodied, are all that Logic as a science need occupy itself with. Truths or knowledge in other fields of mental interest may be useful for illustration, but they are not to be investigated by it. Only the truth, extent, nature, and validity of the laws of rational thinking come under its scientific aspect. But when we wish to know whether other bodies of knowledge, in which reasoning is involved, have been correctly obtained or proved, we subject them to the test of logical laws, and to do this we may be obliged to adopt a large system of rules for practice. The formulation of such rules and the testing of the various material truths of other sciences are left to Logic as an art. The truth and validity of its general laws will be assumed and admitted. The problem will be to find whether individual processes of reasoning have been conducted in conformity with those laws or not. Hence, as an art, it is concerned with a larger system of truths than it is as a science. In both it is concerned with the laws of thought; but as a science it treats those laws as an *end* for its own sake, and as an art it treats them as a *means* to a remoter end. Hence there is no necessity of deciding whether it is one or the other. The old controversy concerning that question may thus be settled by defining the subject as both a science and an art under limitations.* Other aspects of the definition

---

* For a discussion of the nature of Logic, and of the relation between science and art, see Whately: Elements of Logic, Introduction; Fowler: Element of Deductive Logic, Introduction, Chapter II.; Thompson: Laws of Thought, Introduction, Section 6; Hamilton: Lectures on Metaphysics, Lecture VII.; Lectures in Logic, Lecture I.; J. S. Mill: Logic, Introduction; and Examination of the Philosophy of Sir William Hamilton, Chapter XX.

will require separate consideration. But to indicate what these are we shall adopt as the most complete definition for our purposes that which makes Logic "the Science of the Formal Laws of Thought." Three terms of this definition require special examination.

1st. Thought.—"Thought," in common usage, is a very comprehensive term. It is even coextensive with consciousness or mind. We use the expression, "I have such a thing in my thoughts," whereby we mean merely that attention perhaps is occupied with a particular idea. It is not in any such sense that Logic must use the term. To have an idea in consciousness does not necessarily imply that any logical processes are going on, or that the mind is "thinking" logically about the fact. It may denote no more than an act of perception or attention. In the loose sense of the term, therefore, "thought" might denote any conscious act of the mind. But to regard Logic as the science of such activities and their laws would identify it with Psychology. Hence "thought" must denote a more specific act of the mind, and this is supposed to be the act which compares and reasons. The term may denote both the *act* and the *product* of the rational faculty.

The various acts of the mind may be denominated sensation, perception, apprehension or cognition, memory, association, attention, etc. But all of these are comparatively simple acts. They do not require any act of comparison by the mind. "Thought," in the logical sense, does require such comparison. The simple perceptive acts of the mind have but one thing as the object of consciousness, and hence denote either presentations or individual states of mind without taking into account any relations that might be connected with them. "Thought," on the other hand, does explicitly express the consciousness of some relation between two or more objects held together in consciousness at the same time. Thus I may perceive a tree, a house, a man, without performing any mental act which thinks them in relation to other objects of a like or different kind, or without apprehending their *meaning*. But if I think of a tree as a vegetable, of a house as a useful structure, or of a man as rational, I am apprehending the meaning or relation of the several objects, holding two conceptions in the mind at the same time and pronouncing upon their connection or disconnection, as the case may be. Thus "thought," as the subject-matter of logical science is *an act connecting two distinct ideas*, or is the product of such an act. Verbally it is the act, nominally it is the product. Logic does not need specially to distinguish between them for its purposes. But it does require to consider "thought" in its narrower signification as an act of comparison between conceptions in order to distinguish more clearly its own laws from the laws with which Psychology is concerned.

"The term *thought*," says Sir William Hamilton, "is used in two significations of different extent. In the wider meaning, it denotes every cognitive act whatever; by some philosophers, as Descartes and his disciples, it is used for every mental modification of which we are conscious, and thus includes the Feelings, Volitions, and the Desires. In the more limited meaning, it denotes only the acts of the Understanding properly so called, that is, of the Faculty of Comparison, or that which is distinguished as the Elaborative or Discursive Faculty. It is in this more restricted signification that thought is said to be the object-matter of Logic. Thus Logic does not consider the laws which regulate the other powers of mind. It takes no immediate account of the faculties by which we acquire the rude materials of knowledge; it supposes these materials in possession, and considers only the manner of their elaboration. It takes no account, at least in the department of Pure Logic, of Memory and Imagination, or of the blind laws of Association, but confines its attention to connections regulated by the laws of intelligence. Finally, it does not consider the laws themselves of Intelligence as given in the Regulative Faculty;" namely, the Intuitions of pure intelligence, or the ultimate data, facts, and principles which are involved in the primary experiences of mind. But such are the functions with which Logic is *not* conversant. It remains to determine positively what the nature of its object-matter is.

"The contemplation of the world presents to our subsidiary faculties a multitude of objects. These objects are the rude materials submitted to elaboration by a higher and self-active faculty, which operates upon them in obedience to certain laws, and in conformity to certain ends. The operation of this faculty is Thought. All thought is a comparison, a recognition of similarity or difference; a conjunction or disjunction; in other words, a synthesis or analysis of its objects. In Conception, that is, in the formation of concepts (or general notions), it compares, disjoins, or conjoins *attributes;* in an act of Judgment, it compares, disjoins, or conjoins *concepts;* in Reasoning, it compares, disjoins, or conjoins judgments. In each step of this process there is one essential element; to think, to compare, to conjoin, or disjoin, it is necessary to recognize one thing through or under another; and therefore in defining Thought proper, we may either define it as an act of Comparison, or as a recognition of one notion as in or under another. It is in performing this act of thinking a thing under a general notion, that we are said to understand or comprehend it. For example, an object is presented, say a book; this object determines an impression, and I am even conscious of the impression, but without recognizing to myself what the thing is; in that case there is only perception, and not properly a thought. But suppose I do recognize it for what it is, in other words, compare it with, and reduce it under a certain concept, class, or complement of attributes, which I call *book;* in that case, there is more than a perception—there is a thought."

Thought is, therefore, the act or product of the Understanding or Reason as distinct from the various processes of simple Apprehension or Cognition, and consequently Logic is conversant with the laws affecting or regulating this act of comparative knowledge rather than with the laws of perception. It is the science of thought as an act of conception, judgment, and reasoning, or as *the cognition of relations* between conceptions. This act has its own laws distinct from those of other mental acts. But the limitations of the science and its laws can be determined only by comparing its field and functions with those of other sciences. The nature of its general object-matter suffices to define and distinguish it from cognate sciences, on the one hand, and from the physical sciences on the other. "Thought" is, first, a fact quite distinct from physical events, and is, second, in its technical sense, as logically defined, distinct from mental events which are not acts of comparison : hence it implies a double limitation of the subject of Logic ; one its distinction from sciences which investigate the physical laws and causes of events, and the other its distinction from the philosophical sciences which are occupied either with the efficient causes of mental phenomena or with the nature of the being of which they are phenomena. Logic is occupied with the *relations* between those phenomena, as objects of reason or rational thought, or with the laws which attest the *validity* of thought, and which serve, at least, as negative criteria of the truth in so far as it is determined by processes of reasoning. This, however, will be rendered clearer when we examine the relation of Logic to the various sciences.

2d. The Nature of the Laws of Thought.—The laws of thought will be best understood by defining the several uses of the term "law." That there are "laws" of thought we may at present take for granted, unless we are to assume that there are no regulative principles or conditions which determine the uniformity of the mental processes involved in the various acts of thought. But the assumption that there are "laws" of thought neither defines their specific nature nor affords any indication of what they are. This desirable end can be achieved only by an examination of what we mean by "law."

The first of the two general meanings of the term is of little importance to a discussion in Logic, except that it may be serviceable for bringing the true meaning of it into proper relief. But this first conception is its moral and political sense, in which it denotes a command or prohibition in regard to the doing of certain actions. This idea does not imply any absolute necessity of the event commanded, nor does it imply any regularity of such events. It denotes only an injunction to act or not to act. But such a conception is quite the contrary of the idea of "law" as employed by the sciences, where it denotes the regularity with which events occur, or the uniformity of their dependence upon certain conditions which necessitate them. Hence the notion of "law," as applied in the physical sciences, is but an abbreviation for the uniformity of coexistence and sequence, or the uniformity of causation. It describes or implies those conditions which make events regular and inevitable, or which explain why they follow a given order of occurrence. It is thus opposed to chance or caprice. In so far as the idea is synonymous with the notion of cause, it is a convenient term for referring to the explanation of phenomena. In so far as it denotes merely uniformity of events, it is convenient for indicating the general principles of unity that are exhibited or expressed thereby in the multiplicity of phenomenal events.

But this form of discussion is perhaps not quite so clear as illustration. Each of the sciences endeavors to ascertain the laws which prevail in a special class of phenomena. Chemistry tries to find out the laws of affinity and combination regulating the relations between atomic bodies. Thus oxygen and hydrogen must combine in a certain proportion to produce water, and they will combine in no other proportion to produce the same effect. This represents a certain "law" of affinity between the elements. A similar law operates among all elements which combine only in certain definite proportions which can be expressed with perfect mathematical accuracy. In Astronomy we speak of the law of gravitation which expresses the uniform tendency of material bodies to seek the centre of gravity, or to move toward it, when free, with a determinate velocity. In Physics we have the laws of the expansion and diffusion of gases; of contraction and expansion under changes of temperature; of the conservation of energy, and of the transmission of motion. In all these cases there are certain fixed ways of acting which bodies must follow or obey. All facts must be regulated or determined by the conditions which such laws impose.

But there is a science of human reason, and the processes of thought are as much under law as any other phenomena. These laws, like all other laws or conditions of events, are uniformities of nature; in this case, of the nature of the mind. They are embodied in principles or propositions which show how we must think and reason, if we think and reason at all. They are necessary laws of thought, because the mind has no power to evade them. Thus conception, judgment, and reasoning are according to a principle which conditions and validates them, and so is known as their law, or the uniformity of mental action in them which enables us to accept their results. Such laws may be illustrated in the following manner:

If I take the judgment that "All men are mortal," and infer from it, by a process which is yet to be considered, that "those who are not mortal are not men," or, "The immortals are not men," and if I can apply a similar process to all such propositions, it is because of the law *that what is excluded from the conception "mortal" must be excluded from the conception "men."* The assumption of this law is necessary to the making of this inference, and unless it always be true I have no means of guaranteeing the legitimacy of the results. Again, I may take three conceptions which are capable of agreeing with each other when conjoined in the form of a judgment.

Metals.   Iron.   Useful.

By connecting these as subject and predicate I obtain three propositions, one of which is a conclusion or inference from the other two. Thus:

Metals are useful.   Iron is a metal.   ∴ Iron is useful.

Illustrations could be multiplied indefinitely showing the same form and process of thinking, but the one suffices to call attention to the principle upon which the mental act is based. We observe that in the first two propositions there is a common term compared with two others. The first proposition states an agreement between "metals" and "useful," and the second between "iron" and "metals." The common term with which "iron" and "useful" agree is "metals," and this fact affords a reason to suppose that "iron" and "useful" also agree to the same extent and in the same sense. If we find that all such cases of reasoning exemplify the same form of process and conditions we may formulate the law which regulates and legitimates it. It is a logical axiom. Thus, *if two things are identical or agree with a third common thing, they are identical or agree with each other.* This is a fundamental law of thought and sustains the same relation to the validity of reasoning that the law of gravitation sustains to the phenomena of falling bodies or the motion of the planets about the sun. It is a law, because it expresses the uniform way in which the mind acts, and must act, when comparing conceptions and judgments. There are many other such laws, but they do not require examination at this stage of the discussion. One illustration suffices to show what it is that Logic endeavors to investigate and establish. Its laws are generally expressed as principles or assumptions, while those of the physical sciences are usually conceived as causes, conditions, or uniformities of coexistence and sequence. But in both, the essential idea is uniformity of some kind. In the physical sciences it is the regularity with which certain events occur, and may be expected to occur, under given conditions. In logical science, the uniformity is that of mental action when thinking and reasoning, as the fixed modes of comparison and inference which the mind must obey when its action is healthy and rational. In that respect Logic is a science like all other forms of inquiry into the nature and principles of phenomena; only its laws are the laws of *reason*, and not of physical events.

3d. Form and Matter.—The distinction between "form" and "matter," and the definition of the term "formal" as employed in Logic, are two of the most important things to be accomplished, as their peculiar signification appears to influence many of the doctrines of the science and to explain many of the perplexities which are incident to logical processes. The term "formal" we shall find often to be equivalent in many respects to the idea of "law." The "forms" of thought are often called the laws of thought, and they are such in their nature. But why they are so can be more distinctly understood when we have made clear the difference between the "form" and the "matter" of thought, or between the uses of those terms in reference to any subject-matter whatever.

What we mean by "form" when applied to material or physical objects is generally clear enough. It is synonymous with *shape*, the geometrical outline of a body, or the spacial limitations under which a physical object is conceived. The "form" of a body may be regular or irregular, long or short, round, square, rectangular, thick, thin, flat, etc. But no such predicates can be applied to thought, or to states of consciousness. Mental states can neither be said to have "form" in the sense of having the quality of extension, nor to exist in any assignable relation to space. Hence we cannot speak of their "form" as we would of material objects. But inasmuch as the "form" of physical objects does not necessarily depend upon the *stuff* or matter of which they consist, but may remain the same amid all changes of the latter, or the same when the matter may be of different kinds, because extension is independent of material substances, this conception of the case may be chosen to describe, at least by analogy, certain relations between the processes and the objects of thought or consciousness. The fact first suggests, however, the definition of "matter," which in its relation to "form," as applied to physical objects, is merely the physical elements, apart from extension, which make a body or object what it is other than its "form." It is, as already said, the *stuff* of which it consists. But by the same analogy which transfers the use of the term "form" to other than physical objects the term "matter" is also transferred, and the two terms are chosen to express certain relations of uniformity and variation existing between all kinds of phenomena, physical or mental. Hence, as two things may be alike in their "form," while they are of different "matter" or substance, two mental processes may be the same in kind, although occupied about different objects of thought. Thus the judgments, "All men are mortal," and "All metals are elements," are alike in the one aspect of grammatical structure, but are different in respect to the conceptions of which they are composed. Two or more syllogisms consisting of different propositions will illustrate the same truth. The one aspect in which they resemble each other may be called their "form," and the conceptions which compose them, and which may vary indefinitely without affecting the "form," grammatical or logical, may be called their "matter." Hence we may define the "form," at least provisionally, *that which remains constant when the "matter" changes*, and, correlatively, the "matter" may be defined as *that which remains constant when the "form" changes.*

Objections can be made to these definitions, the force of which can be readily admitted. But they are intended to be merely tentative and approximate, and for the purpose of making them general enough to cover all conceivable objects of consciousness. It is a misfortune to be compelled to define the terms reciprocally. But their simplicity and the fact that they are purely relative to each other is an excuse for such a course, and may even make it necessary. All objects, real or ideal, must have both their "matter" and their "form" at the same time. But the fact that either quality may be constant while the other is variable proves the value of defining them in a way to recognize both this variable relation and the permanent coexistence of the two qualities in one way or another at the same time.

But the "form" is usually regarded as what is *essential* to a thing as well as what is constant. This means that a certain quality cannot be dispensed with in reference to a given end, although others may be immaterial or accidental in this respect. Thus a house may serve the same purpose whether made of brick or wood, while the "form" may not only be the same, but must be more or less essential. A judgment must have a subject and predicate, whatever the conceptions constituting it, and so the "form" is essential to its being a judgment, while its matter is not. This idea of what is *essential* must be added to that of constancy in order to complete the notion of "form" and its distinction from "matter."

But it is well to remark again the purely relative character of the two conceptions in order to anticipate and remove a possible objection. The two qualities are so variable that what is the "form" in one relation may become the "matter" in another. It depends wholly upon the nature of the relation in which an object is conceived. The fact, however, does not require development in an elementary treatise of the science. It is noticed only to indicate that the circumstance has not been overlooked in our account of the subject under consideration.

The relation between "form" and "matter," as we have defined it, expresses very clearly how Logic is a science of the *formal* laws of thought. They are the laws which are not only essential to it, but which are the same whatever the subject-matter involved in our reasoning. The laws of thought remain the same in the reasonings of Astronomy, Physics, Politics, or Ethics, but the "matter" changes and does not affect the validity of the process. The "form" of our reasoning in all these cases is essential to its being such a process. Hence Logic, as a science, is "formal," and deals only with the "formal" principles of thought in distinction from the material objects of reason. Logic thus becomes the most general of all the sciences, and its principles are the "formal" conditions of the truth attained in them. What these "forms" of thought are will appear with the development of the science. It is only important at present to know that we do not require to take any account of the particular "matter" of knowledge in order to ascertain these "formal" laws.

II. THE RELATION OF LOGIC TO THE SEVERAL SCIENCES.—Logic we have found to be the science of the formal laws of thought. This fact has led to its definition as the *Science of Sciences* (scientia scientiarum). This is true if considered only in its *formal* functions. Any other conception of it would imply that it was concerned with their material truths.

But it can employ itself with the material knowledge of other sciences only in its function as an Art (ars artium). As a science, however, it has a "formal" relation to all other sciences inasmuch as it determines the laws of thought everywhere affecting the acquisition and legitimation of truth. This is all that it is necessary to recognize in a general way. But inasmuch as it sustains a more intimate and complex relation to the mental sciences, often dealing with precisely the same subject-matter, we require to distinguish between its relation, in some particulars, to the physical sciences, and its relation to the mental. While it formally conditions the mental processes of the physical sciences, it is materially occupied with the laws of thought, and they, the physical sciences, with the laws of things. The other mental sciences, however, to which it has a more peculiar relation beside the general formal one, have quite a distinct object in view as compared with Logic. The two sciences are Psychology and Ethics. With Logic these constitute, properly, the mental sciences.

Psychology deals with *all* mental phenomena; Logic deals with only a part of them. They therefore partly coincide. In so far as they both deal with the thought processes of judgment and reasoning they are occupied with the same field. But they do not deal with these phenomena in the same way, nor with the same object in view. . "Psychology deals with them as laws in the sense of uniformities, that is, as laws in accordance with which men are found by experience normally to think and reason; Psychology investigates also their genesis and origin. Logic, on the other hand, deals with them purely as regulative and authoritative, as affording criteria by the aid of which valid and invalid reasonings may be discriminated, and as determining the formal relations in which the different products of thought stand to one another." These observations of Keynes may perhaps be rendered a little more complete if we remark that Psychology investigates all mental phenomena, including the rational, with a view to ascertaining, first, their *uniformities* as actual events, and second, their *causes*, but does not require to distinguish between normal and abnormal states, valid or invalid ideas. On the other hand, Logic, in addition to being conversant with a more limited field than Psychology, does not deal with the phenomena in common to the two sciences with the same object in view throughout. It deals with the uniformities of *rational* processes, but neither with their causes, nor with any matters regarding the origin of those uniformities. It endeavors to ascertain the uniformities of thought which are the grounds of truth or valid thought, and which exclude all other ideas from recognition, except as facts. Psychology is thus concerned with the *origin* and nature of mental phenomena in general, and Logic with the conditions and the *validity* of the rational processes. The latter, therefore, has to do with the *formal* in contradistinction to the *efficient* principles of thought. Ethics, of course, is the science of the *ends* of conduct, and although it is formally conditioned by Logic, the relation between it and Logic is not so close as between Logic and Psychology. But their respective spheres can be very well defined by the Aristotelian and scholastic formulas expressing the various kinds of causes. This will characterize the difference between them as sciences. Consequently Psychology is the science of the efficient causes (*causa efficiens, ratio fiendi*) of mental phenomena, including the rational processes: Logic is the science of the formal causes (*causa formalis, ratio essendi*) of thought, of valid rational processes: Ethics is the science of the final causes (*causa finalis, ratio agendi*) of human conduct, of the mental phenomena of desire and volition.

It will thus be seen how Logic and Psychology may to some extent overlap each other. This, however, is mainly in regard to the phenomena with which they deal, and not with regard to the objects which they aim to accomplish. They both deal with uniformities of mental phenomena. But Psychology as such does not distinguish between the true and the false; it *explains* mental processes; Logic *validates* those of reasoning and ascertains the principles which condition them. Psychology deals with causes, Logic with principles. This will perhaps show the intimate relation subsisting between the two sciences, and at the same time indicates the difference of method and object pursued by them in their investigations.

III. THE DIVISIONS OF LOGIC.—Logic may be divided according to the particular object which it has in view, or according to the kind of reasoning with which it deals. The first principle of division gives us Pure or Formal Logic, and Material or Applied Logic. The same distinction is observed in the divisions, Theoretical and Practical Logic. These divisions are based upon the distinction between Logic as a science and Logic as an art. Pure or Formal Logic is a science; Material or Applied Logic is an art. The former is conversant only with the pure or formal laws of thought, and does not concern itself with the material truth of any particular proposition; the latter is conversant with the material conceptions of various sciences and endeavors to apply formal laws to the attestation of truth or knowledge. Pure Logic is abstract and theoretical; applied Logic is concrete and practical.

The second principle of division gives us Deductive and Inductive Logic. Deductive Logic is usually defined as occupied with the laws of *a priori* reasoning, and Inductive Logic with the laws of *a posteriori* reasoning. The one assumes general principles or facts in order to elicit into consciousness something which is not explicitly there in the premises, or which, when it is in consciousness, requires a general assumption to validate it as truth; the other assumes facts and endeavors to arrive at general truths beyond them, or ideas not directly deducible from the premises. But this division of the subject can be adjusted to the former, as the following scheme will illustrate:

$$\text{Logic} \begin{cases} \text{Pure or Formal} \begin{cases} \text{Deductive.} \\ \text{Inductive.} \end{cases} \\ \text{Applied or Material} \begin{cases} \text{Deductive.} \\ \text{Inductive.} \end{cases} \end{cases}$$

CHAPTER II.

ELEMENTS OF LOGICAL DOCTRINE

I. GENERAL PRINCIPLES. — The definition of logic teaches us that the science is employed about the laws of thought. Thought, we also found to be an act of comparison, and as the object of Logic and logical doctrine it is usually identified with the processes of reasoning. But all the higher and more complex mental processes presuppose the lower, and the material which they furnish. Thus reasoning deals with ideas, notions, conceptions, etc., which may be called its elements. We cannot understand the nature of that process until we know the nature of the elements involved in it. It may be the function of Psychology to tell us how these elements have originated, but it may not be its function to tell us what they are, or what their relation to reasoning is. Logic must know what the qualities of its elements are, in order to formulate its laws about their relations in the processes of thought. Hence we proceed to inquire what the elements of logical doctrine are.

There are two distinct ways of regarding this question, according as we consider the *objects* of logical science or the *mental processes* concerned in it. Hence we may divide the elements of logical doctrine into two classes : (*a*) The *matter* of logical science which consists of Terms, Propositions, and Syllogisms ; and (*b*) the *form* or process of thought which is the principal object of the science, and which is usually represented as consisting of three subordinate mental modifications ; namely, Conception, Judgment, and Reasoning. The former looks at the question from the point of view of language, and makes no special examination, introspectively or otherwise, into the mental activities of which terms, propositions, and syllogisms are the object. It simply relies upon the normal accuracy of mental processes and develops Logic as a practical doctrine concerning the proper mode of conducting arguments and avoiding error. But from the point of view of the process of thought there is a desire to get a deeper knowledge of the way in which the mind conceives and uses its material, and of the actions involved in understanding the nature and relations of its material objects. Thus we may produce a Logic without entering into any analysis of terms, propositions, etc., and without considering the nature of the mental processes involved. On the other hand, a complete conception of the problem is not possible until we have formed some definite idea of the process as well as the matter of thought. The two aspects may be treated together, as they imply each other. But their peculiarities may be best exhibited by viewing them apart.

II. DEFINITION OF THE MATERIAL ELEMENTS OF LOGICAL DOCTRINE.—**1st. Terms.**—A term may be variously defined. Grammatically considered it is a word which is the name of an idea, or conception. Logically considered it is any word or group of words which is capable of constituting a distinct part of a sentence or a proposition. A grammatical term is single; such as "man," "tree," "with," "from," "walk," etc. A logical term may be either a single word, or a phrase; such as "man," "house," or "the wife of Socrates," "the Queen of England." The singleness of the idea is all that is required to make it a logical term. But terms are the elements, or atoms, as it were, which combine to form a proposition. Logic may deal with them to some extent without considering the processes implied in apprehending their meaning. This fact has given rise to the belief in some schools of thought that the science is exclusively occupied about language. But very little observation is required to perceive that it can deal with language only as a symbol of thought.

2d. Propositions.—A proposition, grammatically considered, is a sentence consisting of words or terms arranged to make sense. Logically considered it is a connection of terms, as subject and predicate, in a way to express their agreement or disagreement with each other. Thus "Iron is an element," and "The security of life is one of the primary objects of government," are propositions respectively of the simpler and the more complex form. They express a relation between the subject and predicate, which cannot be expressed by a term.

3d. Syllogisms.—A syllogism is purely an object of Logic. It is the combination of three or more propositions in such a relation to each other that the last one is an inference or conclusion from the others. It is a simple syllogism if it consists of three propositions, and a complex one when it consists of more than three. It is the material form which all our reasoning assumes, and various characteristics of it which are yet to be examined make it the principal object of the investigations conducted by logical science.

III. EXAMINATION OF THE FORMAL ELEMENTS OF LOGICAL DOCTRINE.—As we have already ascertained, the formal elements are Conception, Judgment, and Reasoning. The last two are more conspicuous objects of logical science inasmuch as their laws are more readily determined and the processes themselves can be more easily regulated. The process of Conception is so spontaneous and follows so closely the psychological laws of mental phenomena that concepts are always found and completed before logical science can take any account of them. There are, therefore, no well-defined laws of Conception which we can regard beforehand in the making of the primary elements with which Judgment and Reasoning have to deal. The qualities of its products, namely, concepts, can be observed and their influence upon different mental processes considered. But it is more difficult to formulate any laws by which Conception is governed. Yet it has its own laws, although they are not usually formulated or discussed, except in Psychology. Logicians are usually content with a statement of the characteristics of various notions, ideas, or conceptions, and hence they omit any detailed account of the processes involved in their formation. But we shall give the subject a little more attention in the present treatise. We proceed, therefore, to a careful examination of each process.

1st. **Conception.**—The term "conception" is an ambiguous one. It is used sometimes verbally, and sometimes nominally or substantively. Hence it may denote either a *process* or a *product*. At present we are concerned with it only as a process, and its object or product will appear for more particular consideration in a moment. But the product of conception as a process is called a *concept*. For the purposes of Logic a concept is the same as a term. In itself it is to be viewed from the mental side, and represents the ideal rather than the symbolical element of thought. A word or term is only symbolical of ideas, an index of thought; a concept expresses the content of the mental act as conceived apart from language, and so is, to some extent at least, ideal as opposed to real. It expresses more clearly than "term" what is connoted or denoted by a word. The distinction between it as a product, and conception as a process, must be kept clear if we are to understand the unity and simplicity of the latter as compared with the diversity and complexity of the former. Hence we must define them as accurately as possible.

1. DEFINITION OF CONCEPTION AND CONCEPT.—*Conception* is the mental act of connecting percepts or individual ideas into a whole. This whole may be of several kinds, as we shall presently see. But the act is illustrated in the process of conceiving "man," for example, as uniting the properties of *animality* and *rationality*. These terms may denote a whole group of qualities. For instance, "animality" may denote material properties, such as solidity, color, form, etc., organic properties, such as peculiar physiological structure, assimilative, circulatory, respiratory organs, etc., and "rationality" may denote sensibility, consciousness, reasoning capacity, etc. All these properties are grouped together in the notion of man. The act of mind which *conceives* them, or *thinks* them as combined in a single object is Conception. The act is the same for all objects comprehending a number of attributes, or for all terms

comprehending a number of individual objects. Thus man besides representing a whole which is a combination of qualities, may denote a whole which is a common concept for a class made up of individual men. It requires, however, the same act of synthesis or comprehension for both kinds of concepts. The nature and properties of these various concepts must be examined later. We can now only take account of the comparing and combining act which groups different qualities or different individuals under the same idea or term. The laws by which the process is governed are the laws of association, of identity and difference, and of necessary connection.

Concepts, as already intimated, are the products of Conception. They denote the ideas themselves instead of their names. Technically a concept may be defined as a *synthesis of percepts*, or a *synthesis of individual wholes* to form a general notion. If the former could be called a simple concept, the latter could be defined as the synthesis of simple concepts, making a complex and general concept of greater extent. A simple concept in this view of the case would be such as "man," "tree," "lion," "institute," "President," etc., in so far as these terms or ideas represented an individual group of qualities. But in so far as they were general names for a large number of individuals of the same kind, the concept would be complex as involving the idea of a group of qualities comprehended in each member of a larger group of individual objects. But the definition of it as a synthesis is perhaps too technical for common use. Hence it may be best to define a concept, in its broadest sense, as *the notion of a group of qualities or individuals which are capable of being thought of at the same time*, the former of which belongs to the same thing and the latter of which represents the same *general* idea, or is applicable to a number of the same kind. But this definition will not throw much light on the subject because it undertakes to do so much. It is an endeavor to define an idea which is supposed to apply equally to *an individual group of qualities* and *a general group of individuals*. The difference between these two things is so great that some logicians use the term "concept" as if it could apply but to one of them and that to the latter, where it expresses a general abstract idea. Some have endeavored to distinguish between "idea," "notion," and "concept." For certain purposes this may be proper and necessary. A complete and accurate logical doctrine may be greatly helped thereby inasmuch as such a distinction is intended to evade the ambiguities incident to the use of the term "concept." But practical purposes may not require any refined niceties in the matter, and as we are at present interested only in throwing light upon the fundamental nature of Conception as a process, we may postpone the immediate consideration of that question. It will come up again when studying the formation of concepts and their several divisions. For the moment we need only to know that there is a mental process by which more than one object of consciousness is perceived as constituting a single whole. This process is one of comparison and synthesis. It is elementary, and conditions all the higher and more complex acts of comparison and synthesis. The complete investigation of its nature and functions may be left either to Psychology or to more elaborate treatises of Logic than the present one professes to be. It is sufficient at present to know enough of it for appreciating its relation to concepts and the kind of knowledge with which Logic deals.

2. THE FORMATION OF CONCEPTS.—I have stated my intention not to draw any important distinctions between "idea," "notion," and "concept." For practical purposes I shall assume them as identical and as denoting objects of knowledge or products of mental action other than the primary experiences of sense and consciousness. They may require to be distinguished from representative states which are the objects of memory. But in a broader sense the products of memory are non-compared data, and may be characterized as a form of ideal percepts. I cannot, however, go into any special questions of Psychology. We must be content to assume that concepts are the result of some form of comparison and of group-

ing either qualities or objects. With this once accepted, at least tentatively, we may proceed to see how they have been formed.

I said that a concept might be a synthesis of percepts. But the term "percept" requires definition and suggests the necessity of considering briefly the elementary processes of knowledge.

A percept is *an individual object of cognition*, such as a color, a sound, a taste, solidity, weight, extension. We may include any state of self-consciousness as such, and as a single object of inner cognition. The consciousness even of a "thought," in the sense of a logical act, so far as it is an individual object of inner perception, may be regarded as a percept. But those illustrations taken from the phenomena of sense perception are the clearest instances, and they suffice to show how a concept is formed. We must be careful, however, in thinking of "a percept as an individual object of cognition," that we do not understand by it any object of sense perception in the ordinary sense of the term, such as a tree, a house, a cloud, matter, etc. We may say practically that we *see* such objects, but in reality we do not. We perceive certain colors and relations of form, and other experiences of the use and nature of objects having these qualities enable us to interpret the meaning of what we see. Thus what appears to be a very simple idea turns out to be very complex. Hence a percept strictly considered can only be a single *presentation* of sense or of consciousness. Sound is given by hearing only, savor by taste, solidity and resistance by touch, color and form by vision. Certain signs also in other senses may be called the percept by which extension or space is derived. But not to go into special problems, we find these illustrations sufficient to show what is meant by "an individual object of cognition." They are the elements out of which concepts are formed.

The process of knowledge which gives us percepts is Perception or Apprehension, which we may denominate internal or external, according as it gives us material or mental percepts. Into its nature Logic does not require us to enter. It is the process which combines them that concerns the problems of Logic. We call that process Conception, as an act, because it involves some form of comparison and discrimination; that is, the holding of two or more elements in consciousness at the same time, and combining or separating them in its own peculiar way. But the process is conditioned by the Laws of Association. By repeated experiences in the conjoint perception of a number of qualities, physical or mental, or physical and mental combined, we come to form a concept of an individual object or whole. By conjoint perception we mean the activity of different senses at the same time, or under such circumstances as enables the mind to think that the qualities perceived *belong to the same thing*. Thus certain experiences of taste, color, and touch enable me to form an idea of an *orange* as uniting the three sets of qualities, so that when I see one of them I may think of the others as associated with it, and requiring only the proper experience to verify them. Again, certain experiences of form, color, solidity, etc., and perceptions of usefulness enable me to form the notion of a *house*. I think of these qualities as cohering in the same object. The same is true of all objects or wholes which we know. The individual percepts, through the agency of association, and perhaps other mental acts, are conceived as constituting one object or mental totality, which is a concept of the first order. All such objects representing a simple synthesis of qualities or percepts I shall call *individual* or *attribute-wholes*. The simplest illustrations of them is that of proper names, such as Plato, Bucephalus, etc. They represent objects which are merely a combination of qualities or attributes, and whose name is not applicable to any other individual. They are only attribute-wholes. Other and general concepts may be more than attribute-wholes. Thus "man," "quadruped," "tree," "circle," "country," "institution," "state," etc., are attribute-wholes, inasmuch as they represent certain combinations of qualities. But they are also more than these at the same time because they may denote a *class* of individuals. This fact makes it necessary to examine into the formation of concepts of the second order. In comparison with this, those of the first order may be best denominated *individual-wholes*. Those of the second order shall be called *class-wholes*.

The process of formation is that of association and comparison. If we perceive objects having like qualities there is a tendency to associate them and to give them a common name. Thus I may see an oak having roots, trunk, branches, leaves, color, form, etc., of a particular kind, and then an elm with the same properties slightly modified. But there may be a sufficient number of qualities alike for me to think that the objects bearing them are entitled to be called by a common name. Hence, instead of calling them both "oaks," or both "elms," I may use the term *tree*, which will apply to the common qualities, while "oak" and "elm" may denote the differences. We may even compare different "oak" trees and use that term as a general one to denote all trees having the essential qualities of "oak," but with slight differences; or to denote all individual oak-trees of exactly the same kind, whether there be any differences or not. The comparison is one based upon the *similarity* or *resemblance* between individual objects, not upon the coexistence of two or more qualities in the same thing, but rather the coexistence or succession of the same quality or qualities in different things. The concept is thus a *class* concept because it applies to more than one individual. It may denote at the same time an individual-whole, which is perhaps what we represent to ourselves in imagination when we think of it. But it also denotes a class of individuals to each of which it is equally applicable. The terms "man," "quadruped," "tree," "circle," "institution," "state," etc., to which reference was made as representing attribute-wholes, are also class-wholes, or *general* concepts, as opposed to individual concepts, although they denote at the same time a totality of cohering attributes or qualities which make them representative of individual concepts. But there are general concepts which represent a class of individuals without, perhaps, indicating that the individuals composing it are complex attribute wholes. These will be such as represent different percepts, qualities, or objects of the same kind, or perhaps of mutually exclusive qualities not belonging to the same object and the same space. Thus "color," "sound," "hardness," may be general or class concepts without denoting individual-wholes. Take the idea of "red," for example. It is a single percept. But this percept may be found in connection with various objects, and so we may form a general idea of "redness" applicable to a number of percepts without denoting their synthesis or combination in the same object or the same space. These will afford the clearest illustration of class-wholes independently of individual wholes, while most concepts may be understood in both relations at the same time. But in all, the same process of comparison, discrimination, and association is necessary to their formation. The difference between the two classes of concepts, and hence between the processes forming them, is, that individual wholes, as a synthesis of percepts, are a combination of different qualities, and class-wholes are a combination of similar qualities in different objects under a common name. To illustrate again, "Lincoln," "Socrates," "Italy," "God," etc., are pure individual wholes; "color," "sound," "redness," "stature," "number," etc., are pure class-wholes, inasmuch as they denote more than one thing without implying a union of various attributes in it; while "man," "vegetable," "oak," "nation," etc., may denote both individual and class concepts.

This distinction between the two kinds of class concepts is the basis of the distinction between the Mathematical or Quantitative Sciences, and the Metaphysical or Qualitative Sciences. The element of the distinction to which I refer is that in which a class term or concept may be used to denote a number of individuals, perceptive or conceptive, *absolutely identical in kind*, or it may denote a number of individuals *essentially different in kind*, but with common properties of a character to justify the use of a common name. The comparison in one case is between individual-wholes, or properties to which the common name applies in the same sense and without a difference. In the other the comparison is between individual-wholes, or properties to which the name applies with a difference. Thus "two," "red," "length," etc., will apply to a class of objects without reference to a difference in kind. These may be called *mathematical* concepts, because each object denoted by them may be *representative* of the whole class irrespective of the distinction between common and accidental, or what are called essential and differential qualities. But there are other concepts, such as "man," "tree," "biped," "Greek," "country," etc., which may be called *logical* or *metaphysical* concepts, inasmuch as they denote individuals with such differences that no one of the individuals is wholly representative of the class. They are concepts connoting only the common or essential qualities and not taking any special account of the accidental or differential properties. It may not be proper to call them "logical" or "metaphysical," and I shall not contend for that name longer than to indicate the distinction which I have defined. The importance of the distinction itself will appear when we discuss the matter of genus or essentia, and differentia or accident. We require at present only to recognize the two acts of comparison, one of which involves an act of *abstraction*, and the other does not. The formation of mathematical concepts does not require any abstraction of special properties, and the bringing of them together to the neglect of certain others, unless it be in a manner which we hardly need to take account of. On the contrary, the formation of what I have called logical or metaphysical concepts, such as "man," "tribe," "race," "animal," etc., requires that we *abstract* certain common qualities and ignore the accidental ones, so as to denote the former by a common name. "Abstraction" is here used merely to denote the thinking away of certain properties from their exclusive application to any particular object. A farther conception of it is not required at present. It is sufficient to remark a process producing two kinds of class concepts with which later problems in logic will have to reckon.

3. The Denomination of Concepts.—The naming of concepts has always been considered an important matter in Logic. It is not especially a process of thought, although it is necessary to make thought an effective instrument in the com-

munication of knowledge, and to give stability and fixity to the various products of mental activity. Some have even considered names and naming the most important aspects of Logic. Into this controversy it is not necessary to enter either on one side or the other. We may be content with one or two remarks about the nature of language and its service to thought.

Language is the symbol of ideas, or consists of the signs by which we can indicate the resemblances and differences between concepts. By it the infinite number of individuals and classes can, to some extent, be tabulated or indexed for use. Without it, perhaps, we should not be able to develop our thinking processes above the grade of animal intelligence. With it we can name an idea so as to keep it by itself, if required, or conceive and speak of a class of ideas, if need be. The same word even may have a double denotation, as we have seen in general concepts denoting either individual or class-wholes. In this way a word may indicate the quality that separates the concept or object named by it from others, or it may apply equally to all members of the class. "Man" may imply or represent the quality or qualities which separate him from a "lion," an "eagle," or all other objects. At the same time it will denote the qualities by which the term may be employed to indicate all individual men. Thus economy of language is obtained, on the one hand, and clearness of conception on the other. But the general service of language is illustrated in the simple fact that where any ambiguity is associated with a single term, our intellectual confusion is completely overcome by the use of two or more terms which may specify the distinct qualities confounded under a single term. The denomination of concepts, therefore, is an important process either in completing the act of thought, or in making it useful when it is completed.*

---

* The student may consult the following references for a discussion of Language and Denomination in their relation to Logic: Bosanquet: Logic, vol. i., Introduction, Sections, 4, 5, pp. 8-30; Thomson: Laws of Thought, Introduction, Chapter on Language ; Hamilton: Lectures on Logic, Lecture VIII.

2d. Judgment.—Like Conception the term Judgment may denote both a process and a product. Its product is a proposition. At present we are concerned only with the act, which can be briefly defined, and which, with a slight modification, will apply to the product. As an act a judgment is *the perception of the relation between two concepts*. It is an act of comparison still more distinct than any involved in Conception. The relation which it expresses is that of agreement or difference between subject and predicate, or of inhesion or non-inhesion between subject and attribute. Much discussion would be required to give a satisfactory theory of that relation, and to decide whether it is wholly one of agreement or difference, or partly of some other character. But larger treatises may be consulted upon this problem. It is sufficient to know that *the act of connecting two concepts* is the fundamental characteristic of Judgment. The laws which determine it may be ascertained without any particular theory of the process, even if such a theory be helpful in the final solution of problems centring about it.

The principal matter of importance to be observed in connection with the nature of Judgment is that the act of comparison involved in it is in its essential elements the same as the act of Conception. The difference between them is only the way in which the result is expressed, or the object-matter with which they deal. Conception is the connecting of percepts; Judgment is the connecting of concepts. The connecting of percepts can be expressed in terms or words: the connecting of concepts must be expressed in propositions. The relation between subject and predicate in Judgment may be variously expressed. But the process will be identical with that which is involved in the formation of concepts. This is perhaps apparent in the fact that the two kinds of judgments correspond to the two general kinds of concepts. There are judgments which express the relation between substance and attribute, corresponding to concepts representing individual wholes; for example, "Iron is hard." Then there are judgments which express the relation of resemblance or difference between subject and predicate, or the relation of classes, corresponding to concepts representing class-wholes; for example, "Iron is a metal." We shall examine the importance of this distinction again. The nature of the mental process of Judgment and its relation to that of Conception is all we require to know at this stage of the discussion.

3d. Reasoning.—Reasoning is a process only a little more complex than Judgment. It may be briefly defined as *inference*. But this will require further explanation. Hence we may adopt Jevons's definition of reasoning as adequate for all practical purposes. It is " the progress of the mind from one or more given propositions to a proposition different from those given. Those propositions from which we argue are called the *Premises*, and that which is drawn from them is called the *Conclusion*." The definition here covers what is known as *Immediate Inference*, and *Mediate Inference*. If from the proposition that "All metals are elements," I infer that "All that are not elements are not metals," I am making an immediate inference; if from the two propositions that "All metals are elements," and "Iron is a metal," I infer that "Iron is an element," I am making a mediate inference. But in both cases my reasoning consists in *the perception of a relation between propositions*, through the medium or agency of concepts whose relation is known, implied, or expressed.

This last definition of Reasoning identifies the process in its nature with that of Conception and Judgment. The difference between them is not in the form of the mental act, but in the matter to which it is applied. In reasoning it is the resemblance or difference of relations between propositions that constitutes the peculiar nature of the matter dealt with, while in Judgment it is the resemblance or difference of relations between concepts. The distinction is thus only a matter of complexity, Reasoning representing in a more complex form only what is found substantially in the earlier process of Conception. At any rate, this way of regarding the question helps to give unity to the mental processes, while it justifies our turning to the differences of matter to which the various mental acts are applied, in order that we may determine the different laws of thought regulating the process according to the changes of matter apparent in the development of the subject of Logic.

CHAPTER III.

TERMS OR CONCEPTS, AND THEIR KINDS

TERMS and concepts have already been defined as denoting ideas. We come next to their divisions, which can be made on several different principles of distinction.

1st. Categorematic and Syncategorematic Terms.—A categorematic word is one which can stand as the subject or predicate of a proposition. Such are "animal" "nation," "excellence," "wise," "beautiful," "perfect." These show that categorematic terms are limited to nouns and adjectives. Verbs ought to be included.

A syncategorematic word is one which cannot stand alone as the subject or predicate of a proposition. Such are "with," "and," "through," "nobly," "very," "indeed," etc. From this we perceive that syncategorematic terms are either *modal* or *relational*: modal, if they are adverbs, relational, if they are conjunctions and prepositions.

A *term* in the logical sense, as the subject or predicate of a proposition, may consist of one or more categorematic words, or of categorematic and syncategorematic words. In the latter case it must consist of a grammatical phrase or clause.

2d. Singular and General Terms.—This division of terms is limited to categorematic words, and perhaps to the class of substantives. But this fact is less important than that the division is a new one and not to be confused with the one already made. It is based upon the differences between the number of individuals denoted by various concepts, and at least *partly* coincides with that between individual-wholes and class-wholes.

1. SINGULAR TERMS.—A singular term is one which can be

affirmed, in the same sense, only of a single object, real or imaginary. *Proper* names are the best illustration. Thus, Napoleon, Paris, Greece, St. Paul, etc., are singular terms because they can apply, in the same sense, only to one object. Expressions like "the Secretary of State," "the Prime Minister of England," "the King of Spain," will be singular when they refer to an individual or particular person. But they are also capable of being general terms. This will be the case when they are used to denote the class of officers by those respective names. Thus when we refer indefinitely to "the Secretary of State" as any man or officer in that station, we use it in its general sense; if we refer definitely to a particular man occupying the place, it is singular. Keynes also indicates how the same expressions may become singular, if an individualizing prefix is added to them: Thus, "the present Prime Minister," "the present Secretary of State," "the reigning Queen of England," etc. Likewise, he thinks such expressions as "the first man," "the pole star," are singular. Perhaps we could add such terms as "the highest good," "the supreme or ultimate end." These are simple cases where the addition to a general name of an adequate prefix transforms it into a singular one. But great caution must be exercised in our judgment of such cases. For example, "the eldest child," is an expression which is applicable only to a particular child, but it is an indefinite particular, and so is general in its import. But the illustrations preceding show that there may be singular terms other than proper names. Some of them, however, are capable of both a general and a singular use. For example, *space*, when it refers to the totality by that name, is singular, but when it is a name for a number of definite spaces, or the different portions of aggregate space, it has all the distinctive characters of a general term. The same is true of the term *universe*, and Professor Bain thinks it true of all aggregate substances which are divided into *parts*, not kinds; as "water," "stone," "salt," "mercury," "flame." But Keynes's remark that we can say "*some* water," "some salt," "some mercury," which cannot be said of proper names or concepts denoting only one object, or an individual whole, seems to me decisive in favor of considering them general terms. Keynes ought, perhaps, to have remarked the reason that such terms are liable to be mistaken for singular ones. It is that general terms apply either to objects which are different in kind, or to objects which have an independent existence, and never unite to form one continuous mass of homogeneous matter, as do "water," "air," "space," and "stone" when denoting rock of the same kind. *Oneness of kind* is a characteristic of singular terms, as well as individuality, and hence it is easy to confuse such terms as "water," "air," "mercury," etc., with the singularity of proper names. But they represent objects which can be divided into individual parts without modifying the qualities of the parts which receive the same names as the aggregates. This is not true of proper names, or of any singular term which is singular in its absolute sense. It would have been more apt if Professor Bain had classified the terms as *collective*, which are still to be examined, for their analogy to such terms is closer than to singular terms. But, nevertheless, I think they can be shown not to be collective. I regard them as abstract.

Proper names may become general terms when used to denote a class of individuals having a given characteristic. Thus "the Napoleons," "the Cæsars," "the Platos," "the Washingtons," are general terms because they denote a class of persons with certain characteristics which belonged to the original person by that name. Again, the name "God" will be singular to a monotheist and general to a polytheist: uncapitalized it is general to a monotheist.

There is, then, no absolute rule by which the mere *form* of a word may be taken to indicate its character. Some general terms by adding a prefix may become singular; some singular terms, used in the plural, or to denote, on certain emergencies, more than one of a kind, become general. It is a case where a change of the *matter*, when the *form* of the term remains absolutely or virtually the same, affects the character of the concept. Hence the rule can be absolute only when the two aspects of the concept, form and matter, remain constant and according to definition.

2. GENERAL TERMS.—A general term or concept is one which can be applied, in the same sense, to an indefinite number of objects, real or imaginary. It is a name applied to class-wholes, such as "man," "vertebrate," "quadruped," "generation," "triangle," etc. Also such terms as "army," "forest," "crowd," "nation," etc., are general terms. But there is so marked a difference between the first class of illustrations and the second that a subdivision of general terms into *distributive* and *collective* has to be recognized. A distributive term is one which applies in the same sense to each individual in the class, such as in the first set of illustrations given; namely, "man," "vertebrate," etc. A collective term is one which applies to an aggregate of individuals composing a whole, and which will not apply to any of the individuals alone. Thus "army," "flock," "bevy," "family," "tribe," etc., are collective terms because they denote a composite or aggregate whole. They are at the same time general terms because they are applicable in the same sense to an indefinite number of such aggregates.

It is important to remark that a collective term may be singular instead of general. Thus the Vatican Library, the American people, the Greek nation, the Seventy-second Regiment, Company B, etc., are singular aggregates because the name will apply to no other objects of the same general kind. Some few terms may be used either distributively or collectively according to the emergency. Thus "people," "the Greeks," "the English," and often the plural of ordinary general terms, as "the trees," "the houses," etc., may be used in either sense, as illustration will presently show. The following diagram will show the relation subsisting between singular and general terms, and their subdivision :

Terms { Singular { Individual.
        General { Collective.
                 { Distributive.

Or,

{ Singular { Individual.
           { Collective.
  General { Collective.
           { Distributive.

Perhaps a doubt about the accuracy of this representation is possible, since a collective term when general may at the same time have the characteristics of a distributive term. This is true of such terms as "army," "herd," "regiment," etc. They are collective in so far as they denote an aggregate whole of individuals; they are distributive in so far as they denote or can apply to a class of such aggregates. But since the last is true only when the aggregates are *treated* as individual-wholes, the general distinction between them may still be observed. It is merely a case again where the difference is due to some variation between the form and matter of the concept. Thus in the sentence, "Mobs are crowds of enraged men," there can be no distinction between the form and matter of the collective term "mobs." But in the sentence, "Mobs are dangerous," the assertion is made of all such aggregates, and hence while the *form* of the term is collective its matter may be both collective and distributive, or distributive alone.

But the distinction between distributive and collective terms is more important for Logic than the distinction between singular and general terms. This is apparent for the reason that the nature of propositions is less affected by the latter than by the former distinctions. We shall learn later that the same logical laws apply to "singular propositions" as apply to "universal propositions," and the distinction between these is parallel with, and determined by, that between singular and general terms. They are not liable to easy confusion. But the collective and distributive uses of terms are often confused and give rise to corresponding fallacies in reasoning. This liability to confusion is illustrated in such propositions as the following: "All the angles of a triangle are equal to two right-angles," and "All the angles of a triangle are less than two right-angles." The two propositions seem contradictory, because the same thing cannot be equal to and less than another at the same time and taken in the same sense. But the first proposition taken collectively in its subject is true, and the second taken distributively is true. Again, "All the trees in the forest produce a thick shade," may be taken in the same double sense. Similarly, "The people filled the hall," and "The people are honest," or, "The Greeks are a nation," and "The Greeks are Caucasian." A case of the plural of an ordinary distributive term becoming collective is the following: "The trees make a forest;" but in the following it is distributive: "The trees are beautiful."

3d. Concrete and Abstract Terms.—It is difficult to give a satisfactory definition of concrete and abstract terms. Scarcely any two authors agree upon the subject, and even if they did agree, observation would teach us that any attempt to apply the definition to an actual classification of concepts would meet with the serious obstacle that the two classes seem to shade off into each other by insensible degrees. It is only in certain cases that the distinction can be made clear, and often, in spite of this clearness, a term may be concrete in one of its applications and abstract in another. The difficulty is largely caused by the unfixed and varied use of the term "abstract," which is sometimes used as the equivalent of "general," and again as denoting the conception of a quality apart from its subject. The confusion occasioned by this usage will appear as we proceed.

The definition of "concrete" and "abstract" terms can be made very simple. The only difficulty we have to encounter after that, is the determination of the particular terms which fall under the one class or the other. But we can at least begin with a definition of them in their purest form, and if subsequent facts require us to qualify it we can do so.

A *concrete* term is a name which stands for a thing, or for an attribute of a thing conceived as an attribute; e.g., "Webster," "Bucephalus," "Parthenon," "white," "clear," "generous," etc. A concrete concept is the same in its meaning, but we do not think of it as a word. It denotes a thing or a quality as the object of consciousness; while spoken of as a term it is the object of Grammar. With the same qualification we may define abstract terms or concepts.

An *abstract* term is a name which stands for an attribute or quality considered apart from the thing possessing it, or conceived and used as a thing, *e.g.*, "redness," "cheapness," "purity," "perfection," "righteousness," "justice," "ability," etc.

Mill's definition is as follows : " A concrete name is a name which stands for a thing ; an abstract name is a name which stands for an attribute of a thing." It is not clear what such a definition will do with adjectives. They are names of attributes and yet Mill includes them among concrete names. On the other hand, they can hardly be included among concrete terms because they do not denote "things" in the strict sense. We must therefore either make a class for them apart from the concrete and abstract, or modify the definition. I have attempted to overcome this difficulty in the definition which I have given. It provides for two kinds of concrete terms, the *substantive* and the *attributive*. I shall also divide abstract terms into two classes, *static* and *dynamic*, or adjectival and verbal nouns. To complete the classification I shall set aside a third class of *mixed* concrete and abstract terms. Thus it appears in the following table :

Terms
- Pure
  - Concrete
    - Substantive = Singular Nouns, *e.g.*, Homer.
    - Attributive = Adjectives, *e.g.*, Pure.
  - Abstract
    - Static = Adjectival Nouns, *e.g.*, Generosity.
    - Dynamic = Verbal Nouns, *e.g.*, Acceleration.
- Mixed = Concrete and Abstract, *e.g.*, Government, Institution, etc.

In this classification no mention of such terms as "man," "animal," "race," "vegetable," "triangle," etc., has been made. Keynes considers them as concrete. The illustrations of concrete substantives were chosen from singular terms, and it is now a question whether general terms are concrete or abstract. The mere fact of being general terms does not make them concrete, as is shown in *verbal* abstracts, which may be general, and also in the *adjectival* abstracts, if they can be considered as general. Jevons thinks them singular. This is extremely doubtful, to say the least. But "man," "animal," "race," etc., denote objects which we are accustomed to speak and think of as "concrete," and so they seem most naturally entitled to be called concrete names. On the other hand, some writers call them abstract, and so regard all general concepts as abstract because the process of forming them is one of *abstraction*. Keynes disputes the legitimacy of this treatment of them. He seems to hold that the process has nothing to do with the nature of the product. This seems to my mind doubtful. But in so far as such terms as "man," "animal," "tree," etc., denote real objects which are definite things, they may with, at least, tolerable propriety, be called concrete. But in so far as they are indefinite and do not seem to represent any clearly conceived individual object, they closely resemble abstract terms in this respect. In fact, as terms become more general, that is, as their *extension* increases they approximate in indefiniteness the abstract concepts which are characterized by this indifference to particular things, and hence they may with some propriety be spoken of as abstract. Perhaps their ambiguity in this respect would justify us in regarding them as abstract in one relation and concrete in another, their abstractness depending upon the proportion of indefiniteness, and their concreteness upon the proportion of definiteness expressed by them. There are other terms which give still greater trouble than these. They are such examples as "color," "sound," "pleasure," "thought," etc. It may be a problem to determine whether they are concrete or abstract. On the one hand, "color" and "sound," as denoting qualities apart from a definite conception of their subject, might be regarded as abstract. On the other hand, as general terms for attributes which are concrete they might be considered as concrete. So "pleasure," "thought," and all names of states of consciousness, as *verbal* nouns, dynamic concepts, might be abstract. But as names of individual facts clearly representable in some way, they might be considered by many writers as concrete. In other words, as general terms denoting facts or individual events, they will be usually conceived as concrete, but as terms denoting attributes, but not definitely denoting their subject, they will appear usually as abstract. This may be true of a large number of terms. If so, they may be treated in two relations at the same time, according to the degree of definiteness and indefiniteness with which they are conceived.

But this manner of speaking about such terms suggests the common usage of the words "concrete" and "abstract," which logical discussion cannot wholly ignore. Ordinarily "concrete" denotes any sensible or real object, which may *represent* the meaning of a term, singular or general. Thus "man," "tree," "biped," denote sensible objects. Even "color," "sound," "odors," denote individual percepts, and so there is always some distinct or definite reality expressed by them, which is supposed to be identical with the concrete. But such terms as "emotion," "reasoning," "thought," "pleasure," "figure," "form," are either not sensible objects or are so vague and indefinite, being often called the "higher abstractions" of the mind, that the common consciousness thinks and speaks of them, perhaps loosely, as "abstract" conceptions. In the same way general terms are often conceived as abstract in proportion to their generalized character, or their remoteness from the individuals which they comprehend. This distinction, then, is mainly that between what is *presentative* or *representative*, and what is merely *thought* in consciousness. Conceptions which call up distinctly to the mind the individual objects which they denote are thus commonly taken for concrete, and those which do not indicate clearly the characteristics named, and which might be called *symbolical* conceptions, after the manner of Leibnitz, are taken as abstract. For this reason it might serve the purposes of Logic much better if the distinction were made between *definite* and *indefinite* concepts, the former taken to denote all terms which clearly denote or connote certain marks, such as *Peter, man, white*, and the latter taken to denote such as do not indicate distinctly any communicable mark of an object or a fact, as "life," "organic," "humanity," "government," "institution," etc. It is this distinction rather than that between the concrete and the abstract, as I have defined them, that is of importance in Logic, because fallacious reasoning is occasioned more by ambiguities of meaning due to indefiniteness than by the question whether a term is abstract or not, unless "abtract" is taken as synonymous with indefinite. As this indefiniteness increases with the *extension* of a concept, general terms will partake of this character as they recede from the individual and concrete, and so will often be called "abstract" in the same degree, although they may not wholly lose a concrete reference. It is possible, therefore, to consider them either as mixed concrete and abstract terms, or as combining definite and indefinite characteristics in an inverse ratio to each other. If the former, we simply adapt them to our definitions; if the latter, we use "concrete" and "abstract" to denote the distinction between the *representable* and the *non-representable* concepts, which is the most important for Logic, as later chapters will show. Wundt has some excellent observations upon this question, which sustain the position I have taken, and should be quoted in this connection:

"It is a necessary consequence of the formation of concepts from the connection of percepts that certain conceptions should stand much nearer than others to the presentations of sense. We have generally expressed this fact by the distinction between *concrete* and *abstract* ideas, but have described the logical process, by which abstract notions have been formed from concrete, as the *process of abstraction*. The influence of this latter procedure, moreover, has essentially changed the meaning of the terms 'concrete' and 'abstract' in the course of history. Scholastic Nominalism, which introduced it into Logic, used the terms for the mere purpose of distinguishing between words. Every substantive noun which denoted an individual object, or a class of objects, was concrete, whereas, on the other hand, a word formed from a concrete term and used to denote a universal property was called abstract. Words like *man, white*, etc., therefore were regarded as concrete, and those like *humanity, whiteness*, etc., were regarded as abstract. In modern Logic this distinction gradually became confused with that between individual and general concepts, since, in depending upon the use of the word 'to abstract,' we became accustomed to characterize as abstract all conceptions whose formation was marked by a distinct process of abstraction. But as this was peculiar to all generic conceptions, there remained finally nothing but singular or individual terms which could represent the territory of the concrete. This confusion is, in fact, logically quite excusable; for, in whatever senses the terms 'concrete', and 'abstract' are applied, it is evident that it is neither a difference in the process of forming, nor the essential constitution of concepts, but a much more external circumstance which is expressed by the terms. Even Mill's proposition to restore scholastic usage of them may be opposed to the practical consideration that, in the modified sense in which they refer to the degree of applying the process of abstraction, they have already obtained, through common usage, a right to a place in the language of science, as well as in practical life, which at the same time indicates a demand for a corresponding logical distinction. If we regard only the present practice of language, it will not appear doubtful that we have here to do chiefly with the *relation of a conception to its representative percept*. So long as the latter consists in a sensible presentation in which the essential elements of the concept are realized, and not merely in the word denoting it, we name it *concrete*. But, on the other hand, so soon as the written or spoken word becomes a single sign or symbol for the concept, it is abstract. In other words, those concepts are abstract to which no adequate representative percept corresponds, and for which, in thought, we can only choose an external and apparently arbitrary symbol. In this sense we should doubtlessly describe such a concept as 'man' or 'animal' as concrete, and such a concept as 'humanity' as abstract. But in opposition to scholastic usage we could call 'the righteous' as well as 'righteousness' by the name of abstract. And further, an individual concept would most frequently be concrete at the same time, while a concrete term would very frequently be general. Also it would certainly happen that, in individual cases, the distinction would remain indeterminate or doubtful. Though words have gradually developed into signs of abstract conceptions, and though, as the history of the change of meaning everywhere shows, abstract conceptions have been developed from concrete, why should we not at times meet a conception which remains in the intermediate stage of development? Conceptions like 'machine,' 'weight,' etc., may, in fact, be completely abstract to one person, and in another be attached to a sense picture, and consequently concrete. From this a single conclusion is evident; namely, that this distinction has little importance for the nature of a conception, but that it has just as great an importance for the development of abstract notions. This development depends upon the constant application of the same process which we find is active in the origin of all conceptions, even the most individual and concrete." *

A pure abstract concept, therefore, I shall consider as defined to be a quality or attribute conceived and treated as a substantive or thing. A mixed abstract and concrete term will then be capable of either reference according to circumstances and the degree of its definiteness. It is general terms that are so frequently of this mixed character and that are the source of confusion in Logic. The pure forms also give rise to a certain order of errors. In the first place, there is the danger of treating abstract conceptions as if they represented independent realities. Thus such ideas as "truth," "beauty," "excellence," "virtue," "nature," "law," etc., are often spoken of by writers as if they represented existences independent of particular objects or persons, of which they are in reality only qualities. The error here is first in the conception, and reasoning is subsequently affected by it as the conclusion is always vitiated or validated by the character of the premises. But modern thought is better provided against the confusion of the abstract and the concrete than ancient and mediæval speculation, when it concerns the pure concrete and the pure abstract concepts. This is not the case with the second form of error which arises from the confusion of the concrete and the

* Wundt: Logik, Bd. I., Zweiter Abschnitt, Cap. I., § 3, p. 97.

abstract in the same terms and propositions. One person may have the abstract and another the concrete aspect in consciousness when using the concept, and unless what is affirmed of it be true of both aspects there will be a difference of opinion between the two, or one will be guilty of error in thought and reasoning. This is less apparent in conceptions than in propositions. Hence it is more frequent that abstract thoughts, propositions, or principles should produce error than the mere concept alone, or the confusion of the abstract and concrete aspects of it, although the error in the use of principles must begin with an error in the use of the concept. But as possible sources of fallacy in the conception of mixed terms we have such general concepts as "religion," "institution," "home," "history," "organism," "socialism," where we may not be assured whether it is the concrete or the abstract form which is prominent in consciousness. But all this is brought out more clearly in propositions than in concepts. Thus if I take the proposition, "Religion is useful," I may mean "religion" in the abstract ideal form, or I may mean in its particular concrete form; that is, all religions and denominations. Or I may say, "Governments are good institutions," and mean either all particular governments, or government in the abstract. In the concrete, governments might be very bad, while in the abstract they might be very good. "Government," as an abstract concept, is only an ideal quality of the aggregate of men or of those in power, and so, in the concrete, can never be better than the men composing it. But in speaking and writing of it, it is often convenient to treat the concept independently of its reference to the men who made it, and then it is considered in the abstract. As long as the men who compose government are bad, I can say that "government is a good institution" only when I consider it abstractly. Concretely government could only be what the men are who compose it. The same observations apply to any other mixed concepts and propositions. This will also be true, but perhaps in a less degree, of such terms as "man," "animal," "vegetable," etc., for we occasionally give them an abstract import, although their concrete reference may be the most frequent. But the significance and importance of these distinctions will be more apparent when we come to consider the *essence* and *accidents* of concepts. For the difference between the abstract and the concrete is closely related or connected with that between essence and accident.

4th. Positive, Negative, Privative, and Nego-positive Terms.—The usual division of terms in this respect is into Positive, Negative, and Privative, but I add the fourth for reasons which will appear in the sequel. I shall first define and illustrate them.

Positive terms or concepts are those which signify the existence, presence, or possession of certain qualities; for example, "good," "pure," "excellence," "metal," "organic," "human," etc. A positive term is thus one which is so both grammatically and logically, or both in form and matter.

Negative terms or concepts are those which signify the absence of certain qualities, as "impure," "inorganic," "unhuman," "non-metal," "ingratitude," "insipid," etc. They are thus negative in both form and matter.

Privative terms or concepts are those which denote the deprivation of certain qualities once possessed or the normal characteristic of the subject. Thus "deaf," "dumb," "blind," "dead," are privative terms. In so far as they denote the absence of qualities they are negative terms. But they differ in their form from negative terms, although materially considered they are only modifications of them. They may be more strictly defined as terms which are positive in their form, but negative in their matter. This will be apparent from the illustrations.

Nego-positive terms or concepts are those which denote the presence of a positive quality, although they appear to be negative. Thus "disagreeable," "inconvenience," "infamous," "ignorant," "displeasure," "immediate," "undone," etc. They are thus negative in form and positive in matter. They can in many cases be most easily distinguished by comparing them with their corresponding positive conceptions. Thus "unhappiness" and "invaluable" have their equivalents in "misery" and "costly," both of which are positive.

Some terms may be taken in either a negative or a nego-positive sense. Thus "uncertain," "unhealthy," "unpleasant," "indistinct," may be conceived as the negatives of "certain," "healthy," "pleasant," "distinct," or as the nego-positive equivalents of "doubtful," "sickly," "painful," "obscure." They are, however, the same modification of positive conceptions that privative terms are of negatives. In fact, privative and nego-positive terms are simply mixed concepts, having an element from each of the other two, but combining them in a reversed relation. In the broader sense, therefore, we can divide all terms into positive and negative. In this broader sense the symbols or indices of negative terms are *in*, *un*, *non*, *less*, *dis*, *a* or *an*, *anti*, and sometimes *de*, and perhaps *mis*.

The following tabular outlines will give the divisions and indicate their nature and relations to each other.

Terms or Concepts.
- Positive = Positive in both form and matter.
- Negative = Negative in both form and matter.
- Privative = Positive in form and negative in matter.
- Nego-positive = Negative in form and positive in matter.

Terms or Concepts
- Pure
  - Positive.
  - Negative.
- Mixed
  - Privative.
  - Nego-positive.

Terms or Concepts.
- Positive
  - Pure = Simple Positive.
  - Mixed
    - Privative.
    - Nego-positive.
- Negative
  - Pure = Simple Negative.

The reason for distinguishing a separate class of nego-positive terms is their liability at times to be mistaken for negatives, and the confusion often incident to the transition, or immediate inference from a purely negative conception to one which is really positive in its meaning, although expressed by the same term. Thus if we were to argue that a thing or an act is *bad* because it is *not-good*, we should be committing a fallacy, considering that "not-good" is a purely negative conception. The same remark applies to the passage from the *not-just* to the *unjust*, and from the *non-moral* or *not-moral* to the *immoral*. For instance, physical acts are "not-moral," but they are also not immoral. In ethics there are three classes of acts—the moral, the non-moral, and the immoral—or in stoical parlance—good, indifferent, and bad. Or we might divide them first into moral and not-moral, and subdivide the latter into non-moral and immoral. Similarly we may recognize just, not-just, and unjust acts. The distinction is here the same. Also in many other departments of thought this triple division is the proper one, although the usual division is dichotomous instead of trichotomous. Where it occurs the negative and nego-positive conceptions do not necessarily coincide. It is true, however, that a negative term is often used as the equivalent of the nego-positive. Thus we sometimes describe an act as "not-just" when we think of it as "unjust." This is when we do not think of the conception as *infinitated*. By an infinitated conception I mean one which comprehends all other possible objects in the world than those denoted by the contradictory positive term. Thus in its strict meaning and extension the negative "not-just" will include all other things in the universe than those expressed by "just." Hence among "not-just" might be found material objects, such as trees, stones, houses, etc. Again, "not-house" would include all other things in the universe than the term "house;" and so on with all similar conceptions. The infinitated concept is simply all else than what is expressed by the positive. But often a negative concept is used either as a euphemism, or as an equivalent of the nego-positive. In such cases they are clearly convertible. Thus wherever "unpleasant" is thought of as "painful," the two concepts can be substituted for each other. Illustrations of this will appear in Conversion. But when the negative term is infinitated this substitution cannot be made, and it is the business of the student and the logician to be on the alert for the confusion incident to such a procedure.

The terms "greater," "less," and "equal" deserve a brief notice. They are all positive concepts, but taken in relation to each other, as they must be, they are *relative* terms, and must be considered in the next section. But being relative concepts we may say that any two of them are the negatives of the third. But their pure negatives are the infinitated forms of their positives, and it is only as relatives that any of them can be considered the negative of the others.

What has been said of the terms "greater," "less," and "equal" can be said of a large number of terms in the language. Indeed, every term may be said to be negative in relation to all other terms, except its own equivalents or synonyms. It will not need to be so in its form, but only in its meaning as related to those terms which do not express the same concept. Hence, while we should call "horse" a positive term, in relation to man it would be negative, because it would be negatively conceived as "not man;" "tree," as not a lion; "external influences" as "not internal influences," etc. Very often, from the mere habit of conceiving different concepts as excluding each other, we assume that particular cases represent contrary concepts, one the negative or contradictory of each other, when in reality they are not so. Many errors of opinion and of reasoning are incident to this mistake. We require always to examine how far the meaning of concepts excludes that of other terms, and not to assume a negative relation from the mere fact that it generally so exists. A very large number of concepts have this relation; but the form of the term will be no adequate criterion of the fact, and we must examine the matter of thought in given cases, in order to decide the question for practical instances.

5th. Absolute and Relative Terms or Concepts.— This distinction, as usually applied, does not have much importance in Logic. Hence it may be dismissed very briefly. There is, perhaps, a more comprehensive sense in which the distinction is a valuable one. But it does not apply in any special way to the few terms known particularly as relative. Most terms are absolute concepts in the ordinary sense of the word.

"Absolute" means literally that which is severed from all

dependence on another. Hence an absolute term is one which expresses what can be thought of by itself, and which does not have its meaning determined by comparison with some correlate object of consciousness. Thus "man," "tree," "earth," "star," "book," etc., are absolute concepts. They have no correlatives implied in them. They may denote things which exist, both in thought and reality, in relation to something else. But this related object is merely associated with the conception and may not be necessary to the meaning of the term, or to its presentation in consciousness. Quite the contrary is true of relative terms.

A "relative" term or concept is one which denotes an object that cannot be thought of without reference to some other object, which is its correlate. Thus "father" and "son," "parent" and "child" are correlatives. Each has its meaning determined in relation to the other. Again, the terms "monarch," "shepherd," "master," "teacher," "subject," "slave," "eldest," etc., are relative. Their correlates can easily be remarked. It will be apparent also how "greater" and "less" may be regarded as relative. But as little confusion is incident to logical processes connected with the use of relative terms as here defined, the subject does not require further consideration.

In a broader sense, as I have remarked, every term is relative. It is first relative to its negative concept, and we often find this means a convenient one for defining a term. But it is never completely satisfactory, because a true definition demands a statement of the positive content of a concept. The negative term, however, is a very convenient one for representing the actual relation sustained by every concept. Further than this, every object may be said to exist in a relation to some other, or all other objects, and the relation between two or more objects may be so close as to mutually affect each other's meaning. Thus "day" and "night," "joy" and "sorrow," "pleasure" and "pain," "true" and "false," are conceived as relatives by contrast. But while they may be so related in thought, the material or real existence of one of the correlates is not implied by the other. Such relation as may exist between other classes of conceptions is too remote and unimportant for logic to take any special notice of it.*

* For discussion of technical problems connected with the qualities of terms the student may consult the following works: Mill: Logic, Bk. I., Chap. II. ; Venn : Empirical Logic, Chap. VII. ; Keynes: Formal Logic, Part I., Chaps. I.-III., pp. 7-50; Jevons: Principles of Science, Bk. I., Chap. II.; Wundt: Logik, Zweiter Abschnitt, Chaps. I.-IV., pp. 86-134.

CHAPTER IV.

THE AMBIGUITY OF TERMS

In this chapter I shall do little more than transcribe the language of Jevons upon the subject. He has said about all that a practical Logic requires to have said upon it. The importance of considering it ought to be apparent to everyone, and will be so to those who know or suspect that the majority of fallacies in our reasoning turn upon the ambiguous use of words. Most of our controversies are logomachies on the same account. We think we are employing the same conceptions when we are only using the same terms; the difference between our conceptions, in spite of the identity in language, may be as great as between different words. Different terms also may be used for the same concepts, and thus give rise to a converse error. But the most frequent source of error is ambiguity. A syllogism will illustrate it:

> No designing person ought to be trusted.
> Engravers are by profession designers.
> ∴ They ought not to be trusted.

It is easy enough to detect in such cases the source of the fallacy. But there are instances where the ambiguity is more subtle, and requires a keener logical insight for its detection. In profounder subjects of speculation it requires a wide knowledge of the use of language and a thorough acquaintance with the laws and processes of the mind. In fact, a reasoner should always be on the alert for the ambiguous use of terms, and he cannot have the possibility of such a source of confusion better indicated than by a few observations upon the simplest words of the language.

We may divide terms into *univocal* and *equivocal*. A univocal term is one with but a single meaning, which is exposed to no mistake of interpretation. An equivocal term is one with more than a single meaning. Very little observation is necessary to show that very few terms come under the class univocal. Proper names, and therefore singular terms, are almost the only conceptions with an unmistakable import, and even some of these are equivocal. A general concept applying only to a single class of individuals of exactly the same kind may be univocal. Thus President Lincoln, St. Paul's Cathedral, Berlin, are univocal. But Washington may be equivocal as applying to a person, a city, or a state. Of common terms that are univocal, Jevons thinks that they are chiefly found in technical and scientific language. He enumerates "steam-engine," "gasometer," and "railway train," and in common life such terms as "penny," "mantelpiece," "tea-cup," "bread," and "butter." "Cathedral" is probably univocal, or of one logical meaning only. But "church" is equivocal, as referring sometimes to a building, and sometimes to a corporate body of men. Compared with the equivocal terms, however, the univocal are very few.

From this point we may simply quote Jevons. He begins with a division of equivocal terms. "We may distinguish," he says, "three classes of equivocal words, according as they are—

"1. Equivocal in sound only.

"2. Equivocal in spelling only.

"3. Equivocal both in sound and spelling.

"The first two classes are, comparatively speaking, of very slight importance, and do not often give rise to serious error. They produce what we call trivial mistakes. Thus we may confuse, when spoken only, the words right, wright, and rite (ceremony); also the words rein, rain, and reign.; might, mite, etc. Owing partly to defects of pronunciation, mistakes are not unknown between the four words *air*, *hair*, *hare*, and *heir*.

"Words equivocal in spelling, but not in sound, are such as tear (drop), and tear, pronounced tare, meaning a rent in cloth; or lead, the metal, and lead, as in following the lead of another person. As little more than momentary misapprehension, however, can arise from such resemblance of words, we shall pass at once to the class of words equivocal both in sound and spelling. These I shall separate into three groups, according as the equivocation arises.

"(*a*) From the accidental confusion of different words.

"(*b*) From the transfer of meaning by the association of ideas.

"(*c*) From the logical transfer of meaning to analogous objects.

"(*a*) Under the first class we place a certain number of curious but hardly important cases in which ambiguity has arisen from the confusion of entirely different words, derived from different languages or from different roots of the same language, but which have in the course of time assumed the same sound and spelling. Thus the word *mean* denotes either that which is *medium* or mediocre, from the French *moyen* and the Latin *medius*, connected with the Anglo-Saxon *mid* or *middle;* or it denotes what is low-minded and base, being then derived from the Anglo-Saxon *gemoene*, which means 'that belonging to the moene or many,' whatever, in short, is vulgar. The verb to *mean* can hardly be confused with the adjective *mean*, but it comes from a third distinct root, probably connected with the Sanscrit verb *to think*.

"As other instances of this casual ambiguity I may mention rent, a money payment, from the French *rente* (*rendre*, to return), or a tear, the result of the action of *rending*, this word being of Anglo-Saxon origin and one of the numerous class beginning in *r* or *wr*, which imitate more or less perfectly the sound of the action which they denote. *Pound*, from the Latin *pondus*, a weight, is confused with *pound*, in the sense of a village pinfold for cattle, derived from the Saxon *pydan*, to pen up. *Fell*, a mountain, is a perfectly distinct word from *fell*, a skin or hide; and *pulse*, a throb or beating, and *pulse*, peas, beans, or potage, though both derived from the Greek or Latin, are probably quite unconnected words. It is curious that *gin*, in the meaning of trap or machine, is a contracted form of *engine*, and when denoting the spirituous liquor is a corruption of *Geneva*, the place where the spirit was first made.

"Certain important cases of confusion have been detected in grammar, as between the numeral *one*, derived from an Aryan root, through the Latin *unus*, and the indeterminate pronoun one (as in '*one* ought to do *one's* duty'), which is really a corrupt form of the French word *homme* or man. The Germans to the present day use *man* in this sense, as in '*man sagt*,' *i.e.*, one says.

"(*b*) By far the largest part of equivocal words have become so by a *transfer of the meaning* from the thing originally denoted by the word to some other thing habitually connected with it so as to become closely associated in thought. Thus in Parliamentary language the House means either the chamber in which the members meet, or it means the body of members who happen to be assembled in it at any time. Similarly the word *church* originally denoted the building (κυριακόν, the Lord's House) in which religious worshippers assemble, but it has thence derived a variety of meanings; it may mean a particular body of worshippers accustomed to assemble in any one place, in which sense it is used in Acts xiv. 23; or it means any body of persons holding the same religious opinions, and connected in one organization, as in the Anglican, or Greek, or Roman Catholic Church; it is also sometimes used so as to include the laity as well as the clergy; but more generally perhaps the clergy and religious authorities of any sect or country are so strongly associated with the act of worship as to often be called the church *par excellence*. It is quite evident, moreover, that the word entirely differs in meaning according as it is used by a member of the Anglican, Greek, Roman Catholic, Scotch Presbyterian, or any other existing church.

"The word *foot* has suffered several curious but very evident transfers of meaning. Originally it denoted the foot of a man or an animal, and is probably connected in a remote manner with the Latin *pes, pedis*, and the Greek πούς, ποδός ; but since the length of the foot is naturally employed as a rude measure of length it came to be applied to a fixed measure of length ; and as the foot is at the bottom of the body the name was extended by analogy to the foot of a mountain, or the feet of a table ; by a further extension, any position, plan, reason, or argument on which we place ourselves and rely, is called the foot or footing. The same word also denotes soldiers who fight upon their feet, or infantry, and the measured part of a verse having a definite length. That these very different meanings are naturally connected with the original meaning is evident from the fact that the Latin and Greek words for foot are subject to exactly similar series of ambiguities.

"It would be a long task to trace out completely the various and often contradictory meanings of the word *fellow*. Originally a fellow was what *follows* another, that is, a companion ; thus it came to mean the other of a pair, as one shoe is the fellow of the other, or simply an equal, as when we say that Shakespeare 'hath not a fellow.' From the simple meaning of companion, again, it comes to denote vaguely a person, as in the question, ' What *fellow* is that ? ' but then there is a curious confusion of depreciatory and endearing power in the word ; when a man is called a *mere fellow*, or simply a *fellow* in a particular tone of voice, the name is one of severe contempt ; alter the tone of voice of the connected words in the least degree, and it becomes one of the most sweet and endearing appellations, as when we speak of a dear or good fellow. We may still add the technical meanings of the name as applied in the case of a Fellow of a College or of a learned society.

"Another good instance of the growth of a number of different meanings from a single root is found in the word *post*. Originally a post was something *posited*, or placed firmly in the ground, such as an upright piece of wood or stone ; such meaning still remains in the cases of a lamp-post, a gate-post, signal-post, etc. As a post would often be used to mark a fixed spot of ground, as in a mile-post, it came to mean the fixed or appointed place where the post was placed, as in a military post, the post of danger, or honor, etc. The fixed places where horses were kept in readiness to facilitate rapid travelling during the times of the Roman empire were thus called posts, and thence the whole system of arrangement for the conveyance of persons or news came to be called *the posts*. The name has retained an exactly similar meaning to the present day in most parts of Europe, and we still use it in post-chaise, post-boy, post-horse, and postilion. A system of post conveyance for letters having been organized for about two centuries in England and other countries, this is perhaps the meaning most closely associated with the word post at present, and a number of expressions have arisen, such as post-office, postage, postal-guide, postman, postmaster, postal-telegraph, etc. Curiously enough, we now have iron letter-posts, in which the word post is restored exactly to its original meaning.

"Although the words described above were selected on account of the curious variety of their meanings, I do not hesitate to assert that the majority of common nouns possess various meanings in greater or less number. Dr. Watts, in his 'Logic,' suggests that the words book, Bible, fish, house, and elephant are univocal terms, but the reader would easily detect ambiguities in each of them. Thus fish bears a very different meaning in natural history from what it does in the mouth of unscientific persons, who include under it not only true fishes, but shell-fish or mollusca, and the cetacea, such as whales and seals, in short, all swimming animals, whether they have the character of true fish or not. Elephant, in a stationer's or bookseller's shop, means a large kind of paper instead of a large animal. Bible sometimes means any particular copy of the Bible, sometimes the collection of works constituting the Holy Scriptures. The word man is singularly ambiguous; sometimes it denotes man as distinguished from woman; at other times it is certainly used to include both sexes; and in certain recent election cases lawyers were unable to decide whether the word man, as used in the Reform Act of 1867, ought or ought not to be interpreted so as to include women. On other occasions *man* is used to denote an adult male as distinguished from a boy, and it also often denotes one who is emphatically a *man* as possessing a masculine character (heroic). Occasionally it is used in the same way as groom, for a servant, as in the proverb, "Like master, like man." At other times it stands specially for husband.

"(c) Among ambiguous words we must thirdly distinguish those which derive their various meanings in a somewhat different manner, namely, by analogy or real resemblance. When we speak of a sweet taste, a sweet flower, a sweet tune, a sweet landscape, a sweet face, a sweet poem, it is evident that we apply one and the same word to very different things; such a concrete thing as lump-sugar can hardly be compared directly with such an intellectual existence as Tennyson's *May Queen*. Nevertheless, if the word sweet is to be considered ambiguous, it is in a different way from those we have before considered, because all the things are called sweet on account of a peculiar pleasure which they yield, which cannot be described otherwise than by comparison with sugar. In a similar way we describe a pain as sharp, a disappointment as bitter, a person's temper as sour, the future as bright or gloomy, an achievement as brilliant; all these adjectives implying comparison with bodily sensations of the simplest kind. The adjective *brilliant* is derived from the French *briller*, to glitter or sparkle; and this meaning it fully retains when we speak of a brilliant diamond, a brilliant star, etc. But by what subtle analogy is it that we speak of a brilliant position, a brilliant achievement, brilliant talents, brilliant style! We cannot speak of a clear explanation, indefatigable perseverance, perspicuous style, or sore calamity, without employing in each of these expressions a double analogy to physical impressions, actions, or events."

Continuing the discussion in a later chapter on "The Growth of Language," the same author goes on to show how these ambiguities have originated. It is a matter of considerable importance to the logician to understand the process, and we reproduce, at length, Jevons's treatment of it:

THE AMBIGUITY OF TERMS

"There are two great and contrary processes," he says, "which modify language, as follows:

"1. GENERALIZATION, by which a name comes to be applied to a wider class of objects than before, so that the extension of its meaning is increased, and the intension diminished.

"2. SPECIALIZATION, by which a name comes to be restricted to a narrower class, the extension being decreased and the intension increased.*

"The first change arises in the most obvious manner from our detecting a resemblance between a new object which is without a name and some well-known object. To express the resemblance we are instinctively led to apply the old name to the new object. Thus we are well acquainted with *glass*, and if we meet any substance having the same glassy nature and appearance we shall be apt at once to call it a kind of glass. The word *coal* has undergone a change of this kind; originally it was the name of charked or charred wood, which was the principal kind of fuel used five hundred years ago. As mineral coal came into use it took the name from the former fuel, which it resembled more nearly than anything else, but was at first distinguished as sea-coal or pit-coal. Being now the far more common of the two, it has taken the simple name, and we distinguish charred wood as charcoal. Paper has undergone a like change : originally denoting the *papyrus* used in the Roman empire, it was transferred to the new writing material, made of cotton or linen rags, which was introduced at a quite uncertain period. The word character is interesting on account of its logical employment ; the Greek χαρακτήρ denoted strictly a tool for engraving, but it was transferred by association to the marks or letters engraved with it, and this meaning is still retained by the word when we speak of Greek *characters*, Arabic *characters*, *i.e.*, figures or letters. But inasmuch as objects often have natural marks, signs, or tokens which may indicate them as well as artificial characters, the name was generalized and now means any peculiar or distinctive mark or quality by which an object is easily recognized.

* For explanation of the terms *intension* and *extension* see Chapter V.

"Changes of this kind are usually effected by no particular person and with no distinct purpose, but by a sort of unconscious instinct in a number of persons using the name. In the language of science, however, changes are often made purposely, and with a clear apprehension of the generalization implied. Thus *soap* in ordinary life is applied only to a compound of soda or potash with fat, but chemists have purposely extended the name so as to include any compound of a metallic salt with a fatty substance. Accordingly there are such things as *lime-soap* and *lead-soap*, which latter is employed in making common diachylon plaster. Alcohol at first denoted the product of ordinary fermentation commonly called spirits of wine, but chemists having discovered that many other substances had a theoretical composition closely resembling spirits of wine, the name was adopted for the whole class and a long enumeration of different kinds of alcohol will be found in Dr. Roscoe's lessons on chemistry. The number of known alcohols is likewise subject to indefinite increase by the progress of discovery. Every one of the chemical terms, acid, alkali, metal, alloy, earth, ether, oil, gas, salt, may be shown to have undergone great generalizations.

"In other sciences there is hardly a less supply of instances. A lens originally meant a lenticular-shaped or double convex piece of glass, that being the kind of glass most frequently used by opticians. But as glasses of other shapes came to be used along with *lenses*, the name was extended to concave or even to perfectly flat pieces of glass. The words lever, plane, cone, cylinder, arc, conic section, curve, prism, magnet, pendulum, ray, light, and many others, have been similarly generalized.

"In common language we may observe that even proper or singular names are often generalized, as when in the time of Cicero a good actor was called a Roscius, after an actor of pre-eminent talent. The name Cæsar was adopted by the successor of Julius Cæsar as an official name of the Emperor, with which it gradually became synonymous, so that in the present day the kaisers of Austria and the czars of Russia both take their title from Cæsar. The celebrated tower built by the king of Egypt on the island of Pharos, at the entrance of the harbor of Alexandria, has caused light-houses to be called *phares* in French, and *pharos* in obsolete English. From the celebrated Roman general, Quintus Fabius Maximus, any one who avoids bringing a contest to a crisis is said to pursue a Fabian policy.

"In science also singular names are often extended, as when fixed stars are called distant *suns*, or the companions of Jupiter are called his *moons*. It is, indeed, one theory, and a probable one, that all general names were created by the process of generalization going on in the early ages of human progress. As the comprehension of general notions requires higher intellect than the apprehension of singular and concrete things, it seems natural that names should at first denote individual objects and should afterward be extended to classes. We have a glimpse of this process in the case of the Australian natives, who had been accustomed to call a large dog *cadli*, but when horses were first introduced into the country they adopted this name as the nearest description of a horse. A very similar incident is related by Captain Cook of the natives of Otaheite. It may be objected, however, that a certain process of judgment must have been exerted before the suitability of a name to a particular thing could have been perceived, and it may be considered probable that specialization as well as generalization must have acted in the earliest origin of language much as it does at present.

"*Specialization* is an exactly opposite process to generalization, and is almost equally important. It consists in narrowing the extension of meaning of a general name, so that it comes to be a name only of an individual or a minor part of the original class. It is thus we are furnished with the requisite names for a multitude of new implements, occupations, and ideas with which we deal in advancing civilization. The name physician is derived from the Greek φυσικός, natural, and φύσις, nature, so that it properly means one who has studied nature, especially the nature of the human body. It has become restricted, however, to those who use this knowledge for medical purposes, and the investigators of natural science have been obliged to adopt the term *physicist*. The name naturalist has been similarly restricted to those who study animated nature. The name *surgeon* originally meant handicraftsman, being a corruption of *chirurgeon*, derived from the Greek χειρουρνός, handworker. It has long been specialized, however, to those who perform the mechanical parts of the sanatory art.

"Language abounds with equally good examples. Minister originally meant a servant, or one who acted as a *minor* of another. Now it often means, specially, the most important man in the kingdom. A chancellor was a clerk, or even a doorkeeper, who sat in a place separated by bars or *cancelli* in the offices of the Roman Emperor's palace; now it is always the name of a high or even the highest dignitary. Peer was an equal (Latin *par*), and we still speak of being tried by our peers; but now, by the strange accidents of language, it means the few who are superior to the rest of the Queen's subjects in rank. Deacon, bishop, clerk, queen, captain, general, are all words which have undergone a like process of specialization. In such words as telegraph, rail, signal, station, and many words relating to new inventions, we may trace the progress of change in a lifetime.

"One effect of this process of specialization is very soon to create a difference between any two words which happen from some reason to be synonymous. Two or more words are said to be synonymous (from the Greek σύν, with, and ὄνομα, name) when they have the same meaning, as in the case, perhaps, of teacher and instructor, similarity and resemblance, beginning and commencement, sameness and identity, hypothesis and supposition, intension and comprehension. But the fact is that words commonly called synonymous are seldom perfectly so, and there are almost always shades of difference in meaning or use, which are explained in such works as Crabb's 'English Synonyms.' A process called by Coleridge DESYNONYMIZATION, and by Herbert Spencer, DIFFERENTIATION, is always going on, which tends to specialize one of a pair of synonymous words to one meaning and the other to another. Thus wave and billow originally meant exactly the same physical effect, but poets have now appropriated the word 'billow,' whereas wave is used chiefly in practical and scientific matters. Undulation is a third synonym, which will probably become the sole scientific term for wave in course of time. Cab was originally a mere abbreviation of cabriolet, and therefore of similar meaning, but it is now specialized to mean almost exclusively a hackney cab. In America car is becoming restricted to the meaning of a railway car.

"It may be remarked that it is a logical defect in a language to possess a great number of synonymous terms, since we acquire the habit of using them indifferently, without being sure that they are not subject to ambiguities and obscure differences of meaning. The English language is especially subject to the inconvenience of having a complete series of words derived from Greek or Latin roots nearly synonymous with other words of Saxon or French origin. The same statement may, in fact, be put into Saxon or classical English; and we often, as Whately has well remarked, seem to prove a statement by merely reproducing it in altered language. The rhetorical power of the language may be increased by the copiousness and variety of diction, but pitfalls are thus prepared for all kinds of fallacies.

"In addition to the effects of generalization and specialization, vast additions and changes are made in language by the process of *analogous or metaphorical extension* of the meaning of words. This change may be said, no doubt, to consist in generalization, since there must always be a resemblance between the new and old applications of the term. But the resemblance is often one of a most distant and obscure kind, such as we should call analogy rather than identity. All words used metaphorically, or as similitudes, are cases of this process of extension. Thus the old similitude of a ruler to the pilot of the vessel gives rise to many metaphors, as in speaking of the Prime Minister being at the helm of the state. The word governor, and all its derivatives, is, in fact, one result of this metaphor, being merely a corrupt form of *gubernator*, steersman. The words compass, pole-star, ensign, anchor, and many others connected with navigation, are constantly used in a metaphorical manner. From the use of horses and hunting we derive another set of metaphors; as, in taking the reins of government, overturning the government, taking the bit between the teeth, the government whip being heavily weighted, etc. No doubt it might be shown that every other important occupation of life has furnished its corresponding stock of metaphors.

"It is easy to show, however, that this process, besides going on unconsciously at the present day, must have acted throughout the history of language, and that we owe to it almost all, or probably all, the words expressive of refined mental or spiritual ideas. The very word *spirit*, now the most refined and immaterial of ideas, is but the Latin *spiritus*, a gentle breeze or breathing; and inspiration, *esprit*, or wit, and many other words, are due to this metaphor. It is truly curious, however, that almost all the words in different languages denoting mind or soul imply the same analogy to breath. Thus, *soul* is from the Gothic root denoting a strong wind or storm; the Latin words *animus* and *anima* are supposed to be connected with the Greek, ἄνεμος, wind; ψυχή is certainly derived from ψύχω, to blow; πνεῦμα, air, or breath, is used in the New Testament for Spiritual Being; and our word ghost has been asserted to have a similar origin.

"Almost all the terms employed in mental philosophy or metaphysics, to denote actions or phenomena of mind, are ultimately derived from metaphors. Apprehension is the putting forward of the hand to take anything; comprehension is the taking of things together in a handful; extension is the spreading out; intention, the bending to; explication, the unfolding; application, the folding to; proposition, the placing before; intuition, the seeing into; and they might be almost indefinitely extended. Our English name for reason, the understanding, obviously contains some physical metaphor which has not been fully explained; with the Latin *intellect* there is also a metaphor.

"Every sense gives rise to words of refined meaning; sapience, taste, insipidity, goût, are derived from the sense of taste; sagacity, from the dog's extraordinary power of smell; but as the sense of sight is by far the most acute and intellectual, it gives rise to the larger part of language; clearness, lucidity, obscurity, haziness, perspicuity, and innumerable other expressions, are derived from this sense.

"It is truly astonishing to notice the power which language possesses by the processes of generalization, specialization, and metaphor, to create many words from one single root. Professor Max Müller has given a remarkable instance of this in the case of the root *spec*, which means sight, and appears in the Aryan languages, as in the Sanscrit *spas*, the Greek σκέπτομαι, with transposition of consonants in the Latin *specio*, and even in the English *spy*. The following is an incomplete list of the words developed from this root: Species, special, especial, specimen, spice, spicy, specious, specialty, specific, specialization, specie (gold or silver), spectre, specification, spectacle, spectator, spectral, spectrum, speculum, specular, speculations. The same root also enters into composition with various prefixes; and we thus obtain a series of words, suspect, aspect, circumspect, expect, inspect, prospect, respect, retrospect, introspection, conspicuous, perspicuous, perspective; with each of which, again, a number of derivatives is connected. Thus from suspect we derive suspicion, suspicable, suspicious, suspiciousness. I have estimated that there are in all at least two hundred and forty-six words employed at some period or another in the English language which undoubtedly come from the root *spec*."*

Jevons's discussion suffices to illustrate quite fully the fact of ambiguous conceptions and the laws of their development, but it does not indicate those instances which are likely to

---

* For a more complete study of the ambiguity of terms and the origin of it, the student may consult the following works: Locke: Essay on Human Understanding, Book III., Chapters IX. and X.; Mill: Logic, Book IV., Chapter V.; Trench: On the Study of Words · Max Müller: Lectures on the Science of Language.

give trouble in the study of Logic. It is perhaps just as well that we should first understand the wide extent to which words vary in meaning, and to which they are modifiable by generalization and specialization, before considering more particular classes affecting logical problems. We can in that way have the general laws regulating or causing their ambiguity most distinctly impressed upon our mind, inasmuch as they affect the meaning of terms even where logical confusion may not be the consequence of the change. Most of the conceptions chosen by Jevons in illustration would very seldom give rise to any serious fallacy in reasoning, as, in spite of a certain kind of ambiguity, they are well enough understood to prevent serious logical mishaps. But this is not the case with a very large set of conceptions current in science and philosophy, general conceptions whose technical meaning varies with the schools arguing for or against certain doctrines involved in them. They belong largely to the class of terms which are either very abstract in their import, or are liable to the confusion of their abstract with their concrete conception. This is a source of ambiguity already touched upon, and which Jevons does not treat of. It is perhaps a source of more logical errors than all the ambiguities his discussion illustrates, although it comes under the same laws as those which he does notice. Pure abstracts and pure concretes, as adjectival nouns and proper or singular names, are not likely to give much difficulty. Fallacies are much more incident to mixed abstract and concrete conceptions, or those concrete terms, usually so-called, whose different meanings are either closely allied, or connected with closely resembling objects having nevertheless important differences between them. Where the separate concrete meanings of the same term have no natural affiliation, they will not easily give rise to error. Thus "spirit," denoting *mind* on the one hand, and *alcohol* on the other, will not be used in connections where fallacy will be the consequence of such a possible ambiguity. But "spirit," denoting mind or intelligence in one case, and something immaterial in another, may be the source of error, because the two conceptions are closely related. The process of generalization gives rise to a tendency away from the concrete and may produce confusion as long as the new and old conceptions are not clearly defined. An opposite source of error is that of the process of specialization, which is from the more abstract to the concrete ; in so far as the term "abstract" is taken to denote the common qualities denoted by the general concept apart from the particular, concrete, and differentiated individual to which it may also apply. The generalization and specialization indicated have to do, almost exclusively, with the terms that may be mixed abstracts and concretes, and the source of confusion increases as the indefiniteness of a conception increases with the process of generalization.

A highly illustrative set of ambiguous terms, more important to scientific and philosophic study than any which Jevons has stated, and which yet come under the laws which he enunciates, are such as the following : *Motive, intuition, experience, idea, cause, God, religion, faith, feeling, knowledge, sensation, reason, first principles, a priori, government, law, nature, moral, right, justice, nation, church, authority, origin, freedom,* etc. Thus motive may denote in mechanics the cause or force producing motion, and by analogy the cause of volition ; then it may denote the idea of an end, which, in so far as it is the object of the mind, is rather the effect than the cause of volition. Intuition may denote immediate perception, or again it may denote universal perception, the first indicating only a direct process of knowledge without reference to time or the number of persons involved, and the second indicating that all individuals experience it. The term has, further, four or five other meanings, but these two suffice to illustrate an ambiguity which might well give rise to all the controversies waged about certain doctrines in philosophy. Add to this the still greater ambiguity of the terms "experience" and "idea," and we can well imagine why so much dispute has centred about the theory of "innate ideas." On the one hand, experience has meant, first, simple sensations ; second, any realized state of consciousness ; third, a series of mental states giving as a resultant a new component not found in the primary state ; on the other hand, idea has meant, first, simple presentations of sense (Locke) ; second, general conceptions which are the product of the higher intellectual faculties; and third, mere opinion. These are sufficient to make a perfect labyrinth of complexities in argument. Again, cause is sometimes merely the condition of anything, and at others the active agent producing a phenomenon. In this latter sense it may be either some influence external to the thing affected, or, as in the case of the mind and its volition, or of any agent which contributes of its own energy to the effect, it may be internal. God sometimes denotes the first cause without indicating whether it is more than force ; then it may denote the agent effecting the order of the world, whether regarded or not as the creator of matter ; again the term may denote a supreme intelligence, with various attributes of perfection and power, not necessarily implied in the former conceptions, but not excluded by them. Religion may denote certain beliefs about God and the world, or it may denote a certain attitude of mind toward these things, or it may further denote the belief and practice of certain moral doctrines, with various subordinate meanings. Faith has at least four distinct meanings : First, intellectual assent to propositions above the attestation of reason, and thus equivalent to intuition, if it gives immediate assured knowledge, or to mere probability, if it can only produce less than absolute conviction or certitude ; second, it may denote the acceptance of truth on authority ; third, fidelity of disposition in living up to a promise, treaty, or a law ; and fourth, trust in a person. Feeling has a similar application, now denoting a tactual sensation, now a firm and ineradicable conviction = intuition ; again, a variable mental state of excitement which can easily be eradicated from an influence upon knowledge, and so is equivalent to emotion, and lastly, a general conception for the primary elements of knowledge. The remaining terms and many others possess similar multiplications of meaning, which are the source of all the controversies and their incident fallacies in the world of speculative knowledge. The few that we have specified may serve to illustrate the importance of being on the alert for them, and of first giving a term that analysis or definition of its meaning which is at least a partial provision against the contingency of error.

To recapitulate. First, words are ambiguous when they are capable of more than a single meaning. Second, they become ambiguous through the process of transition from a generalized to a specialized form, or the reverse. Third, this process causes confusion in Logic mainly when it results in the liability to mistake the abstract for the concrete, or one concrete conception for another closely allied to it.

CHAPTER V.

THE INTENSION AND THE EXTENSION OF CONCEPTS

1st. Nature of Intension and Extension.—Allusion has been made to the intension and the extension of conceptions without explaining the meaning of those terms. We come now to determine this meaning, which is a very important matter in Logic. Various terms have been employed to express the same fact and relation as are expressed by extension and intension. Thus comprehension, depth, connotation are frequently taken as synonymous with intension, and extent, breadth, denotation for the extension of terms. Much controversy exists about the true use of the terms denotation and connotation, which I shall consider later in the chapter. The only matter of real importance is the meaning of the concepts which they are supposed to describe, and this can be determined, in a large measure at least, without complicating ourselves with this controversy.

Nearly all, if not absolutely all terms have a twofold meaning or application, which is expressed by their intension and extension. A simple example will make this apparent. The term "man" may apply to all the individual men represented by the word. It is thus a name for the individuals in the class, and we should call each one by that name when asked to define what he is. On the other hand, the term also expresses a certain number of qualities or marks which make up the individual. "Man" is thus not only a name for the individuals of the class, but for a certain conjunction of qualities, which may be thought of without regard to the range of application possessed by the term. Tree has a similar application. It may be a name for a class of objects, or it may denote a certain number of vegetable qualities. Metal, quadruped, biped, vertebrate, triangle, figure, nation, city, custom, etc., are only other instances of the same fact. There are some terms, however, which are not class concepts; for example, singular or proper names. They seem only to denote a combination of qualities, and it is true that they cannot denote more than one individual. But this does not hinder it from having a numerical application to that extent, and this is all that is necessary to justify the application of extension to it. In all these terms *the intension refers to the quality or qualities possessed by an object* having a given name; *the extension refers to the number of objects* included under the name. These will at least serve as approximate definitions of the two terms until completed. We require at present merely to know that the extension indicates the objects to which a name *applies*, and the intension, the attributes which it *implies*.

The clearest illustration of terms with extension will be class concepts, and especially all concrete general concepts, as "man," "quadruped," "Caucasian," "European," "tree," "book," "animal," "house," etc., and perhaps the clearest illustration of intension will be pure abstract terms and singular or proper names, which latter, although they denote an individual in a class, more particularly indicate a certain quality or union of qualities that are thought of rather than the *range* of the term, as "Lincoln," "Berlin," "Rome," "Europe," "Plato," "Declaration of Independence," "Magna Charta," etc. But all terms have both references at the same time, and differ only in the degree of their intension and extension in relation to each other. That is, every term has both intension and extension. The extension of "man" is the number of individuals to which it is applicable as a name; the intension is the number of qualities which it implies, and so with other terms.

The application of extension and intension to concrete conceptions, singular or general, affords no difficulty. But is it possible to apply them to abstract terms? The answer to this question involves the previous questions whether abstract terms are singular or general, and whether singular terms can be said to have extension. Inasmuch as I have claimed singular concepts for extension, merely indicating that the extension is at its *minimum* in them, there remains only to settle whether abstract terms are singular or general. Jevons thinks them singular. This may possibly be the case in some instances. But it is certainly not the case in such terms as "sweetness," "justice," "ability," "color," etc. For there are several kinds of "sweetness," several divisions of "justice," and various forms of "ability" and "color." Perhaps some writers would make "color" a concrete concept. I hold that it may be either concrete or abstract, according to its reference. But it is not important to settle this question, whether it may be both, or is only one of them. If it be concrete it certainly has extension and intension together, and if it be abstract and general, as it must be, both extension and intension are characteristic of it; so that the two qualities can be denied of it only on the supposition that it is abstract and singular, and that all singulars have only intension. But with the qualification already mentioned these properties are possessed by both concrete and abstract terms, so that the reduction of abstract concepts to singulars would be no obstacle to their simultaneous possession of intension and extension.

It remains to consider whether *adjectives* can possess intension and extension. In so far as they are the names of qualities they indicate some degree of intension. But are they class terms? There are two ways of answering this question. The first is Mill's view that attributive terms of this kind always imply a subject. Thus "white" implies all white objects, and so must represent a class. The second is the view that, even if they do not necessarily imply anything definite about things, they may connote different kinds or degrees of the quality expressed by them. Thus there are many shades of "white," "blue," "green;" many kinds or degrees of "pure," "noble," "benevolent," etc. In this sense they will also have extension. And also they would possess this quality if they were singular, according to the principle I have asserted for the minimum extension of a term, and it is not necessary, perhaps, to take account of any other degree of it.

It may also be a question whether any great importance attaches to either the affirmation or the denial of extension to abstract terms and attributives. The chief interest of the logician, probably, is in the possibility of representing them by the usual symbols, the logical circles, in illustrating the relation between the subject and predicate of a proposition. But as this symbolization is only one of analogy, and as the legitimacy of it even is disputed by some, we may disregard the question whether it be strictly applicable to abstract nouns and to adjectives, and, if we desire, leave the whole matter undecided, so far as practical Logic is concerned, because fallacies do not, to any extent, turn upon the question whether abstract terms and adjectives are capable of extension or not. In purely theoretical Logic it may be somewhat different. There, we may be interested merely in the truth about this special case, and undoubtedly we shall find the matter of extension less clear in its application to abstract terms and adjectives, even if it be granted as possible, than to concrete substantives. For, being terms which merely imply a subject, but do not express it definitely, they will most likely represent either a single quality with its various degrees, or only such as the substantive from which they are taken represents. Of course some adjectives are not taken from substantives. But many of those that are so taken, as "manly," "human," "animal," "personal," "heavenly," undoubtedly connote as many qualities as the substantive, and so equal it in its intension at least, and it is possible to say that their extension is also equal to that of the terms from which they are derived; although it might be proper to regard it as a sort of *relative* or *derivative* extension. Certainly they do not as terms indicate individuals so distinctly as substantive concretes. But even if they do not, Logic does not require us to give them more than the minimum extension, and this will apply to all concepts representing any idea whatever. This is to say that extension must not be confused with the notion of plurality.

After this discussion we may conclude the section with a more accurate definition of extension and intension. The former is usually described as referring to the individuals of a class, and this conception of it leaves the impression that in order to have extension at all a term must be a class concept. But this is only an incident of the disproportion that may exist between the two properties. But to cover the case more distinctly they may be defined as follows:

Extension is *the quantitative power of terms or concepts*, and so indicates their numerical application. It may refer to individual- or class-wholes, whether substantive or attributive, real or conceptual.

Intension is *the qualitative power of terms or concepts*, and so indicates their denotation of qualities. It may refer to a single quality, a group of qualities in an individual-whole, or the common quality or qualities of a class-whole, whether substantive or attributive, real or conceptual.

Conceptual-wholes are such as denote thought products which are not conceived properly as either substantive or attributive. Illustrations are such concepts as "proposition," "word," "syllogism," "science," "botany," etc. Some terms describing mental states may be regarded as attributive; *e.g.*, "sensation," "memory," etc.

2d. The Relation between Extension and Intension. —The relation between the intension and the extension of concepts is determined by comparing the broader with the narrower term. Thus, if we take the term "metal" and compare it with the term "iron," we shall find that "metal" is a name for a larger number of objects than the term "iron," because it includes all that is denoted by "iron," and all other metals besides. The extension of metal is, therefore, said to be greater than the extension of "iron." But the intension of "iron" is greater than that of "metal," because it contains all the qualities necessary to regard it as a "metal," and in addition the quality or qualities necessary to make it "iron" and to distinguish it from other metals, such as gold, silver, lead, etc. Again, "matter" will have greater extension than "metal," and "steel" less than "iron." But the intension of "steel" will be greater than that of "iron," and the intension of "matter" less than that of "metal." The same comparison can be instituted between any set of related terms, such as biped, man, European, Frenchman, Louis XIV., or vertebrate, quadruped, horse, Bucephalus; or figure, quadrilateral, square.

From this we deduce the general law that *as the extension increases the intension decreases, and vice versa.* The same law is sometimes expressed in a different way. Thus, *extension and intension vary in an inverse ratio to each other*, or they are *inversely related to each other*. Hamilton represented the relation by a cone or pyramid, in which the apex indicated the least intension or extension, as the case might be, and the base the greatest. We may thus symbolize the relation for a series of terms, and we may indicate both the order of greatest and least extension, the greatest and least intension, and their reciprocal relation inversely considered. The following figures will represent them:

Fig. 1 represents the order of least and greatest extension, beginning with the apex of the triangle or cone. "Plato" has the least extension of the series and applies to only one individual. "Vertebrate" has the greatest extension of the series, including all the others and all other beings having a certain anatomical structure. In Fig. 2 the order is reversed. "Vertebrate" has the least intension because it stands for the fewest qualities, and "Plato" the greatest because it denotes the largest number of qualities. Fig. 3 represents the reciprocal relation of the two properties. "Plato" has at the same time the greatest intension and the least extension, and "vertebrate" the greatest extension and the least intension. The intermediate terms vary between these extremes by an indeterminate ratio, but are presumably in the same relation.

We must bear in mind, however, that the formula for this relation is not strictly accurate. It is not true that the extension *always* increases as the intension decreases. Some logicians, in their objections to the law thus enunciated, go so far as to say that the relation in some instances may be reversed, so that the intension would increase as the extension increases. This, however, is an exceptional state of the case and may be dismissed from consideration of the general rule. The law is, perhaps, not meant by any logician to be absolutely and universally true in the strict sense in which it is sometimes expressed. The formula is a convenient one for indicating a relation sometimes strictly true, and sometimes true with qualifications. It is enunciated in mathematical terms for the sake of clearness rather than because it is literally true in all instances of comparison, although in the ideal logical world it might be so. Nevertheless, the absolute and universal application of the law, as we have formulated it, is subject to the following limitations :

1. The law cannot be interpreted in any strict mathematical sense. It is, for instance, not true that when the intension is doubled the extension is halved. The number of individuals may even be increased without decreasing the intension of either the class term or of the individual under it. Thus I may increase the number of persons to whom the term "man" is applicable, and still not alter the quantity of intension represented by the term.

Some logicians might reply that, in fact, this never takes place ; that every individual added to a class differs by some mark from all others, and so may decrease the intension of the general term in the same proportion, at least, that its extension is increased. This is conceivably the case, and, if a fact, would sustain the law in its general sense, although it might not prove its strict or definite mathematical interpretation in terms of any specified ratio. Besides, the law is perhaps defensible as a general formula in the same sense in which psychologists speak of the inverse ratio between sensation and perception, or between the consciousness of feeling and the consciousness of an object, where we wish merely to express the fact that the two do not vary together in the same way, but that as one becomes more distinct the other becomes less so. This mode of expression may be applicable to the variations, at least as a general rule, between extension and intension, and be relatively true when one increases without the other decreasing, and absolutely true, although not by any assignable ratio, when the increase of the one is accompanied by a corresponding decrease of the other. Nevertheless, the variation is so irregular that the law has only a conventional value in the form in which it is usually enunciated.

Perhaps a distinction between two kinds of general terms would enable us to formulate the law in different ways, one of them to suit its simple mathematical conception, and the other to suit a less definite conception of it. It is a fact that *general* terms or concepts are of two distinct kinds, although the distinction is not explicitly recognized by logicians. The two divisions I shall call *mathematical* generals and *logical* generals. The latter term is perhaps an unfortunate one, because all generals are logical in the broadest sense. But I defend its use in the technical sense to be defined as the only resort at my command for expressing the notion I have of the terms described by it. By mathematical general concepts I mean those which are absolutely alike in their content or intension, or such as are grouped together under the same name solely on account of their numerical value. In such cases the extension may be increased either without altering the intension, or in connection even with an increase of the intension, supposing, of course, that the increase of the intension was the same in all individuals of the class. Thus if I assume a number of gold coins exactly alike, I may add to the number denoted by a particular name any number of like coins with exactly the same qualities, and here I increase the extension without increasing the intension. And again, if in adding a new individual to the class, I discover a quality not known before, but which is found to belong to all members of the class, I have increased the intension of the concept (not of the thing) in the same proportion as I have increased the extension. If the new individual has a new property not in the others of the same class, the intension of the general term or genus may remain the same, or fixed, while the extension is increased. Hence wherever the addition to the general concept is a purely numerical or mathematical one the law of inverse ratio between extension and intension does not hold in its strict sense, and can only be taken as a general statement of the independence of each other in their variations.

But in the case of the second class, namely, *logical* concepts, the ratio of variation between extension and intension may more definitely accord with the statement of the law. By a logical general concept I mean one which does not apply in exactly the same sense to the individuals or species which it comprehends, or which strictly connotes only the *common* qualities of a class, and does not denote the differences. The difficulty, however, in strictly defining them is that they are usually applied in a mathematical sense at the same time, and hence we may have no concepts which denote only the common qualities of a class without enumerating the individuals mathematically at the same time. This is the case with the mixed concrete and abstract terms. But such a general concept, if it actually exist in its purity, will be such as comprehends individuals and species not taken merely in an additive sense, but as denoting certain common qualities allowing great variations and differences in all other characteristics. Thus "vertebrate" is a general concept comprehending individuals and species, with such differences as do not distinguish one of them from another as comprehended in the general concept. It is the same with the general term "animal." It denotes a certain common quality or qualities which permit all sorts of other differences without disturbing or confusing the application of the term. In all such cases the increase of the extension of the concept may be either mathematical, or both mathematical and logical. In case it is mathematical, or the mere addition of an individual exactly like those already denoted by the term, the law is subject to the limitations mentioned in the instances of mathematical generals. But if the increase is due to the addition of a new species, presenting new characteristics compared with those denoted before, the term is *generalized* logically as well as mathematically, and its intension is decreased with its increase of extension. Thus the term "crow" for a long time denoted a certain class of birds with a black, glossy plumage. The color of their feathers was a common quality, being thus involved in the intension of the class. But as soon as "crows" with a white or gray color were discovered, this intension was decreased by throwing the quality of color out of regard as a distinctive characteristic, and the extension was increased in proportion to the new additions to the class. The generalization is accomplished by taking into account fewer common qualities, and hence the ratio between the intension and the extension varies inversely in all such cases. Again, the term "Frenchman" ordinarily denotes a person with certain race characteristics, and born in France. But we often find it used to denote persons having only the race qualities and born in any other locality. "Hebrew" once denoted the inhabitants of Palestine. It now denotes a race without any implications as to country. "Book" was once synonymous with "beechen boards," having printed or written matter upon them. It is now broadened to denote any mass of paper bound in a particular way, whether containing printed and written matter or not. The matter, indeed, is an entirely unessential characteristic, as it was once the essential. Here again the law is strictly illustrated, as the intension decreases with an increase of the extension. The same fact is clearly illustrated in all cases of generalization and specialization already discussed. Here the intension and extension vary inversely.

But it is important to remark the limitation with which this is true. The addition to the general term numerically may be manifold as great as the deduction from its intension. A hundred, more or less, individuals may be added to the extension and only one or two qualities subtracted from the intension. And even if only one individual with a difference is added to the class, and two or more qualities subtracted from the intension, the law is still true in the sense it was intended, although not true in the mathematical sense that the ratio of variation is the same for both intension and extension. In its proper logical meaning the law simply indicates, without any numerical implications, that as you extend the application of a term to new species you decrease the number of generic qualities comprehended by it. This will be quite uniformly true of *logical* extension, but variable and subject to modifications in *mathematical* extension.

The importance of this distinction and discussion will appear when we come to consider the question of genus and species, or essentia and differentia. At present it suffices merely to specify the qualifications under which the law is true.

2. The second limitation of the formula regarding the relation between extension and intension is that it is not applicable to all conceptions independently of the relation between genus and species. Thus the extension of "man" cannot be compared with that of "bear;" of "horse" with "lion;" of "government" with "science." Nor can any measure of the relation between the extension and the intension of such terms be determined. Hence the law cannot apply to terms taken promiscuously. It can apply only to conceptions which represent a superordinate and a subordinate notion, or a genus and a species. Neither genera nor species can be compared in this relation with their own kind, but only with each other. The comparison must be limited to terms representing different degrees of generalization. All concepts may thus be brought under one comprehensive head, but apart from such a relation their extension and intension will not be an object of determination at all. Hence this limitation of the law. But it should be remarked that this does not set aside the formula. It only qualifies it and its application. In its proper conception and under appropriate conditions it remains valid, although it may be possible to exaggerate its importance.

INTENSION AND EXTENSION OF CONCEPTS 79

But even those who criticise the doctrine admit considerable importance for the meaning which they assume is latent at the basis of it. Bosanquet, who subjects the formula to a somewhat searching criticism, virtually admits all that the law ever meant to express by saying that conceptions may vary their import and range of application in a way much as the doctrine of an inverse ratio asserts. He says: "It is certain that to abstract and to distinguish—to know what belongs to one relation, and what, again, though conjoined with that relation, yet does not arise out of it, but out of some other condition or cause—is the first duty of the scientific intelligence. In consequence of this activity arrangements of individual objects under a series of abstractions, each applying to a wider aggregate than the last, meet us on every hand and most obviously of all in family relationships as estimated among civilized nations."* When we take into account, therefore, that the diminution of intension, in any case, does not involve, necessarily, a decrease in the number of qualities in the individual or sub-class, but only the number of *common* qualities constituting the genus, as before determined, and so with the increase of the same, there will be nothing to seriously object to in the law, except its precise mathematical and promiscuous application. Its importance is therefore vindicated.

3d. The Denotation and Connotation of Terms.— Some logicians use this distinction as identical with that between the extension and the intension of terms. Others employ it in a somewhat different sense. Indeed, it is only this variation of usage and its frequent coincidence, or close connection with the application of extension and intension, that makes it necessary to consider the matter at all. I shall notice briefly the doctrines of Mill, Fowler, and Keynes.

Mill's distinction is between *connotative* and *non-connotative* terms. "A non-connotative term," he says, "is one which signifies a subject only, or an attribute only. A connotative term is one which denotes a subject and implies an attribute. By a subject is here meant anything which possesses attributes. Thus

* See Bosanquet's Logic, Vol. I., Introduction, § 8, pp. 46-71.

John, or London, or England, are names which signify a subject only. Whiteness, length, virtue, signify an attribute only. None of these terms, therefore, are connotative. But *white, long, virtuous*, are connotative." The distinction here made Mill holds to be a restoration of scholastic usage. As far as can be determined from his language, "denotative" merely means "significative;" for he does not define it. He practically identifies it with "non-connotative." "Connotative" he defines, according to scholastic usage, as equal to "*connotare*, to mark along with; to mark one thing *with*, or *in addition to* another." This identifies it with *imply*, or the implicative power of a term.

Objections can be produced against his classification of connotative and non-connotative terms on the basis of these definitions themselves. For instance, he says that proper names "do not indicate or imply attributes as belonging to individuals." This can be directly challenged. Bucephalus, The Secretary of State, Mont Blanc, quite distinctly imply attributes. It does not suffice to make a proper name that we merely capitalize a word. It is such only by virtue of its application to a single individual that is a group of attributes. But if, besides denoting a subject, it implies an attribute or attributes, it becomes connotative, according to the definition, though classified as non-connotative. On the other hand, adjectives, though classed as connotative, are as much the names of attributes alone as abstract nouns, and therefore should be non-connotative. Besides they do not *denote* a subject or *imply* an attribute. On the contrary, they denote an attribute and imply a subject. This would prevent them from being connotative, according to the definition. But if "connotative" be regarded as synonymous with implication, as seems to be the case with Mill, adjectives would be connotative, but so would abstract nouns for the same reason. Hence Mill must either classify these together, or change the definition of connotative. There is certainly, in this respect, an inconsistency between his definition and his classification, while at the same time the definition given does not describe the nature of adjectives.

It will be apparent in Mill's usage that the terms cannot be identical with that of extension and intension, and hence his doctrine either has a diminished importance or is too confusing in its implications to be of much service in the problem under consideration. Fowler and Keynes take the liberty to modify their application or meaning so as to coincide with extension and intension, "denotation" being identical with extension and "connotation" with intension. In this application they require no further discussion. If I might be allowed to introduce an innovation, I would prefer, since we already have the terms extension and intension, to employ "denotative" to indicate the application of a name to individuals without any reference to their qualities, and hence only to denote them as concrete wholes, numerically or mathematically considered. "Connotative" I would employ to *connote* or comprehend a union of attributes, or a class of individuals. It would thus apply to individual and general wholes, one as an aggregate of qualities and the other as an aggregate of individual objects. Denotative might apply to the names of singular attributes or singular individuals as such. In this way I should remain by the strict import of the word *connote*, and gain an economical term for a logical synthesis of any kind, and retain a separate one for all concepts representing an individual unit of any kind. In this sense I have used the terms where I found it necessary to employ them technically at all. But as no particular logical doctrine is dependent upon this usage, I do not urge either the adoption or the importance of it.*

*For general discussion of this question and the problem relating to the extension and intension of terms the student may consult the following references: Mill: Logic, Book I., Chap. II., § 5; Fowler: Deductive Logic, Chap. II.; Keynes: Formal Logic, Chap. II. ; Hamilton: Lectures on Logic, Lects. XI. and XII. ; Venn: Empirical Logic, Chap. VII. ; Bosanquet: Logic, Vol. I., Introduction, § 8, pp. 46-71 ; De Morgan: Formal Logic, Chap. XII.

CHAPTER VI.

DEFINITION AND DIVISION

1st. The Predicables.—Definition and Division are complicated processes, the former especially, since there are several kinds of definition. But to understand them and their place in logical science and disquisition we must examine into the nature and meaning of the so-called predicables. They are usually stated as five in number, as follows:

> Genus   (γένος)   = Genus.
> Species   (εἶδος)   = Species.
> Differentia (διαφορὰ)   = Difference.
> Proprium   (ἴδιόν)   = Property.
> Accidens   (συμβεβηκός) = Accident.

The most natural order of considering the predicables would be that in which they are stated. But as the meaning of the first three is dependent upon understanding the fourth, and perhaps the fifth, I shall begin with the fourth and return to the others.

1. PROPERTY.—This term is, for all practical purposes, synonymous with *quality* and *attribute*. Some writers endeavor to distinguish between them, but the distinction serves no important logical purpose. Hence by a property of a thing I mean *any quality, mark, characteristic, or attribute of it* which goes to make it what it is. Thus whiteness is a property of snow; hardness, of iron; yellowness, of gold; brilliancy, of a diamond; instability, of a liquid, etc. It is the same with any namable quality of an object. But qualities or properties are not all of the same value or importance to the existence of a thing. Hence they are usually divided into *essential* and *non-essential*. Essential properties are those which are necessary to the particular nature of an object, which would not be what

it is except for these qualities. For example, the essential property of a pen is that it be fit for writing ; of man, that he have life and consciousness ; of a lamp, that it be able to give light; of a tree, that it be of wood, have a trunk and branches, etc. ; of iron, that it have a certain density, metallic lustre, and molecular cohesion. In a great many cases it is difficult to assure ourselves that a given quality is essential rather than non-essential. This is because of the indeterminate extension of the term under notice. Thus we might consider sweetness as an essential property of sugar, and so it is in the common use of the term. But to the student of chemistry this is not necessarily the case, as we have seen that science will often generalize a term without carrying along with the increased extension the essential property of its narrower import. An instance of this is the term "metal." Originally it was supposed that a specific gravity greater than water was a necessary property of metals. But on the discovery that potassium, sodium, and lithium were metals because of their metallic lustre and structure, this property became one of their essential characters, and their specific gravity was disregarded by the term metal obtaining a more general application. We have, therefore, always first to determine the extension of a concept before indicating its essential properties. Fluidity will be an essential property of water, if we do not include the term ice in it. Animal fibre is an essential property of "meat," if we do not use the term synonymously with the term "food." But all this merely indicates how difficult it is to name *absolutely* the essential property denoted by a given term. We have first to settle what extension is given it, and this will be determined by the limits assigned to the presence of a given quality, and this quality will be an essential one so far as it is identified with the extension of the term, or made to determine that extension. This last modifying clause has to be added because some *universal* properties are regarded as non-essential. But the essential properties will determine the limits of extension for any given meaning, but will not stand in the way of a higher generalization eliminating that quality as essential, and substituting another in its place. But in this case the concept is materially changed, and is not the same as before. In particular cases, therefore, before naming the essential quality we must see that the term is clearly understood or that its extension is not greater and cannot be greater than the property to be specified.

A non-essential property is one which is not necessary to the concept or existence of a thing. For instance, whiteness is not an essential property of man; redness, of an apple; a specific length, of the sides of a triangle; iron material, of a ship, etc. The non-essential properties are called *accidents*, which are admirably defined and illustrated by Jevons. "An accident," he says, "is any quality which may indifferently belong or not belong to a class, as the case may be, without affecting the other qualities of the class. The word means that which *falls* or happens by chance, and has no necessary connection with the nature of the thing. Thus the absolute size of a triangle is a pure accident as regards its geometrical properties; for whether the side of a triangle be one-tenth of an inch or a million miles, whatever Euclid proves to be true of one is true of the other. The birthplace of a man is an accident concerning him, as are also the clothes in which he is dressed, the position in which he rests, and so on. Some writers distinguish between separable and inseparable accidents. Thus the clothes in which a man is dressed is a separable accident, because they can be changed, as can also his position, and many other circumstances; but his birthplace, his height, his Christian name, etc., are inseparable accidents, because they can never be changed, although they have no necessary or important relation to his general character."

Accidental properties are almost as indeterminate in particular cases as the essential. This is because the term may, like that, be relative in its import. That is, what is accidental in one relation may be essential in another. Thus muddiness may be an accidental quality of rivers, but an essential quality of a certain river. A better illustration is that of red hair. It is a purely accidental property of a man, as a man, but it may

DEFINITION AND DIVISION

be essential to him as an individual. There are instances, however, where the accident will hardly be considered as in any way essential even to the individual. But this may be controverted, and it is not necessary to sustain the position for our purposes. All we require to remember is that, usually at least, the accidents are relative to certain qualities regarded as essential, and that by narrowing the extension of a concept, we may make them essential in that sphere. The only place where an accident will be absolutely such is in the case of the individual or singular concept, and even here it may be disputed whether the distinction between essence and accident can be drawn. It may require for its existence a comparison with some other individual having common attributes and certain differences, in which case the common attributes would be the *essentia* of a common name and the difference would be the accidents of it.

If this be the case, however, it only shows the ambiguity of the term "accident." In fact, it has more than one signification. Sometimes it is used synonymously with any quality of an object, and in this case it is equivalent to property or attribute. Again, it is used synonymously with *differentia*, which is the property defining a species. In this sense it is accidental in one relation and essential in another. A third meaning is that which denotes a property in no way essential to either the genus, the species, or the individual. This is perhaps the sense in which it is usually taken by logicians. But it is sometimes relative in this usage. Thus seven feet stature might be an accident of a giant as a man, but necessary to his being a giant. But a case where this relative import is not apparent, if it exist at all, would be the size of the hand, the presence of a mole on the skin, liability to blush, etc., which would be regarded as accidents of the individual. This is the proper sense of the term, although it may be disputed whether it is ever absolute in its meaning even under the circumstances just considered. But it is not necessary to decide this point. Its relative import suffices to set aside certain properties which may be disregarded in determining the essentially important meaning of a concept. It only remains to remark that some *accidentiæ* are regarded as *universal* or coincident with the genus or species, as curly hair of the negro, and others are merely *casual* or contingent, as sickness, flatness of the nose, shape of the head, etc.

Some logicians use the distinction *peculiar* property to denote what belongs to a whole class and to that class only, as risibility in man. But this comes under the general head of universal accidents, as one kind of them, and needs no further consideration.

2. DIFFERENTIA.—Differentia, or difference, is the name of that particular property which distinguishes one species from another. For example, *bipedality*, or two-footedness, is a differentia of man as compared with quadrupeds; the possession of feathers, a differentia of birds as compared with horses; cloven-footedness of cattle is a differentia as compared with horses; redness of core or pulp is a differentia of "blood-oranges" as compared with ordinary oranges, etc. In all such cases the difference, or differentia, is a quality or property in addition to the generic or common qualities, and merely determines the species or individual under the class or genus. In this respect or relation it may even be spoken of as *essential*, although it is essential only to the species or individual.

There is no word used by logicians as the counterpart or complement of differentia except the term *genus*. But this is also contrasted with *species*, and species is the name for an individual or narrower class rather than of a quality or group of qualities expressed by the term differentia, although it is determined by this characteristic. Sometimes the term *essence*, however, is used as the opposite of difference, and denotes the quality or qualities essential or necessary to the existence of a class or individual. It is identical with essential property. For example, vertebrate anatomy and rationality are essential to the class "man;" four-footedness to the class quadrupeds; cloven-footedness to cattle; a certain pungent taste, color, etc., to oranges; woody structure, trunk, and branches to trees, etc. But here, as in the case of the term differentia, it is relative in its import. Besides, it is somewhat ambiguous in that it is sometimes used to denote the common qualities of the class or genus, and at other times is applicable to the differentia as an essential quality of the species. Usage is not so uniform as is desirable in this matter. Indeed, Whately calls the differentia the *formal* essence of a thing, and the genus the material part or essence of a concept.

The confusion incident to the discussion may be avoided if we adopt a new term corresponding to differentia, as genus corresponds to species, and retain essence or essentia as contrasted only with accident or accidentia. There has been some difficulty in selecting a proper term that would express the characteristic meant to be indicated by the term. But the best I have been able to accomplish in this matter is the selection of *conferentia* (from *con* and *fero*, to bring together, as differentia is from *dis* and *fero*, to separate). I had seriously thought of using *communia* (neuter plural of the Latin *commune*, what is common), because it would etymologically import the common qualities. But the fatal objection was that no singular of it could be used without confusion with the English *commune*. I have therefore chosen "conferentia," as a good logical and etymological complement of differentia. It has the objection to contend with that the English equivalent, "conference," is not associated with any logical usage of the kind here wanted. But we can confine ourselves to the form "conferentia," with the distinctive meaning assigned to it.

By "conferentia" I shall mean the common quality or qualities which "bring together" like individuals, or constitute the application of a term to a class. It is therefore the essence of the genus, as the differentia is the essence of the species. It is also a relative term, as what is conferentia in one class may be the differentia in relation to a co-ordinate species. But in it we have a convenient expression for the *common* qualities uniting individuals in a class, in contrast with those important qualities that are not common to all under it. Generic and general, as well as "conferential," may be the corresponding adjectives. In this way we may reserve essence, or essentia, to denote the essential properties of either genus or species, and accidentia for the non-essential. The following table summarizes the discussion and classifies properties :

Property- { Essentia { Conferentia = Common properties, or essence of the Genus. Differentia = Distinctive properties, or essence of the Species. Accidentia { Universal Casual } Non-essential to either Genus or Species.

4. GENUS AND SPECIES.—These are important terms to define. The concepts to which they apply play a very large part in all logical discourse. They are closely related to the extension and the intension of concepts. They differ from them only in implying a distinction between the general or conferential and differential properties of concepts, which is not necessarily involved in the difference between extension and intension, although when the variation between these is in an inverse ratio it coincides with, and perhaps in a measure determines, the distinction between genus and species. But this relation is a matter for more advanced Logic to consider. The question of importance in all practical reasoning is the influence exerted upon it by those various conceptions known as genera and species, and the logical relation between them. It will be seen most clearly in the fallacies of Equivocation and Accident, when we come to consider them. It will also be valuable in interpreting the relation between subject and predicate in one class of judgments, the process of Conversion, and the relation subsisting between essential and accidental properties, on the one hand, and conferential and differential properties, as concomitants, on the other. We now proceed to define and illustrate the two terms very carefully.

A conception which applies to a whole class of objects is called a *genus*. Thus " man " is a genus-concept because it is a name which applies to the various kinds of individuals, tribes, or nations of men. " Substance " is a genus because it includes iron, clay, brass, gold, silver, water, etc. It is a general name for various species and individuals under it.

A conception which applies to a narrower class, or an individual under a genus, is a *species*. Thus "Caucasian" is a species compared with the genus man; "iron" is a species compared with the genus substance; triangle is a species compared with the genus figure. A species, therefore, has the less, and a genus the greater, extension.

But it must be farther observed that the terms "Caucasian" and "iron" may also be genera in relation to a lower order of concepts. Thus "Caucasian" includes Germans, Frenchmen, Englishmen, etc.; iron includes steel, malleable iron, wrought-iron, cast-iron, etc. On the other hand, "man" is a species in comparison with the higher orders "biped," "vertebrate," "organic being," etc. To take a single term illustrating both relations, the concept "metal" is a genus compared with "iron," "gold," "silver," "platinum," etc., but a species compared with the term "substance."

From these illustrations it is apparent that *genus* and *species* are relative terms wherever they are convertibly applicable to the same concept in different relations. It will be noticed that a term is always a genus in relation to a narrower extension, and a species in relation to a wider extension. We may thus proceed in either direction until we reach the limits of farther progress, as Being, organized being, vertebrate, man, American, Lincoln. Here all intermediate terms are either genera or species according as they are conceived in relation to a higher or a lower order; according as they include a lower class, or are included in a higher. But the two extreme terms cannot be viewed in this twofold relation. "Being" is a genus, but not a species. On the other hand, the term "Lincoln" is a species, but not a genus, supposing, of course, that we can speak of an individual as a species. The former is not a species, because there is no higher genus under which objects can be brought, and the latter is not a genus because it cannot be divided into individuals or species. All singular terms are species, but not genera. They are called *infima species*, or lowest species. They are always individuals, so far as Logic is concerned. On the other hand, the highest genus is called the *summum genus*, or *genus generalissimum*. It is represented by some such term as "being," "thing," "something," "ultimate reality," but in all cases must be represented by a single concept. It is thus worthy of remark that there is only *one* absolute summum genus, while there may be an indefinite number of infima species. All intermediate concepts are sometimes called *subalterns*, as being either genera or species, according to the relation in which they are viewed.

It is necessary to notice the use of the terms genus and species as used in natural history. A species is there "a class of plants or animals supposed to have descended from common parents, and to be the narrowest class possessing a fixed form ; a genus is the next higher class." This is Jevons's definition, but he does not illustrate it. Perhaps in natural history the term "tree" would represent a *genus*, and oak, elm, maple, etc., a species, while the distinct kinds of oak, elm, and maple would be varieties. But the peculiar use of the term spècies here is that it is supposed to be fixed, and not relative as in Logic. In this conception of the term its meaning is quite distinct from that of Logic, where, as we have seen, any but the *summum genus* and the *infima species* may be either a genus or species, according to its relation to a higher or lower order. In natural history it is supposed to represent certain fixed characters and relations to a common progenitor. But the acceptance of the doctrine of evolution prevents any such determinate line of distinction from being drawn. The conception of "species" becomes an indistinct one, and does not imply any necessary assumption about a particular common ancestor. It denotes only a certain aggregate of characteristics with differences less marked and distinct than between genera. This change of meaning, therefore, makes the term approximate its logical import, if it may not ultimately identify the two. But it is necessary to remark the differences that have hitherto prevailed between the logical use of the term and its use in natural history.

It is important to notice an ambiguity in the logical use of the term genus. This ambiguity is apparent, as I have already remarked in the contrast between genus and species, on the one hand, and the contrast between genus and differentia, on the other, where species and differentia are not identical. The fact determines a double use of the term genus. Differentia, as we have seen, denotes certain properties determining the species, but it does not determine the whole content or intension of the species. It denotes only the distinctive qualities. The species is therefore a concrete thing, combining the differential and conferential qualities, or the common qualities expressed or implied by the "genus" and the added differential quality which makes it a species distinct from some other species having less or different qualities. But strictly taken, a specific term is concrete and *denotes indifferently an individual or an aggregate of qualities.* It *implies no special relation between extension and intension.* But the genus, or terms that represent genera, are not so distinct in their meaning. On the one hand, they denote numerically, that is, *extensively,* all that is denoted by the various species under them; logically, or *intensively,* they denote less than any given species. In relation to species, therefore, genera are greater in extension and less in intension, but in so far as they denote the common qualities they apply equally to individual species and to the whole number of species, while specific terms will not apply to the whole of the genus. The relation between genus and species in this respect can be illustrated as follows: "Man," as a genus, applies equally to Germans, Frenchmen, Englishmen, or Caucasians, Mongolians, Negroes, etc. The only difference between them is quantity or quality of intension. Germans contain certain qualities which Frenchmen do not. But the difference is not great enough to prevent the use of the term "man" to denote both species. In this sense the genus, or man, *is identical with all the species taken together and extensively considered.* The contrast, therefore, between genus and species is not a complete one. Indeed, strictly speaking, they cannot be contrasted at all. They differ only in regard to the number of individuals denoted by them, and in this respect are alike *concrete* in their signification. As applying to all the species the genus is but a common name for a number of individual-wholes, and applies mathematically to all alike. The fact that genus and species are not wholly distinct is apparent in all common judgments involving purely class terms. Thus, "Man is a biped," or "All Germans are men." I cannot reverse the subject and predicate because the difference of extension between them will not permit of it. There is, however, a connection always subsisting between genus and species which allows a statement in the form of a simple proposition. But no such connection exists between the different species under the same genus. Thus I can never say "the Germans are Frenchmen." Of them I can only form a negative proposition, where no relation of extension exists. The distinction, therefore, between species is absolute. Between genus and species it is not absolute, but is only of *quantity, whether of extension or intension.*

But when the term genus is contrasted with the term differentia the matter is quite different. Genus and differentia are contrasted not as larger and smaller classes of individual-wholes, but as different qualities or groups of qualities. In this sense genus, or a generic concept, denotes or connotes certain common qualities which characterize the whole class, but which are quite distinct from the differential qualities which form the differential essence of the species. In this meaning the genus is as distinct from the differentia as one species is from another. It cannot apply as a name to any of the qualities represented by the differential elements. It is only a name for the common qualities of a class. Thus "man," as a genus contrasted with differentia, does not denote individual-wholes at all, but only a certain group of common qualities. It is, therefore, not only an abstract conception, thus conceived, but denotes only *the intension* of the concept, exclusive of the differentia, and so serves as the basis for determining its extension. But the extension is not the matter in thought when conceiving it as contrasted with the differentia. It denotes or connotes only the common qualities, and anything affirmed of it in this sense will not agree with the species or differential concept. For example, I may say, "Pine wood is good for lumber." This is not true of every specific *form* of pine wood, but only of its substance or generic qualities. Matches might be made of pine wood and yet not be good for lumber. The affirmation, therefore, can be true only of the genus as contrasted with the differentia, and not of the genus including all the species. The distinction, therefore, between the two kinds of genera, as here drawn, is the same as that between the two kinds of general terms, the *mathematical* and the *logical*. Hence I distinguish the mathematical and the logical genus: the first, or mathematical, to denote the genus as applicable numerically to all the species, and the second, or logical, to denote or connote simply the conferentia or common essence, and not affirmable of specific characters.

The importance and meaning of this distinction must be brought out by further illustration. To effect this requires a brief explanation of what is meant by the connection between subject and predicate. Usually we suppose, or are told, that the predicate is more or less identical with the subject. Thus if I say, "Man is a biped," I mean that two-footedness is a quality of man. But I may mean by a similar judgment that one quality invariably and universally accompanies another. Thus to say, "Man is intelligent," may mean that along with a certain representative quality or qualities standing for man will be found the equally universal and necessary quality intelligence, but which is not *analytically* represented to consciousness in the mention of the name "man." In other words, I affirm the agreement or concomitance of certain conferential qualities, and this will be true of their connection, generically considered, wherever found. To take the former illustration, "Pine wood is good for lumber," goodness for lumber is connected with the generic qualities of pine wood, but it is not connected with every particular or specific form of it. The statement undoubtedly assumes a particular form and quantity of the pine wood as essential to its making lumber. But this only shows that it is not the *mathematical* genus of which it is affirmed, but only of the *logical*, and hence the difficulty when we come to compare the predicate with the species, or the differential qualities.

A negative illustration will bring out the same truth. For instance, I cannot say that "All men are white," but I can say, "The Caucasians are white," because "whiteness" is true of the species, and not true of the genus, taken either logically or mathematically. But now if I turn around and say, "The Caucasian race is the most intelligent," this may not be true of all individual Caucasians numerically considered, but only *generally;* that is, their intelligence is an accidental characteristic connected with such essential qualities as make them men, and with which whiteness is found frequently enough to make the statement of the race in general. We therefore speak of the logical and not the mathematical genus in our proposition.

The importance of the distinction I have drawn will appear when I come to consider the doctrine of Judgments and Fallacies. But it is fully justified in the ambiguity remarked in the use of the term genus, contrasting it, on the one hand, with the species, with which it differs only *quantitatively* in regard to extension, and on the other with the differentia, to which it is opposed and with which it differs *qualitatively* in regard to intension. We have, therefore, to keep constantly in mind that all *generic* concepts have a double capacity, the mathematical, and the philosophic or logical. If this fact is closely guarded the student will be saved many a misstep in reasoning.

2d. Analysis of Concepts.—By the analysis of concepts I mean here the breaking up of them into their parts and subordinate elements. There are two forms of analysis, *Division* and *Partition*, which must be considered in their proper order.

1. DIVISION.—Division is the analysis of the *extension* of a concept, the separation of a genus into its species. The process is usually called Logical Division. Thus we are said to divide the genus "tree" when we indicate the species of tree to which the term applies, as, for example, into oak, elm, maple, willow, ash, pine, etc., some of which at least are still

DEFINITION AND DIVISION

farther divisible into subordinate species or individuals, as oak into white, black, and red oaks; pine into white and yellow pines; willow into white, weeping, and swamp willows, etc. They may be divided by some other principle if we so desire. Thus we might divide "tree" into those of deciduous leaves, and those of evergreen leaves, etc. It is not necessary in every case to proceed upon the same principle. But whatever principle is used is called the *Fundamentum Divisionis.* In order to be a principle of division the quality or circumstance, taken as such, "must be present with some and absent from others, or must vary with the different species comprehended in the genus. A generic property, of course, being present in the whole of the genus, cannot serve for the purpose of division." The principle of division, therefore, must be some differentia, or differential quality, which is the distinctive feature of the species. Thus if I divide apples into red, green, and yellow, color is the principle of division, and each specific color is the differentia of its class. An accidental property will not suffice for any permanent or scientific division.

Hamilton enumerates seven rules governing the process of division. They are:

(*a*) "Every division should be governed by some principle."

(*b*) "Every division should be governed by only a single principle."

(*c*) "The principle of division should be an actual and essential character of the divided notion, and the division, therefore, neither complex nor without a purpose."

(*d*) "No dividing member of the predicate must by itself exhaust the subject."

(*e*) "The dividing members, taken together, must exhaust, but only exhaust, the subject."

(*f*) "The divisive members must be reciprocally exclusive."

(*g*) "The divisions must proceed continuously from immediate to mediate differences."

These are not all of equal importance, and might be reduced to a smaller number. For instance, the first is practically the same as the second. The fourth and fifth (*d* and *e*) might be summarized in one. The third and the seventh (c and g), although important in a complete enumeration, are more likely to be observed naturally than some of the others. But they require to be kept in mind. Jevons reduces them to three, which serve for most all practical purposes.

The importance of the rules is seen in what is called *Cross Division*, which is the naming of species that interpenetrate or overlap. Thus, if I divided trees into tall trees, green trees, pine trees, and dead trees ; or books into octavos, histories, theoretical books, dictionaries, etc., I should be using more than one principle of division, and indicating species that were not mutually exclusive. An illustration of the proper form of division is the following table, or outline:

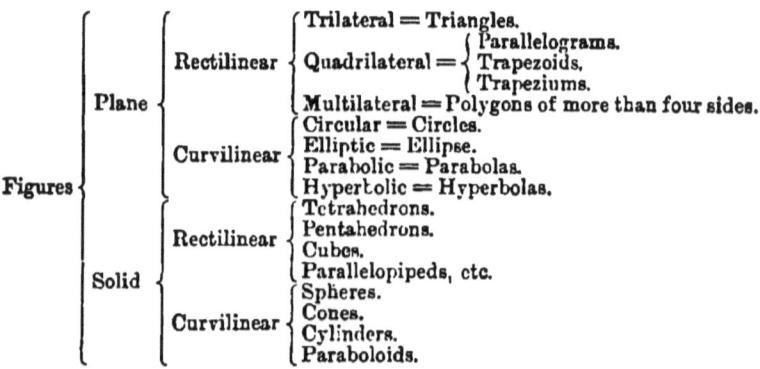

In division a genus, in relation to a species, is said to be *superordinate;* a species in relation to a genus is said to be *subordinate;* and a species in relation to a species is said to be *co-ordinate*. Thus, in the division of man into Caucasians, Mongolians, etc., "man" is superordinate in prior relation to Caucasians, Mongolians, etc., and they are subordinate to man in ulterior relation, while Caucasians, Mongolians, etc., in relation to each other, are co-ordinate.

We may adopt, as is apparent in this outline, a new principle for each successive process of division. In the division of plane and solid figures, however, it is the same. In the first division it is the *form* in general ; in the second division it is the kind of bounding lines ; in the third division it is the different relative positions and relations of lines. An illustration of a completely new principle in each division will be the following. I do not pretend that it is perfect, but only that it illustrates the point under consideration :

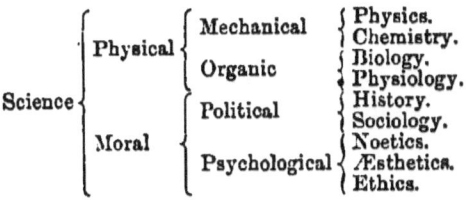

$$\text{Science}\begin{cases}\text{Physical}\begin{cases}\text{Mechanical}\begin{cases}\text{Physics.}\\\text{Chemistry.}\end{cases}\\\text{Organic}\begin{cases}\text{Biology.}\\\text{Physiology.}\end{cases}\end{cases}\\\text{Moral}\begin{cases}\text{Political}\begin{cases}\text{History.}\\\text{Sociology.}\end{cases}\\\text{Psychological}\begin{cases}\text{Noetics.}\\\text{Æsthetics.}\\\text{Ethics.}\end{cases}\end{cases}\end{cases}$$

The simplest form of logical division is called *Dichotomy*, which is the continual division of a genus into two species, a positive and a negative. This is the simplest mode of making the division exhaustive. A threefold division is called Trichotomy; but there is no technical name for the forms after that. Dichotomy is very useful in certain kinds of discussion, but in other circumstances is not so convenient. An example of it is found in what is called the *Tree of Porphyry*, named after the Greek logician who originated it. It may be represented thus :

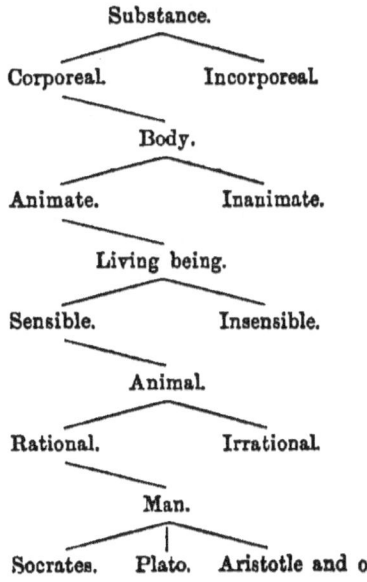

Substance.
Corporeal. Incorporeal.
Body.
Animate. Inanimate.
Living being.
Sensible. Insensible.
Animal.
Rational. Irrational.
Man.
Socrates. Plato. Aristotle and others.

Man could be dichotomously divided into Caucasian and non-Caucasian, the former into Greeks and not-Greeks, and so on. But the process must be terminated in the last analysis with a mention of the individuals, and there may be several points where this may be legitimately done.

The usefulness of dichotomy has its limitations. Thus it would be useless to divide Europe into France and not-France, the British Empire into England and not-England, or America into Rhode Island and not-Rhode Island. "Dichotomy is useless and even seems absurd in these cases, because we can observe the rules of division certainly in a much briefer division. But in less certain branches of knowledge our divisions can never be free from possible oversight unless they proceed by dichotomy. Thus, if we divide the population of the world into Aryan, Semitic, and Turanian, some race might ultimately be discovered which is distinct from any of these, and for which no place has been provided; but had we proceeded thus:

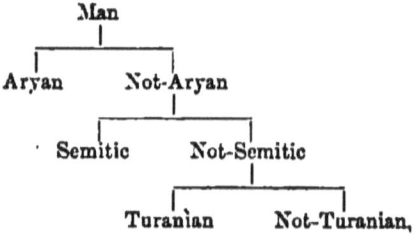

it is evident that the new race would fall into the last group, which is neither Aryan, Semitic, nor Turanian. All the divisions of naturalists are liable to this inconvenience. If we divide Vertebrate Animals into Mammalia, Birds, Reptiles, and Fish, it may happen at any time that a new form is discovered which belongs to none of these, and therefore upsets the division."

Jevons might have remarked that dichotomy can be best applied where our knowledge has not been exhausted, or where changes of boundary are likely to take place. Where there are distinct and known limits to a body of knowledge, the genus can be more clearly exhausted by the usual method of division. Dichotomy is needed where there is an indistinct and uncertain field of ideas.

2. PARTITION.—Partition is the analysis of a concept by a statement of its *intension*. It is simply a process describing the concept by its qualities, or parts constituting it. The concept may be viewed either as an individual or a class-whole, and partition merely defines it by its properties. Partition may be *mathematical or quantitative* and *logical or qualitative*. It is mathematical when the division or analysis is into its parts expressed in terms of space or time. Thus the concept *tree* is partitioned mathematically into roots, trunk, branches, and leaves; logically, it would be divided into its vegetable properties—color, woody fibre, raising of the sap by capillary attraction, etc. The concept "life" (a person's age) would be partitioned mathematically into childhood, maturity, old age, etc.; logically, it might be divided or partitioned into its length, goodness or badness, mode of spending it, etc. But in all cases we should disregard the questions of genus and species, and merely endeavor to consider the properties, essential or accidental, which constitute a concept. To make the partition exhaustive we should be obliged to state all the properties that make up an object; but as this is often prevented by the limitations of our knowledge of them, we have to be content with such as we know. A more complete illustration of the analysis than those given may better show the extent to which it may be carried. Take the case of "gold." The qualities of it are that it is material, metallic, solid, elementary, yellow, malleable, precious, useful, conductor of electricity, etc. "Man" again may be partitioned into animality, rationality, color, weight, sociality, etc.

The usefulness of the process is not so apparent in common conceptions. But if it were carried out carefully with such conceptions as "virtue," "thought," "mind," "religion," "cause," "intuitive," "law," "nature," etc., many a controversy would be modified in its incidents. Thus the notion "cause" may be partitioned into uniformity of sequence and co-exist-tence, and efficiency of power, and perhaps other qualities. Controversies about our knowledge of it would be materially affected by the presence or the absence of the second quality. Usually the partition implies that an object has more than one, or that a concept represents more than one quality. But we may regard every simple descriptive or declarative proposition as a case of partition. It is a descriptive definition, which partition aims to give. But it is not necessary to carry the idea of partition so far, except to intimate the broad distinction between it and true logical definition. Partition, where it serves any useful purpose, assumes a multiple of qualities which require recognition as well as the relation of genus and species. It is opposed to division as intension is opposed to the extension of concepts. Hence it is a process complementary to division.

3d. Definition of Concepts.—The definition of anything in practice is undertaken in several ways. But Logic has strictly to do with only one of them. This is called *Logical Definition*. It is necessary, however, to notice the several modes of definition in order to distinguish the logical from them in the proper way. They may be called *Etymological Definition*, *Descriptive Definition*, and *Logical Definition*.

1. ETYMOLOGICAL DEFINITION.—This is the definition of a concept by the word-roots from which the term originated. For example, "inquisition" would be etymologically defined by saying it was from two roots or words denoting "to inquire into;" "playfulness," from a root and suffixes denoting the quality of being full of sport, etc. But while this form of definition is very valuable in some circumstances it is of no importance in logical doctrine, or for elucidating any of the laws of thought.

2. DESCRIPTIVE DEFINITION.—This is the definition of a concept or a thing merely by describing it, and is essentially the same as partition. It most frequently occurs in the mention of accidental properties of objects when distinguished from complete partition and definition proper. It is, in fact, imperfect definition in the omission of one or the other of the

two essential conditions of the logical form of it. Thus, a descriptive definition of a triangle would be, that it is composed of straight lines and a symbol very frequently used in geometry. But such an account of it might as well apply to a rectangle or a parallelogram. A descriptive definition, however, may be made to approximate very closely to the logical. The above illustration wants but little modification to become that. Indeed the true definition is descriptive, but it is completely descriptive, involving a relation not expressed in this imperfect form. The difference between them can perhaps be technically expressed by saying that the ordinary descriptive definition depends upon the *accidentia* or the *conferentia*, and logical definition upon the *differentia*.

3. LOGICAL DEFINITION.—Logical definition is the statement of the *genus and differentia* of a concept, and is thus occupied with the whole of its *intension*, as Division is occupied with the whole of its *extension*. To illustrate, I may define "man" as a "rational animal." In this statement "animal" is the genus to which "man" is supposed to belong as a species, and rationality is the differentia which distinguishes him from other species. Again, I may define a circle as "a curved line everywhere equally distant from a point within called the centre;" or tree as "a vegetable with woody fibre, root, trunk, branches, leaves, and a certain magnitude;" or a house as "a building used for a place of residence," etc., and in each case fulfil the requirements of a logical definition.

It is important to make two remarks in regard to these and all logical definitions. *First*, they do not state the mathematical genus and species, which would identify the process with division in its principles, but the conferentia or logical genus, and the differentia. *Second*, it is always the species that is defined, never the genus.

The difference between definition and division, in their treatment of the genus, is, that the division of a concept proceeds *progressively* to the species under it, and may continue on down to the individual, but definition proceeds *regressively* to the proximate genus, or some appropriate genus which may serve as such, and goes no farther unless called upon to define that in the same way. Its progressive movement, if it can be said to have any, is a statement of the differential property or properties of the species defined, not a development of its extension or subordinate species. Since it is the species, therefore, which is always defined, and never the genus, unless it be also a species in relation to a higher genus, the ultimate or *summum genus* can never be defined. A logical definition, as we have shown, must state the genus; but as the *summun genus* is never a species it cannot be logically defined. Hence the impossibility of defining ultimate truths or principles, or the simplest concepts. They can be dealt with only descriptively or partitively, or divisively. Everything below them can be defined.

But it is important to observe the meaning of stating the "genus and differentia" in a logical definition. We observe, first, that it is not the mathematical "genus and species," as already remarked. Hence the term genus is used in its meaning as contrasted with differentia, and so denotes the logical as opposed to the mathematical genus. It therefore denotes the common qualities, or *conferentia* of the concept defined, in comparison with co-ordinate species. This is to show that logical definition is a statement of the conferentia and differentia of a conception, and not of the genus and species mathematically considered. Hence we see how it is exclusively occupied, if not explicitly, then implicitly, with the *intension* of concepts. The genus, as contrasted with the differentia, denotes only the common qualities of objects in the same class, and the differentia those which separate the individual or species defined from others in the same class. Thus to define a "bed" as "a piece of furniture for reclining upon," is to intimate that "bed" has qualities in common with other things known as furniture, and it is distinguished from a chair, which is used for sitting, by the property of being used for reclining. Instead of specifying the common qualities *partitively*, however, a general name suffices to imply them. But not having any corresponding abstract term for differentia, these properties have to be distinctly indicated. The genus and differentia, therefore, as stated in the definition, are simply all the qualities that make up the species defined. These are the conferentia and the differentia.

The conclusion from this must be that the predicate of a logical definition is always equal to, identical, and convertible with the subject. The importance of this will be apparent in the doctrine of Conversion and of Reasoning. All that we require to observe here is the difference between the ordinary simple proposition and the proposition which is regarded as a definition. The statements that "man is a biped," or "man is mortal," are not definitions. One states merely the genus of man, and the other a property of him. The first implies a property, that of two-footedness, and the second may be said to imply a genus. But neither of them specifies any differentia that would distinguish man, on the one hand, from other bipeds, and, on the other hand, from other mortals. They are not definitions because they do not state the whole intension of the species. Hence a true definition expresses the full meaning of the species defined, and can be used convertibly with it. Thus we can say equally that "man is a rational animal," or "rational animals are men," a process which cannot be performed with the simple proposition unless it is considered a definition. The fact that the two forms of proposition are the same to all appearances, and the fact that the mind uses the subject and predicate of definitions convertibly with each other, often lead to confusion, by inducing the treatment of simple propositions as definitions. This source of error will be treated in its proper place. At present it suffices to call attention to the fact.

In regard to the rules regulating correct definitions it will suffice to state Jevons's account of them, and it will always be important for the student to keep them in mind. They are five:

(a) "*A definition should state the essential attributes of the species defined.* So far as any exact meaning can be given to the expression 'essential attributes,' it means the proximate genus and difference."

(b) "*A definition must not contain the name defined.* For

the purpose of the definition is to make the species known, and as long as it is not known it cannot serve to make itself known. When this rule is not observed, there is said to be a '*circulus in definiendo,*' or 'circle in definition,' because the definition brings us around again to the very word from which we started. This fault will usually be committed by using a word in the definition which is really a synonym of the name defined, as if I were to define a 'plant' as 'an organized being, possessing vegetable life,' or 'elements' as 'simple substances,' vegetable being really equivalent to plant, and simple to elementary. If I were to define 'metals' as 'substances possessing metallic lustre,' I should either commit this fault or use the term metallic lustre in a sense which would admit other substances and thus break the following rule."

(c) "*The definition must be exactly equivalent to the species defined.* That is to say, it must be an expression, the denotation of which is neither narrower nor wider than the species, so as to include exactly the same objects. The definition, in short, must denote the species, and nothing but the species, and this may really be considered a description of what a definition is."

(d) "*A definition must not be expressed in obscure, figurative, or ambiguous language.* In other words, the terms employed in the definition must be all exactly known, otherwise the purpose of the definition, to make us acquainted with the sufficient marks of the species, is obviously defeated. There is no worse logical fault than to define *ignotum per ignotius*, the unknown by the still more unknown. Aristotle's definition of the soul as 'the entelechy, or first form of an organized body which has potential life,' certainly seems subject to this objection."

(e) "*A definition must not be negative when it can be affirmative.* This rule, however, is often not applicable, and is by no means always binding." *

---

* The following references may be consulted on matters pertaining to this chapter: Mill: Logic, Book I., Chaps. VII. and VIII.; Venn: Empirical Logic, Chaps. XI. and XII.; Hamilton: Lectures on Logic, Lects. XXIV. and XXV.; De Morgan: Formal Logic, Chap. XII.; Whately: Elements of Logic, Book II., Chap. V.; Supplement to Chap. I., §§ 2–6.

CHAPTER VII.

PROPOSITIONS OR JUDGMENTS

1st. Definition.—Words or terms unconnected express only concepts outside of any distinctly affirmed relation. In this way they do not convey truth, but only ideas or conceptions. Logic has to deal with the connection of concepts and their implications. The manner in which terms and concepts are joined together determines what a proposition shall be. It is not every combination of terms that forms a logical proposition. Some combinations may be mere phrases or ejaculations. But those combinations expressing a certain kind of relation, namely, a declarative relation between two terms, are the propositions with which Logic is concerned.

A proposition in Grammar is called a *sentence*; in Logic, it is most frequently called a *judgment*. A proposition or judgment, therefore, in Logic, is *the affirmation or denial of agreement between two conceptions*. It involves a comparison between them and a perception of this relation. Thus the proposition, "Gold is a metal," expresses a certain agreement between the concepts "gold" and "metal," an agreement which implies that the same quality is common to both, or that "gold" is a species of "metal." On the other hand, the proposition, "Man is not a quadruped," expresses a disagreement in a certain particular between the two concepts—a disagreement which implies that "man" is not in the class "quadrupeds," or does not possess the particular quality which distinguishes quadrupeds. This agreement or disagreement is not limited to single concepts or terms, but may include the same relation between groups of concepts constituting phrases. Thus, "The City of Washington, in the District of Columbia, is the Capital of the United States of America," is a proposition, only a little more complex in its elements than the former illustrations.

The terms between which the relation is asserted or denied are called the *subject* and the *predicate*. The subject is that of which something is affirmed or denied ; the predicate is that which is affirmed or denied of the subject. "Subject" (subjectum, ὑποκείμενον) means *underlying* thing ; "predicate" (prædicatum, κατηγορούμενον) means that which is asserted. The subject and predicate may be either grammatical or logical. The grammatical subject or predicate will be a single term ; the logical subject and predicate will consist of the grammatical subject with all its modifiers. Taken together they express in thought a single idea or conception, and hence Logic may treat them accordingly. All complex propositions are thus reduced to a single form.

The term expressing the connection between the subject and predicate is called the *copula*, and is always some form of the verb *to be*, or its equivalent. In many, perhaps the largest number of propositions, the verb *to be* is not found, and hence they appear to be wanting in a copula. Thus the proposition, "Napoleon ruled France," contains no expressed copula. In all such propositions, however, the predicate is said to include the verb and its dependent terms, and so to include the suppressed copula. Thus in the illustration given, "ruled France" is called the predicate, and the proposition seems to consist of only subject and predicate. But if we resolve the expression "ruled France" into its exact logical equivalent, "was the ruler of France," we have the copula and the predicate in the simple form. Hence the term "France" will not be the predicate alone, but "ruled France" must represent it with the copula implied or included in it. It is necessary to so consider the matter in order to deal logically with all such propositions. This logical treatment of them depends upon such a conception of the relation between subject and predicate as can be reduced to a general or universal law. A more complete discussion of the nature of this relation will be appropriate after we have considered the divisions of judgments.

2d. Divisions.—Propositions can be divided in a great many ways. The first division into *Indicative, Interrogative,* and *Imperative,* with perhaps the *Optative* and *Exclamatory,* as recognized by some, is grammatical, and it is only with the first class that Logic has to do. The essential meaning of the others, so far as the relation of concepts is concerned, can be reduced to the first form, the declarative, or indicative in such emergencies as require a logical use of their matter.

1. LOGICO-GRAMMATICAL PROPOSITIONS.—There is a second division which is both grammatical and logical, but which has not been uniformly the same with logicians. Sometimes it has been into *Categorical* and *Conditional,* with a subdivision of the second into *Hypothetical* and *Disjunctive.* Sometimes Conditional and Hypothetical simply interchange places in this division, and in a third form they are made synonymous with each other, giving us a co-ordinate division of three kinds, into Categorical, Conditional *or* Hypothetical, and Disjunctive. This last division I much prefer to all others, for the reason that their relation to each other in structure and meaning can be more easily determined than in any other classification.

There are, however, two classifications which may be given and that are of considerable convenience in explaining the meaning and relations of various kinds of propositions. The first one proceeds in the order of increasing complexity, and is intended to mark the nature of the additions made to determine the more complex forms. The following diagram exhibits the classification with illustrations:

Propositions
- Categorical
  - Declarative = A is B.
  - Disjunctive = A is either B or C.
- Conditional
  - Hypothetical = If A is B, C is D.
  - Dilemmatic = If A is B, C is either D or E.

It is apparent in this division that the simplest form is the declarative proposition, where the assertion is absolute and definite as regards both subject and predicate. The disjunctive form is equally assertory in its form of expression, but differs from the first in allowing some doubt or choice about the predicate, there being one alternative which excludes the connection of the other with the subject. The hypothetical expresses a definite dependence of one proposition, a declarative proposition upon a condition. Hence it adds a declarative assertion to a conditional one. The dilemmatic proposition simply adds a disjunctive one to a conditional proposition.

But a second classification is much preferable to this because it conforms to the three forms of reasoning, and, in a measure, determines them. In this classification we make the disjunctive appear as co-ordinate with the other two, although it is in reality a combination of them, and to which we apply the distinction between form and matter. The categorical and conditional propositions are regarded logically as pure and unmixed. The disjunctive we make categorical in its form of expression, but conditional in its meaning. This the following diagram will show:

Propositions { Categorical = Assertory in form and matter.
Conditional = Hypothetical in form and matter.
Disjunctive = Categorical in form, but Conditional in matter.

In regard to the disjunctive proposition, under this conception of it, it need only be said that its form of expression is undoubtedly assertory. It is positively affirmed that "A *is* either B or C." But the meaning of the disjunction, or the alternative expressed, can be understood only as implying "*if* A is B, it is not C," or "*if* A is C, it is not B." We shall discover later on in the discussion of reasoning that this is the only interpretation of the case which will enable us to reduce disjunctive reasoning to the regular form, or to understand it as a mode of the usual process of reasoning.

A Categorical proposition is one in which a statement is unconditionally made; as, "A is B," or, "Man is mortal." A Conditional or Hypothetical proposition is one in which the assertion is conditional or dependent upon a supposition of some kind; as, "If A is B, C is D," or, "If a stone be released from support it will fall to the ground." The first clause of the conditional proposition is called the *antecedent*, the second the *consequent*. The symbols of such propositions are *if, even if, provided that, although*, sometimes *when*, or any form of expression denoting a condition. A Disjunctive proposition is one which implies or asserts an alternative in the relation between the subject and predicate; as, "A is either B or C," or, "Metals are either hard or soft." The symbols of the disjunctive proposition are *either* and *or*. Some ambiguity is connected with their meaning, which will have to be considered when discussing the Disjunctive Syllogism. But as it does not affect the form and general meaning of the proposition by that name, the matter need not be discussed at present. We have only to remark what the disjunction means when it is complete, and that is, that the alternatives expressed by the terms *either* and *or* should be exhaustive. It means that the connection between the subject and predicate must be one or the other of two things. In the proposition A is either B or C, the question whether A is B or A is C is indefinite or undecided, but it is definitely one or the other, and hence the proposition either means that A is B and not C, or it means that A is C and not B. Hence, although the proposition stands as a direct assertion, it means that if A is B, it is not C, or if A is C, it is not B, or if it is not B, it is C. This is the reason that it is usually classed as a form of conditional judgment. But if it be closely examined it will be found to contain both assertory and conditional elements. It is categorical in its *form*, and conditional in its *matter* or meaning.

2. PROPOSITIONS ACCORDING TO QUALITY.—Propositions may be divided into *Affirmative* and *Negative*, according as they affirm or deny the agreement between the subject and the predicate. This relation is called or determines their *quality*. An affirmative proposition asserts an agreement between subject and predicate; as, "Gold is yellow," or, "Doves are birds." A negative proposition is one which denies an agreement between subject and predicate; as, "Men are not trees," or, "Gas is not heavy."

3. PROPOSITIONS ACCORDING TO QUANTITY.—Propositions according to quantity are divided into *Universal* and *Particular*. The distinction between them is determined by the question

whether the predicate is affirmed or denied of the whole of the subject. Hence a universal proposition is one in which the predicate is said to be affirmed or denied of the *whole* of the subject; as, "All men are mortal," or, "No men are trees." A particular proposition is one in which the predicate is said to be affirmed or denied of a *part* of the subject; as, "Some men are wise," or, "Some snow is not black." But the difficulty with this definition is that there is a sense in which the predicate is affirmed or denied of the whole of the subject in the particular proposition. For according to what has been said of the nature of the subject it may include what is known in grammar as the "logical subject," which consists of all the terms constituting a complex conception and standing in the relation of "subject" to the proposition. In this sense the predicate of a particular proposition is affirmed or denied of the *whole of its logical subject*, but of only a part of the grammatical subject. If therefore we could say that a universal proposition affirms or denies the predicate of the whole subject, grammatical *and* logical, and a particular proposition, of a part of the grammatical subject only, the difficulty would be removed. But it returns again in such propositions as "All good men are respected," which would be particular according to the definition. For the predicate is affirmed of only a part of the grammatical subject.

It would, therefore, be better for the purposes of definition either to divide propositions into *Definite* and *Indefinite*, or define universal propositions as affirming or denying the predicate of the whole of a *definite* subject, and particular propositions, of an *indefinite* subject. This is what is really meant by universal and particular propositions, and hence, with the proviso that they shall be identical in meaning with definite and indefinite, we shall adopt them as expressing the division of propositions according to quantity.

But this twofold division is the result of a reduction from a division which is frequently *fivefold*. Propositions are frequently divided, according to quantity, into *Universal, Singular, General, Plurative,* and *Particular*. The first and the last have been adequately defined and illustrated. The intermediate three may be reduced to one or the other of the first and the last, as their definition will prove. A singular proposition is one in which the subject is a *singular* term, and hence definite in its meaning ; as, "Louis XIV. was king of France." Here the predicate is affirmed of the whole of a definite subject, and hence for all logical purposes the proposition is universal. That is, the same laws of reasoning, mediate or immediate, will apply to singular or apply to universal propositions. A general proposition is one in which the distribution of the subject is ambiguous; as, "Metals are useful," "Man is intelligent." It is not stated whether "*All* metals are useful," or "*All* men are intelligent," or whether some are so. The propositions are capable of either interpretation, and according as we think of *all* or *some*, are universal or particular. Plurative propositions are undoubtedly particular. They are introduced by the word *most*, or its equivalent ; as, "Most ruminants are horned," and require mention only because of a peculiar syllogism which is valid in spite of its composition from particular premises ; of which again. But they affirm or deny the predicate definitely of more than half the subject, but indefinitely in regard to which of the two halves is exhausted in the term *most*. They are, therefore, classed as particular propositions. A summary of this reduction appears in the following table :

Propositions	Definite	Universal Singular General	Universal.
	Indefinite	Plurative Particular	Particular.

The mark of a universal proposition usually consists of some adjective denoting quantity, such as, *all, every, each, any* (meaning *all* individually), and *whole*. But wherever we find the predicate referring definitely to the whole of the subject we may treat the proposition as universal. This merely implies that some propositions may be universal in their matter, but indefinite in their form. The signs of particular propositions are also certain adjectives of quantity, such as *some, certain, a few, many, most, any* (meaning an indefinite individual), or such others as denote *at least a part* of a class.

The signs of a negative proposition are *no, not,* and *none,* the first and last being prefixed to the subject, and the second joined to the copula. Examples of them are, "No metals are animals," "None of the rebels were punished," and "Men are not quadrupeds." The term "no" is one which denotes both universal quantity and negative quality in propositions.

The quality and quantity of propositions may be combined in classifying them, and we shall have universal affirmative propositions, universal negatives, particular affirmatives, and particular negatives. It has been usual to choose an abbreviated symbol to denote each of these classes. The first four vowels of the alphabet—A, E, I, O—have been chosen for this purpose. A is the symbol of a universal affirmative, I of a particular affirmative, E of a universal negative, and O of a particular negative. Henceforth we shall employ them with this denotation whenever it is most convenient. It will be interesting to remark that A and I occur in the Latin *affirmo,* and E and O in the Latin *nego.* There is no significance in this, save perhaps as a mnemonic aid. The following table summarizes results:

Propositions
- Universal
  - Affirmative = A.
  - Negative   = E.
- Particular
  - Affirmative = I.
  - Negative   = O.

4. ANALYTIC AND SYNTHETIC PROPOSITIONS.—Another division separates propositions into *Analytic, Essential* or *Explicative,* and *Synthetic* or *Ampliative.* An analytic proposition affirms of its subject a predicate, which is implied in the very conception of the subject. Thus, "Matter is extended," "Water is moist," "Living beings are organic," "Wood is a substance," are all analytical judgments because the subject cannot be represented to the mind without thinking implicitly or explicitly of the notion expressed by the predicate. The use of the term "essential" to describe such judgments means that the property expressed by the predicate is an essential one, which is neces-

sary to conceiving the subject. The term "explicative," describing the same judgment, means merely that the predicate develops or unfolds what is involved in the thought of the subject. On the other hand, a synthetic proposition or judgment is one in which the predicate conveys information not necessarily implied in the conception of the subject. Examples of them are, "Water is a conductor of sound," "Plato was aristocratic," "Some men are honest," "The Popes were patrons of art." In these instances the predicate is not necessarily associated with the subject. It is no part of our conception of water that it conducts sound, nor of the Popes that they should be patrons of art. It would seem from this, therefore, that analytic judgments assert essential qualities of the subject, and synthetic judgments accidental qualities of it. If so, the distinction is a very clear, and perhaps a very useful one.

But the division of propositions into analytic and synthetic has little or no importance for Formal Logic. Its chief importance is in the domain of psychology and philosophy. The laws of reasoning, mediate or immediate, are not affected by it. Besides this there is often a great difficulty in distinguishing between the two classes of judgment so named, because of the confusion to which we are liable in distinguishing between an essential property which is universal, and a universal property which is accidental. Indeed it may be gravely doubted whether any universal property can be accidental. At least some would doubt it, and it may be a mere matter of our knowledge as to whether a given property is essential or accidental. If so, the distinction between analytic and synthetic propositions will only express the difference between our mode of representing a concept uniformly and the accidental association of some other property with it, less frequent in our experience. That is, the proposition "Body is extended " may appear analytic to the mind who has always or most frequently experienced it in connection with the idea of extension, while the want of frequent experience in connection with its sonorousness might make the proposition "Body is sonorous " a synthetic proposition. On the other hand, the limitation of experience to hearing might make the latter proposition analytic, and the former synthethic.

It will be seen, therefore, that the distinction is not only a relative one, but is mainly of psychological importance. We may, consequently, dismiss it from further consideration.

5. MISCELLANEOUS PROPOSITIONS.—There is a species of propositions called *Tautologous* or *Truistic*. They are those which affirm the subject of itself, and so may be regarded, in form at least, as a kind of analytic judgment. They are such as " A is A," " Whatever is, is," " Man is man," " A beast is a beast," etc. Some of them, after all, are synthetic, and although the predicate is the same word as the subject, it conveys a slightly, or even wholly, different meaning ; as, for instance, " A man's a man," " The king is king," etc. They are tautologous in form, but instructive in matter. In reasoning we require to be on the alert for such ambiguity. Otherwise the consideration of truistic propositions has no logical importance.

There is another division of propositions into *Pure* and *Modal*. " The pure proposition simply asserts that the predicate does or does not belong to the subject, while the modal proposition states this *cum modo*, or with an intimation of the mode or manner in which the predicate belongs to the subject. The presence of any adverb of time, place, manner, degree, etc., or any expression equivalent to an adverb, confers modality on a proposition. ' Error is always in haste,' ' Justice is ever equal,' ' A perfect man ought always to be conquering himself,' are examples of modal propositions in this acceptation of the name. Other logicians, however, have adopted a different view, and treat modality as consisting in the degree of *certainty* or *probability* with which a judgment is made and asserted. Thus, we may say, ' An equilateral triangle is *necessarily* equiangular,' ' Men are *generally* trustworthy,' ' A falling barometer *probably* indicates a coming storm,' ' Aristotle's lost treatises may *possibly* be recovered ; ' and all these assertions are made with a different degree of certainty or modality." But this does not affect the nature and relations of the copula, and if we remain by the definition of the predicate, we shall find that modal articles and terms simply modify *attributives*, verbal or adjectival, and no special significance should be attached to them when they do not affect the quantity of the proposition.

Some logicians distinguish propositions into *True* and *False*. But this has to do with their *matter* as valid, and not their *form*, as a mode of thinking, and as Logic is of formal laws it is not concerned with the material truth or falsehood of propositions. A system of pure and formal Logic, correctly illustrating the laws of thought, could be constructed upon materially false propositions as well as upon true ones. It is not a science of truth in general; but only of the formal laws of thought. It is, therefore, not concerned whether propositions be true or false.

3d. Ambiguity of Propositions.—Judgments are rendered ambiguous in three ways: *First*, by the ambiguous use of certain terms; *second*, by the inverted position of certain terms and clauses; and *third*, by the double meaning of certain propositions even when there is no ambiguity in any of the terms composing it. The first and the third of these influences affect propositions in the same way, giving them a double import, in which one of the implied propositions is the complement of the other. They may be called *Duplex* propositions because they are susceptible of analysis into two distinct judgments. Those due to the second cause may be called *Inverted* propositions.

1. INVERTED PROPOSITIONS.—These are of two kinds, according as the inversion is of the subject and predicate, or of some relative clause. In regard to the first, an example, such as may frequently be found in poetry, is, "Full short his journey was," or, "Great is Diana of the Ephesians." In such cases the order of subject and predicate must be reinverted before the proposition can be dealt with logically according to the formal rules of conversion and reasoning. In regard to the second class, the subject may sometimes be mistaken for the predicate when it is described by a relative clause standing at the end of the sentence; as, "No man is honest who

cheats his neighbor," or, "No one is fit for a king who cannot rule himself." The real subjects in these propositions are, "No one who cheats his neighbor," and "No one who cannot rule himself," and unless we keep this fact in mind such instances would give trouble in determining the *Figure* of a syllogism, as will appear when that subject is to be discussed.

2. DUPLEX PROPOSITIONS.—A duplex proposition is one which is capable of a double meaning and can be analyzed into two distinct judgments. There are three kinds: *Partitive, Exclusive*, and *Exceptive*. The chief characteristic of these propositions is, *that the complementary proposition implied by them is of the opposite quality of that which is asserted in the given instance*. This will be very important to keep in mind, because the process of reasoning will be affected by the question whether one or the other of them is the real one in the thought of the reasoner, as will be illustrated.

(*a*) *Partitive Propositions.*—These express a part of a whole, of which the implied proposition is a complementary part, and are determined by the ambiguous use of the terms "*All—not*," "*Some*," and "*Few*." "All—not" is often conceived as the same as "Not all," and hence when the proposition seems to be universal it is really particular. As an illustration we have, "All metals are not denser than water," or, "All men are not red-haired," where we may mean that "Not all metals are denser than water," and "Not all men are red-haired." Strictly construed the original propositions are E in *form*, but in *matter* they are either I or O, with the other of the two implied when one of them is distinctly intended. When I say that "Not all men are red-haired," or "All men are not red-haired," in the sense of the former, I mean that "Some men are red-haired," and that "Some men are not red-haired." Whichever of the two I have in thought, the other is implied as its complement.

Again, the term "some" is subject to a similar ambiguity, denoting *some but not all*, and *some at least, and it may be all*. Thus the proposition "Some metals are precious," especially if, in speaking, the emphasis be upon the word "some," may mean that "Some metals are precious," and "Some metals are not precious." This is when the term is equivalent to *not all*, or *only a part*. In such instances it implies its complementary opposite, so that it means I and O at the same time. If it be I, it implies O; if it be O, it implies I. The strict and proper import of the term, however, when describing particular propositions is that in which it denotes "*some, and there may or may not be all.*" The importance of this will appear in considering the matter of Opposition. But in actual reasoning we must be on the alert for the ambiguity to which the term is incident, and be ready to detect the fallacy which it may occasion.

A third proposition of a partitive and duplex nature is that introduced by the term "*few;*" as, "Few cities are as large as Vienna," or, "Few men can be President," etc., in which we mean that "Most cities are not as large as Vienna," and "Most men cannot be President." Such propositions imply a complementary opposite, because the term "few" denotes *some, but not all*, or *a few, but not all*. The expression "a few" taken alone does not imply any complementary conception, but is equivalent to the unambiguous use of the word "some." "A few," therefore, introduces a proposition which will be either I or O alone, unless "A few—not" be regarded as ambiguous like "All—not." But "few" introduces a proposition which has the meaning of I and O together, as the illustrations given very clearly prove. The confusion to which such propositions may give rise will be seen in those forms of reasoning where the validity of the conclusion turns upon the question whether they are to be interpreted as I or O. Thus the danger can be illustrated:

All philosophers are mortal.
Few representative men are philosophers.
∴ Few representative men are mortal.

Taking "few" in its duplex import, the conclusion would mean that "Most representative men are not mortal," when we know that they are all so. Hence we cannot treat the minor premise as a simple unambiguous proposition, but must interpret it as meaning both I and O, in which the conclusion would be valid with I, where "few" is equivalent to "some" (unambiguous), but vitiated with O, for reasons that will appear when discussing the doctrine of fallacies.

(b) *Exclusive Propositions.*—They are introduced or have their meaning determined by such particles as *only*, *alone*, and *none but*. They, therefore, limit the predicate to the subject, and are illustrated by such propositions as "Only Caucasians are white," "Giants alone can be seven feet tall," "Elements alone are metals," "None but honest men can be trusted," etc. When I say that "only elements are metals," I do not necessarily mean that "all elements are metals," but that the class "metal" belongs exclusively to the class "element," and that it can be affirmed of no other class. Hence the meaning of the proposition is either "All metals are elements," or, "All compounds (not-elements) are not metals." The first of these is what is called the *simple converse*, and the second may be called the *complementary opposite* of the exclusive proposition in question. It is with one of these, or the conception expressed by one of them, with which we have to deal in reasoning, or in testing any case of reasoning involving an exclusive proposition. Thus, "Only Caucasians are white" must be reduced either to "All white men are Caucasians," or to "Those who are not Caucasians are not white," when testing the formal process of the syllogism. The error to which we are liable in using them without considering their duplex meaning is illustrated in the following argument:

    Only elements are metals,
    Oxygen is an element.
    ∴ Oxygen is a metal.

Now we know that oxygen along with a number of other substances is not a metal. Or again, in a better case:

    Only men are allowed to vote.
    Criminals are men.
    ∴ Criminals are allowed to vote.

But we know that criminals are not allowed to vote, and hence to test the character of the reasoning we must use either the converse or the complementary of the proposition in the major premise. Thus if I say:

All who are allowed to vote are men.
Criminals are men,

it will be seen that I can draw no conclusions, for reasons to be noted when we study the syllogism; and so with the complementary proposition; as,

Those who are not men are not allowed to vote.
Criminals are men.

There are cases also where there might appear to be a formal error in the reasoning, but which is perfectly correct when we consider either the converse or the complementary of the exclusive proposition expressed. Thus:

Only elements are metals.
Gold is a metal.
∴ Gold is an element.

This is correct, but the reason for it will appear again.

It is important to say a few words about the *quantity* of exclusive propositions, because the question may be asked whether they are universal or particular. The answer cannot be made without an explanation of this peculiarity. Exclusive propositions are of two forms. An illustration of the first form is, "Only citizens can vote," and of the second form, "Only some men are wise." The first may seem to be universal, and the second particular. The second, which is a proposition in I, is a particular proposition, and differs from the ordinary instance only in implying the complementary opposite, O. The first form is the most frequent, and has the peculiarity that the terms "only," "alone," and "none but," have the effect of distributing the predicate while the subject is left undistributed. The meaning of distribution will have to be ascertained in its proper place, but we may say regarding it at present, that it denotes that the whole extension of the concept distributed is taken into account. When undistributed, a part of that extension is taken into account, and it may or may not be that all of it is considered. This makes the exclusive proposition, as it stands in the first form, a sort of inverted universal. But it is neither a universal nor a particular in that form. Hence we can determine its quantity only by taking either its simple converse, or its complementary, and in either of these cases the result is a universal. Consequently, the first form of the exclusive proposition must be treated in terms of what it implies, namely, a universal proposition.

The second form is the ordinary particular proposition and must be treated accordingly, except that the signs "only," "alone," etc., do not distribute the predicate. They have only the effect of implying the complementary opposite proposition O.

One exception, however, must be made in regard to these observations. It relates to the negative instances of exclusive propositions, such as, "Only bad men are not wise." They have a complementary opposite but no converse, while the affirmative cases have both. Negative exclusive propositions are in reality particular propositions in O. The exclusive particle does not distribute the subject and being negative the predicate is distributed. Consequently they cannot be converted without violating the rule that no term shall be distributed in the converse which is not distributed in the convertend. In the reduction of reasoning containing them, therefore, we can use only the complementary opposite, and not the converse as our substitute.

(c) *Exceptive Propositions.*—These are such as are introduced or modified by the terms *All but, All except, All save,* etc. For example, "All except those under twenty-one years of age are citizens," "All the planets, except Venus and Mercury, are beyond the earth's orbit." Such propositions appear to be universal, and simple at the same time. But they really consist of two particular propositions; namely, I and O; as, "Some men are citizens," and "Some men are not citizens," or,

"Some planets are beyond the earth's orbit," and "Some planets are not beyond the earth's orbit." If the class "men" or the class "planets" were divided into two species with a name according to the two portions indicated by the nature of the subject, exceptive propositions might be resolved into two universals, A and E, instead of two particulars, I and O. Thus, "All who are twenty-one years of age and over are citizens," and "All who are below twenty-one are not citizens." But in either case we have an illustration of the complementary nature of the two propositions developed from the duplexity of an exceptive judgment. This class, however, is not so important in Logic as the two previous classes, because fallacies are less frequently incident to the use of them. We require only to observe the peculiar nature of the conception involved in such judgments, and to be on the alert for any disturbing influence it is likely to exercise.*

The following outline is a *résumé* of the chapter:

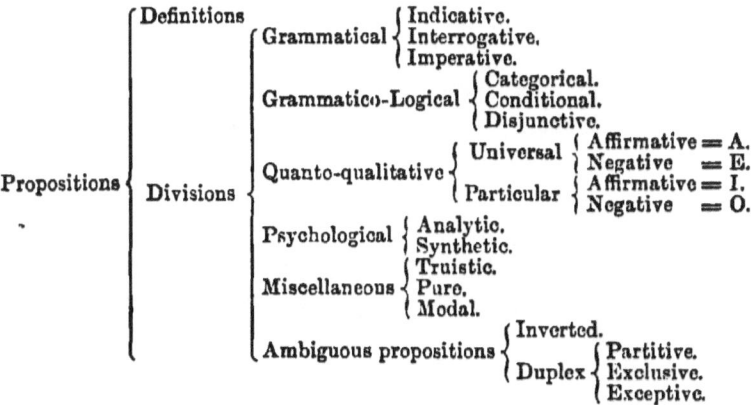

* General references on Propositions are the following: Mill: Logic, Bk., I., Chaps., IV., V., and VI ; also Examination of the Philosophy of Sir William Hamilton, Chap. XVIII ; Hamilton: Lectures on Logic, Lect. XIII. ; Venn: Empirical Logic, Chaps. IX. and X. ; De Morgan : Formal Logic, Chap. IV. ; Wundt: Logik, Dritter Abschnitt, Cap. II. ; Keynes: Formal Logic, Part II., Chap. I.

CHAPTER VIII.

THE RELATION BETWEEN SUBJECT AND PREDICATE

1st. Nature of the Relation Between Subject and Predicate.—We have briefly alluded to the relation between subject and predicate without explaining it. This was in the case of drawing a distinction between the judgments "Man is a biped" and "Man is wise." But now this relation must be examined more carefully and illustrated by symbolic diagrams which will be convenient for testing visibly certain forms of reasoning and inference.

There has been much dispute about the nature of this relation. One set of logicians has claimed, and the other has denied, that the relation can be expressed in terms of quantity; in other words, that the subject and predicate are considered as expressing merely a relation of quantity. Thus the judgment "Men are bipeds" is supposed to have its meaning indicated in the notion that the subject expresses a number of individuals not greater and possibly less than the number indicated by the predicate. This conception refers us to the matter already discussed; namely, that of *extension*. In formal Logic it has been customary to deal with all propositions as if we had to take no other relation into account, and owing to the peculiarly different character of such propositions as "Man is wise" the correctness and accuracy of the general practice has been impeached by some writers, and more particularly the mode of representing the relation by geometrical figures. It is the merits of this question which we wish to examine with some care.

The student must remember what has been said about the extension of terms and the laws regarding the relation between intension and extension. In the present problem it has been customary to deal only with their relations in extension and to choose geometrical figures to represent them in order to show some of the characteristics affecting the process of reasoning. If propositions express a relation of quantity, of equality, of more or less, between terms it would be natural to represent them mathematically. But as the proposition "Man is wise" does not seem to indicate a relation of quantity or extension between subject and predicate, but a relation of attribute to its substance; and as judgments like "Man is a biped" imply the connection of attribute and substance, whatever else is thought of, it has been maintained that symbols of quantity representing the relation are misleading. This is the problem to be considered.

In examining the relation between subject and predicate, we may adopt a division of propositions or judgments which has not been mentioned, but which is based upon the distinction between the *extension*, and the *intension* of concepts. Accordingly all judgments may be divided into *judgments of extension*, and *judgments of intension*. An example of the former is, "Man is a biped," or "Horses are animals;" of the latter examples are, "Man is wise," and "Trees are tall." A good way to represent the difference is found in the following propositions, the first extensive and the second intensive: "Man is a mortal," and "Man is mortal." The great difference between them is remarked in the nature of the predicate and its relation to the subject. In all propositions the predicate is either *substantive* or *attributive*. In judgments of extension the predicate is substantive; in judgments of intension it is attributive. As we have already explained, such propositions as "John struck James," or "The king rules his subjects," the verb expresses a function or attribute of the subject, or an attribute that is attributive, and hence the presence of a substantive *object* does not affect the attributive nature of the predicate, or the intensive nature of the judgment. But with this uniform relation of a substantive predicate to extensive and of an attributive predicate to intensive judgments, we may remark another important distinction which is its corollary. *In the extensive proposition the subject is contained or comprehended in the predicate or excluded from it ; in the intensive proposition the predicate is contained or comprehended in the subject or excluded from it.* The mode of comprehension, however, is distinct in each case. In the former it may be called that of *inclusion* or *exclusion* ; in the latter that of *inhesion* or *non-inhesion*. By inclusion of the subject in the predicate I mean that it is contained as a species in the genus, or as one class in another class term of equal or greater extension. The *exclusive* proposition is no exception to this, because we found that its logical meaning was the converse of its grammatical form. Hence the predicate of the extensive judgment is a *class* concept. By the inhesion of the predicate in the subject I mean that it is contained, or inheres, in the subject as an attribute in a substance, or rather expresses that relation. In the extensive judgment, therefore, the number of individuals denoted by the subject can never be greater than the number denoted by the predicate. This establishes a *relation of quantity* between them, and hence they may be called *quantitative* judgments. The relation being quantitative can be represented in some mathematical way, if not to indicate the nature of it, certainly to indicate an accident quite uniform with the essential qualities and proportionally variable with them. In the intensive judgment the predicate is an attribute or quality of the subject, and is comprehended in it rather than the reverse, as in the extensive proposition. Intensive may, therefore, be called *qualitative* judgments, because the connection between subject and predicate is a *relation of quality*. It is, therefore, a question whether they can be represented or symbolized by any figures expressing relations of quantity.

The answer to this question will be found in the fact to be shown that both kinds of judgment can be conceived in both a quantitative and a qualitative form at the same time, and qualitative in a double sense. If this can be proved, the symbolization of one will be that of the other also. We proceed, therefore, to the examination of this question.

First, in extensive judgments, although the predicate is a substantive and class concept, it connotes certain attributes which belong to the subject in the same way as in the intensive judgment. Thus, when I say "Man is a biped," I not only mean that the class of individuals or species "man" is included in the equal or larger class "biped," but I also mean that "man" is characterized by the quality of *two-footedness*, common to the whole class of bipeds. I therefore affirm or imply this attribute of him, and the extensive judgment becomes at the same time an intensive one.

Second, in intensive judgments, although the predicate is an attribute term, it is generally assumed that it cannot stand alone in thought, but must qualify some substantive. Thus, when I say "Man is wise," I not only mean that wisdom is an attribute of the subject, but I equally mean that he is a *wise something*. I do not mean that "Man is a wise man," for this is tautological. But I mean that he is a "wise creature," in which case I have a substantive predicate, as implied in the simple intensive form, and the intensive judgment becomes an extensive one at the same time. It is true that we do not ordinarily, perhaps never, think of this class relation in such a judgment as "Man is wise," but the fact that the extensive conception of it coincides and is perfectly compatible with the intensive conception, is sufficient to give it that double logical construction, as in the case of the extensive judgments. This is especially the fact when we reflect that in the extensive judgments we may not ordinarily represent to our thought the attributive relation between subject and predicate any more than in the intensive judgment we represent the class relation, and yet no one questions that the attributive relation is implied in the extensive judgment. The extensive relation is equally involved, or implied in the intensive judgment, although it may not be thought of. At any rate, it is possible to represent it so, in perfect compatibility with the attributive relation more particularly expressed by it. This will enable us to represent the relation quantitatively or mathematically as in the extensive judgment.

Another fact sustains the same conclusions. When I say "Man is wise," there is nothing in the nature of this form of statement to prevent my affirming wisdom of other beings as well. Take the proposition "Trees are tall," and we may also say, "Houses are tall." "Tallness" is not exclusively an attribute of "trees," and hence "trees" belong to a larger class of objects having the same quality, "tallness," as "man" belongs to a larger class of beings having the quality "two-footedness," expressed in the proposition "Man is a biped." Hence, so far as form of statement is concerned, the judgment "Man is wise," may admit that other beings are "wise" as well, and as long as this is the case, *formally*, the judgment is extensive as well as intensive.

It may be important to consider the relation between the judgment "Man is a wise creature" and a definition. I have said that the predicate "wise," in the simple intensive proposition, may be affirmed of other individuals and species than the subject, and that this constitutes a significant resemblance to extensive propositions, because their fundamental characteristic is precisely this fact, that it is affirmable of other individuals besides the given subject. But this is not the case with a definition. In a definition the subject and predicate are identical in extension, and convertible with each other. Thus, if I define *man* to be a *rational animal*, I can as well say that "Rational animals are men." This is merely because I regard the property "rational" as belonging exclusively to man, and so make it the *differentia*, while the word "animal" refers to the *conferentia*. It is, therefore, the *total* predicate which is identical with the subject, while the generic term, with its conferentia, indicates an extensive relation numerically greater than the subject, and it is only the differentia that can make the total equal to the subject. Now it is to be remarked that the conversion of a simple intensive judgment, such as "Man is wise," into its corresponding extensive judgment, such as "Man is a wise creature," looks very much like a definition, and hence the wisdom might not be predicable of anything else than the subject. It might be argued that this is the possible case with all intensive judgments, and if so their resemblance to extensive judgments is modified. But the reply to this is very clear.

In the first place, extensive judgments do not require that the predicate be greater in extension than the subject. It may be equal to it and supply all the conditions necessary. A definition, therefore, may be an extensive judgment, and can always be treated so. Farther, it is important to remark that, *formally*, a definition has to be treated as all other propositions, and it is only *materially*, that is, when we consider the modifying attributive as a *differentia*, that we can treat its predicate as convertible with the subject. In the second place, the attribute "wise" in this particular judgment, and the qualifying term in any other proposition, may indicate either the conferentia or the accidentia, and in either case involve a predicate of broader extension than the subject, when a substantive is modified by them. It is only when the modifier expresses the differentia that the predicate can ever be equal in extension to the subject, assuming it to qualify a substantive. Hence all judgments which are not definitions represent predicates of greater extension than the subject, and as even definitions cannot be *formally* distinguished from them as such, they must be treated logically in the same way, although when *materially* known to be definitions we may consider their extension as reduced in reality to that of the subject. But even this does not prevent them, as we have shown, from being extensive judgments, and therefore, whenever an intensive proposition is assumed to imply a substantive element in the predicate, it possesses quantitative properties identical in character with those of the so-called extensive judgment and may be represented accordingly. The figured symbols representing the quantity, or relations of quantity in extension between subject and predicate, shall be illustrated presently, and as soon as another interesting feature of both kinds of judgment has been considered.

What we have considered up to this point in the two forms of proposition is *quantity of extension*, expressed or implied.

What we have still to consider is *quantity of intension;* for this is as marked a characteristic of judgments as any other property, although it is not usual to regard it in symbolic Logic. But take the simple intensive judgment, "Man is wise." We say the predicate is here contained in the subject, or denotes a property belonging to it. Not only is it to be remarked that the same predicate *may* belong to other subjects also, but this one is not the *only* property of the subject. As in the extensive proposition the subject can never represent a greater number of *individuals* than the predicate, excepting in negative propositions as explained, so in the intensive proposition the predicate can never represent a greater number of *attributes* than the subject, except in negative propositions, as before. The quantity of intension, therefore, of the subject in intensive judgments must be equal to, or greater than, that of the predicate. Consequently the mode of symbolizing the mathematical relation would be the reverse of that in the extensive proposition; that is, with the same figures, but with a reversed position for the signs of the subject and predicate. Now, as we have shown that the extensive proposition has its intensive interpretation; thus, " Man is a biped " equals " Man is two-footed," denoting qualitatively what is implied by the extensive form, the quantity of intension between subject and predicate is the reverse of the quantity of extension between the same terms, and hence the mode of representing it symbolically will be the reverse again. This we proceed now to illustrate in full. But we must first explain how it is done, and shall then represent the quantity of extension, as if it applied only to extensive judgments. Afterward we can extend the principle to intensive propositions.

We should, perhaps, remark a connection between subject and predicate which is in some cases different from the two we have discussed, and which would be considered preferable to them by certain schools of thought known as Empiricists and Positivists, or such as oppose all Metaphysics. Instead of supposing that subject and predicate expressed real objects, or things in which all predicates affirmed inhere as qualities or attributes, they would say that judgments expressed the connection of coexistence or sequence between subject and predicate. This view would get rid of the necessity of supposing the subject to always express a substantive, and the predicate a substantive or attributive conception, the latter distinctly indicating the inhesion of a quality in a substance, and the former implying it while affirming a class-whole to which the subject belonged as a numerical part. In this way they would not mean to imply any necessarily metaphysical connection, but only one of coexistence or sequence between the two terms. Thus in the proposition "Man wears clothing," or "Man is a clothed being," we mean, they would say, that the essential qualities or characteristics of man are accompanied by the accidental one of wearing clothing, not that a being has this as an attribute. This is only to say that certain facts or phenomena, say bipedality, bimanousness, rationality, etc., are accompanied by the other quality of being clothed. The same thing, perhaps, could be said of any proposition, such as "Gold is yellow," "Iron is hard," when we suppose that "gold," and "iron" are names for certain qualities, among which "yellow" may be found in one case and "hardness" in the other.

It is true that many propositions seem to express, or to be resolvable into, this kind of connection. But it is not opposed to the kinds of connection we have previously investigated, and may be said to be always coincident with them. Hence, while we admit that such a connection is most apparent in many cases, it does not exclude the idea that the relation between subject and predicate is that of a subject and attribute, in which we include, but may not *expressly* think of, the relation of mere concomitance or non-concomitance. Besides, in extensive judgments this relation of mere connection by coincidence or sequence is not easy to imagine, unless we resolve the predicate into its attributive meaning. But it is just as easy to conceive, and more suitable to the traditional forms of logical discussion to admit, that we think of other relations than of mere coincidence and sequence. Judgments of extension mean to indicate, when conceived attributively, as they in reality always are, the *conferential* qualities of the subject, and to ignore the *differential*. But judgments of intension intend to express the inhesion of a quality in the subject without distinction of *essential* or *accidental*. These connections are simultaneous with that of mere coincidence or sequence, and hence no important end is served by a controversy about the question.

In order to represent the relation between subject and predicate, mathematically, Euler chose circles whose area could correspond to the suppositions already made about the equal or greater extension of the predicate as compared with the subject. As we have said, we shall first limit the symbolic representation to judgments of extension, because there can be no question about their quantitative nature and their mathematical representation accordingly. All the figures we shall employ may apply to the extensive proposition, "Metals are substances." We shall use the letter S to denote the subject, and the letter P to denote the predicate.* Hence S will stand for "metals" and P for "substances." Now as the extension of "metals" cannot be greater than that of "substances," the area of the circle representing it must not be greater but *may* be smaller. Hence proposition A may be represented in the following manner:

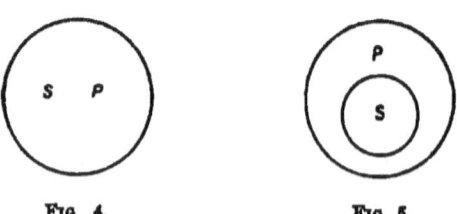

Fig. 4.   Fig. 5.

Proposition I, which would be "Some metals are sub-

* We must not confuse this with the later use of the same symbols to indicate the *minor* and *major* terms of the syllogism. They are respectively the subject and predicate of the conclusion, but are not always such in the premises.

RELATION BETWEEN SUBJECT AND PREDICATE 131

stances," will be represented in the following manner, to be explained again:

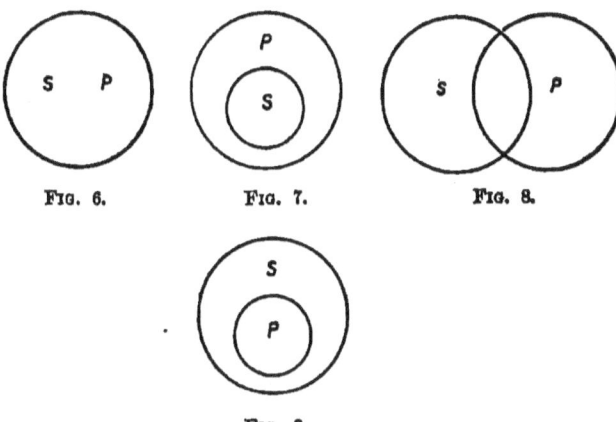

FIG. 6.   FIG. 7.   FIG. 8.

FIG. 9.

Proposition E, "No metals are substances," would be represented in only one form, as follows:

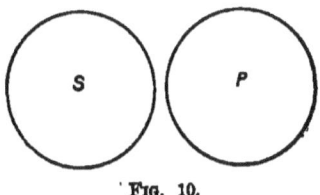

FIG. 10.

Proposition O, which would be "Some metals are not substances," would require three distinct figures for its symbolization, as follows:

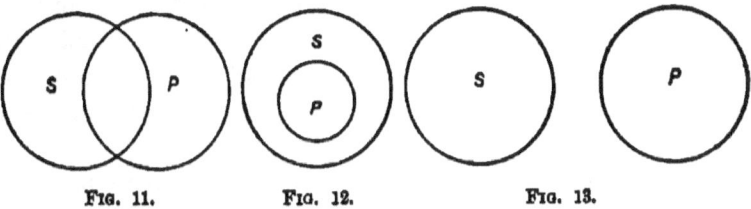

FIG. 11.   FIG. 12.   FIG. 13.

In Fig. 4 the two circles are supposed to coincide and make one, representing an equal extension between subject and predicate. This, as remarked, is true of definitions, and might be true of all other propositions in A, *so far we know positively* from the assertion. But the extension of the predicate may be greater, as we happen to know it is in many cases, and hence for that conception Fig. 5 has to be employed. It means that the area or number of individuals denoted by "metals" is less than that denoted by "substances," and it *may* always be so in A propositions, so far as we can determine from the form alone.

In proposition I, represented by Figs. 6, 7, 8, and 9, it is interesting to remark that two of them are identical with Figs. 4 and 5, and a third differs from Fig. 5 only in the position of the letters S and P. But this resemblance is due to the character of particular propositions. We have remarked that the sign "some" properly denotes in Logic *a part, and it may or may not be all*, and hence Figs. 6 and 7 very clearly indicate this possibility. For if "All S is P," it is evident that "Some S is P," although the former does not follow from the latter. But the conditions of a particular proposition being what they are, it is *possible*, so far as the statement is concerned, that "All S is P," when "Some S is P." Figs. 6 and 7 provide for this possibility. In Fig. 8 some portions of S and P are excluded from each other, and hence when it is compared with Fig. 11, it is found to represent both I and O, and hence it might be taken to symbolize the duplex proposition, where "some" implies its complementary opposite. But it does symbolize I, whether we regard it as an ambiguous representation or not. Fig. 9 has no ambiguity about it, if we regard carefully the relation expressed by the position of the letters S and P. But it has the fault of not admitting the possibility that "All S is P," at the same time, which is one contingency in proposition I. So far as we know from proposition I, proposition A is also true, and Fig. 9 does not indicate this.

We must remark, before passing to the negative propositions, that affirmative judgments must express *inclusion*. A must express total, and I partial inclusion at least. But when we come to negative propositions the relation expressed must be one of *exclusion*: E must express total, and O partial exclusion at least. Hence in Fig. 10 we have the only possible symbol of E. The two excluding circles denote that subject and predicate are not connected in a given respect, and hence no part of the extension of one can be included in that of the other.

In proposition O the exclusion must be at least partial. Fig. 11 represents it, and is like Fig. 8 for I, although different arcs or portions of the circles must be chosen to represent the exclusion, as compared with the arcs in Fig. 8 to represent inclusion. Fig. 11, however, does not indicate the possibility that E may be true, which is the case, so far as we know from the proposition. Hence for the same reason that Fig. 6 may represent I, although also the symbol of A, Fig. 13 may represent O, although the symbol of E. Fig. 12 explains itself as indicating that some of the circle S is not included in the circle P; but it is defective in not admitting the possibility of E at the same time. What is desirable in all this symbolization is that the figures shall properly represent the differences between A, E, I, and O, and at the same time represent the possibility that A is true when I is, and that E is true when O is. We desire also, at the same time, to represent the equal possibility that A shall not be true when I is, and E when O is. In other words, we require a symbol which will represent our entire ignorance as to whether there is *total* or *only* partial inclusion implied when I is affirmed, or whether there is *total* or *only* partial exclusion when O is affirmed. If this be possible the number of figures or symbols might be reduced. As it is at present we escape confusion only by carefully observing the relation between the various positions of S and P.

Ueberweg has a representation which will simplify matters very much. He reduces them all to four figures, representing respectively the propositions A, E, I, and O. He employs a system of dotted lines in order to express the various possibilities involved in the expressed relation of subject and predicate, but neither affirmed nor denied by the form of the judgment. Thus in proposition A, so far as we know, the extension of the predicate may be equal to or greater than that of the subject. The form of the proposition does not say which it is. We only know that it *cannot* be less. Fig. 4 implies that it is equal when it may be greater, and Fig. 5 implies that it is greater when it may be equal to that of the subject. Hence not only the incompleteness of the symbol in each case, but also the liability to confusion with those for other propositions. If, therefore, we can find symbols quite distinct from each other, and yet expressing all the possibilities of the propositions, we can greatly simplify the problem. This Ueberweg has done in the following manner, Fig. 14 standing for proposition A, Fig. 15 for E, Fig. 16 for I, and Fig. 17 for O:

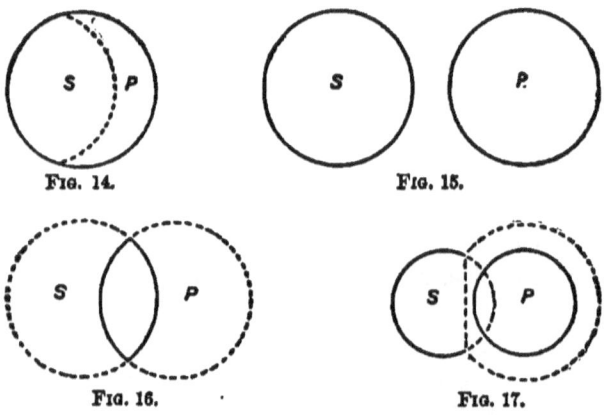

Fig. 14.   Fig. 15.
Fig. 16.   Fig. 17.

In Fig. 14, which we see is quite distinct in form from all others, the dotted line means that so far as we can tell from the proposition A, "All S is P," the extension of P may be either equal to or greater than S. Fig. 15 leaves no room for doubt. It always expresses total exclusion and nothing else. Fig. 16 makes it unknown whether there be a difference of extension between subject and predicate, and yet allows the possibility of A being true when I is true. That is, it is stated and indicated by the undotted lines that "Some S is P," while it may also be true that "All S is P," as is apparent, if S be exhausted within the area of the undotted lines. It is equally possible that O be true if S includes the area of the dotted lines. The same is true of the extension of P. It must be equal to or greater than S, if proposition A be possible, and must express some exclusion if O be possible. The doubt in both cases is expressed by the dotted line.

Fig. 17 is much more complicated. We have to express the partial exclusion of S and P, the possibility of their total exclusion; that is, proposition E; the possibility of their partial inclusion; that is, proposition I; and the possibility of the total inclusion of P in S, as the larger dotted circle implies, and all these at the same time. A little observation will show that this has been done. Thus if S contain only what is represented by the undotted curved line and the straight dotted line, O is symbolized and E is also possible. But if S contains the whole circle represented by the dotted and undotted arcs, O is true and I is possible. On the other hand, supposing that S has or can have a larger extension than P, the larger dotted circle represents again the possibility of both O and I. Taken altogether these four symbols are the only complete ones, and are the only representations which are not liable to confusion with each other.

An important observation in the quantitative relations of subject and predicate is, that in negative propositions, E and O, no comparison of equal, greater, or less can really be made. When subject and predicate exclude each other, we do not indicate anything about the relative extension or number of individuals denoted by them, and hence in negative propositions relations of extension are not commensurable. As a consequence of this, negative propositions either compare *species* which always exclude each other, or compare *genus and species* in the inverted order of universal affirmative judgments. E compares co-ordinate species, which are represented by total exclusion because of their distinct differentia. O either compares co-ordinate species and admits the possibility of E, or total exclusion in regard to the same species, or it compares a genus with a species, taking the genus for the subject and the species for the predicate. Thus, under the genus vertebrate, we can say, "No men are horses," = E, or "Some men are not horses" = O, with the possibility of E; or again, "Some vertebrates are not men" = O. But in the last case it is not implied that the extension of "vertebrates" can possibly be greater than that of the predicate, "men." Hence, so far as the form of judgment is concerned, no definite comparison in the quantity of extension can be made between subject and predicate in negative judgments.

It is otherwise in affirmative propositions, because they express a relation of inclusion, and hence always a relation between genus and species; and never between species and species; as, "All men are vertebrates" and "Some men are negroes."

Thus far we have symbolized only extensive judgments, and it remains to see whether a similar representation can be employed for judgments of intension. It is a very simple matter to solve this problem. We have only to recall the previous reduction of intensive propositions to the extensive form, involving quantitative relations as well as qualitative between subject and predicate, in order to see that the symbolization we have already employed will apply equally to intensive judgments. Hence the same figures will represent the quantitative relations of extension in such propositions as "Men are wise" and "Man is mortal," as in "Men are bipeds," or "Trees are vegetables." At any rate, such a representation is very convenient for testing certain forms of reasoning, and is not incompatible with the relation of intension expressed by such judgments. But assuming that all judgments are both quantitative and qualitative, we may represent the quantitative without interfering with the qualitative relation, and this is all that the symbols of Euler are intended to express. They compare subject and predicate only in the *quantity of their extension*, marking either their inclusion or exclusion.

But when we come to compare subject and predicate in respect of their *quantity of intension*, the matter is somewhat different. Here we mark the relation of inhesion or non-inhesion, and it is a question whether we can symbolize it in any mathematical manner or not. But since we have established the fact that the quantity of extension in the relation between subject and predicate is the direct reverse in the relation expressed by the quantity of intension, we may simply reverse the positions of the letter S and P in order to symbolize the mathematical relation in the intensive judgment as such. In the proposition "Man is wise," we have seen that the predicate can never be greater in intension, or quantity of intension, than the subject, and that it is contained in the subject. This quantitative relation is very clearly represented by Figs. 4 and 5, only in Fig. 5 the letter S will be placed in the larger, and the letter P in the smaller circle, as in Fig. 9. As Fig. 5 stands S is contained in P, as representing quantity of extension, or the comprehension of the subject in the predicate. But as the relation must be reversed for the quantity of intension, or the comprehension of the predicate in the subject, P must be included in S. This is the case for the proposition A, which Figs. 4 and 5 represent. For the quantity of intension in propositions E, I, and O, the same figures will serve as for the quantity of extension, only we must, as in proposition A, reverse the positions of the letters S and P. Of course, in Figs. 4, 6, 8, 10, 11, and 13, this reversal of S and P is not necessary. We require only to keep in mind the reversed order of inclusion as compared with extensive judgments. It is to be remarked, however, that in *negative* propositions the quantity of intension is no more determinate than the quantity of extension, relatively considered, as the same principles are applicable here as there. In the Ueberweg scheme we require to reverse the position of the letters S and P only in Fig. 14, and understand the reversed relation in the others, although in all cases no harm will be done by actually changing them in order to mark the contrast between quantity of intension and quantity of extension.

It will be evident that the scheme for the quantity of intension will apply also to judgments of extension, so far as they can be reduced, as we have shown, to judgments of intension. "All men are bipeds," conceived as meaning "All men are two-footed," can be represented in the same way as "All men are wise." Consequently we not only have the same symbolization for the two kinds of judgment, but a double system according as each judgment is considered in the quantity of its extension or the quantity of its intension. But it will not be necessary for practical purposes to consider more than one of them. The scheme for the quantity of extension is the one usually employed, and as it avails to represent and test all practical cases of reasoning, and the relations between terms, we shall confine our method to it alone. It is the symbolization for the quantity of extension that is used to explain *the distribution of terms*, which is the next topic for consideration.

2d. The Distribution of Subject and Predicate.—By the distribution of a term we mean that something is said about the whole of what it contains; that is, about the whole of its extension. Thus in the proposition "All men are mortal" we state something about the whole class of men, and hence the subject in this proposition is said to be *distributed*. An undistributed term, therefore, is one in which we do not say something about the whole class denoted by the term. Thus in proposition I, "Some men are negroes," we do not assert something of the whole class of men, and hence the subject is said to be undistributed. This appears very clearly in the symbolization we have adopted. For instance, in Fig. 5 the circle S represents that something is said about the whole of the class it denotes, and so in any other figure symbolizing a universal proposition, while the undistributed character of the subject in particular propositions is equally evident when allowance is made for their proper implications. But no ambiguity in regard to this matter will be noticed in Figs. 14, 15, 16, and 17.

In regard to the predicate, it may not be so easy to determine its degree of distribution from the figures without careful explanation. But this, perhaps, may make it clear. In proposition A, "All metals are elements," we perceive without difficulty that something is said about the whole of the subject, and that it is therefore distributed. But nothing is said or implied about the whole of the predicate, except as represented in Fig. 4, where we assume that the proposition is a definition. But as this can never be assumed *formally*, and as all propositions must be formally treated in Formal Logic, Fig. 5 and Fig. 14 are the only proper symbols of the quantitative relation between subject and predicate. Nothing, then, is definitely said about the whole of the predicate in proposition A, because other substances besides "metals" may be included in "elements." If we said anything about the whole of the class "elements" in the proposition "All metals are elements," we could reverse the order of subject and predicate and say, "All elements are metals." But if other substances besides metals are elements this latter proposition could not be true, and hence as long as we can possibly say that other things besides metals are elements, or as long as the proposition "All metals are elements," is entirely silent about the extension of the term "elements," we have asserted nothing about the whole of it, as indicated by the larger circle P in Fig. 5, and hence the predicate is *not distributed*. In proposition I, "Some metals are elements," the same conclusion is apparent, and the predicate is *undistributed*.

In proposition E, "No men are trees," for instance, not only is something said about the whole of the subject, but *something is also said about the whole of the predicate*. It is definitely excluded in its whole extension from the subject, as Figs. 10 and 15 indicate : That is, "men" are not any part of the class "trees," so that the whole of the class "trees" is excluded from the subject, and we can as well say "No trees are men," as "No men are trees." Hence in the negative proposition E the predicate is *distributed*, since something is said or denied about the whole of it. In the negative proposition O the same conclusion will be apparent, if we merely observe that a part of the subject is definitely excluded from the whole of the predicate, as is clear in Figs. 12 and 17. The predicate of O, "Some elements are not metals" is therefore *distributed*. We therefore summarize the rules for the distribution of subject and predicate as follows. We give two forms of statement, and the student may adopt the most convenient:

		Subject.	Predicate.
Propositions { Universal {	Affirmative A.	Distributed.	Undistributed.
	Negative E.	Distributed.	Distributed.
Particular {	Affirmative I.	Undistributed.	Undistributed.
	Negative O.	Undistributed.	Distributed.

All *Universal* propositions, A and E, distribute the *subject*.

All *Particular* propositions, I and O, do not distribute the *subject*.

All *Affirmative* propositions, A and I, do not distribute the *predicate*.

All *Negative* propositions, E and O, distribute the *predicate*.*

The symbol which I shall adopt to indicate the distribution of a term will be a small circle placed around the subject or predicate, as the case may be. Thus the distribution and non-distribution of terms in A, E, I, and O may be represented as follows, the cross indicating a negative proposition:

A, Ⓢ = P.   E, Ⓢ × Ⓟ.   I, S = P.   O, S × Ⓟ.

* General references on the relation between subject and predicate are the following: Venn: Empirical Logic, Chaps. VIII. and IX.; Symbolic Logic, Chaps. I. and VII., inclusive; Bosanquet: Logic, Book I., Chaps. I. and VII., inclusive; Keynes: Formal Logic, Part II., Chap. VI.; Wundt: Logik, Dritter Abschnitt, Cap. I. and II.

CHAPTER IX.

OPPOSITION

1st. Meaning of Opposition.—Opposition treats of the relations between the propositions A, E, I, and O, growing out of their quantity and quality; that is, out of the fact that they are universal and particular on the one hand, and affirmative and negative on the other. It has not to do with the subject and predicate, or the elements of the proposition as such, but with the propositions as a whole. When they contain the same matter they have certain relations of agreement or conflict which it is the business of the logician to exhibit. Unless they do contain the same matter no such relation can be determined. We can, in such cases, only decide upon their quantity and quality, and so merely treat them as universal or particular, affirmative or negative. But the various relations of agreement and conflict between conceptions give rise to corresponding relations between propositions containing them, and hence we require to ascertain the rules which regulate the extent to which any given proposition is true or false when another is known to be true or false. Some propositions, if true, interfere with the truth of others, and some do not. On the other hand, some, if false, necessitate the truth of others, and some do not. The sense and extent to which this is true remains to be determined.

If "All horses are animals," it cannot be true at the same time that "No horses are animals," or that "Some horses are not animals." This we express by saying that if A is true, E and O cannot be true at the same time; it is *inconsistent* with both of them. Also, if it be true that "No men are quadrupeds," it cannot be true that "All men are quadrupeds," or that "Some men are quadrupeds." This we again express by saying that if E be true, A and I cannot be true at the same time; it is inconsistent with them. But it is important to observe that if "All men are quadrupeds" is false, it follows that "Some men are not quadrupeds." It may be true, also, that "No men are quadrupeds," but the falsity of the universal affirmative with the same terms does not prove that fact. It can only prove the truth of the particular negative, and it remains entirely unknown from that proposition whether its universal negative is true or false. Hence if A be false, it follows that O must be true, but it does not follow that E is true or false. Now if it be false that "Some men are not mortal," it must follow that "All men are mortal," and, as we have shown in the first case, the negative of this, namely, "No men are mortal," is false. This we express by saying that if O be false, A is true and E is false. Similarly, if I be false, E must be true and A must be false. In this way we find that if A be true, O is false, and if A be false, O is true; again, if E be true, I is false, and if E be false, I is true. On the other hand, if O be true, A is false, and if O be false, A is true; and if I be true, E is false, and if I be false, E is true. This kind of inconsistency between A and O, on the one hand, and E and I, on the other, we call *contradiction*. In a loose sense the words "contradiction" and "contradictory" are used to express any kind of inconsistency which prevents two things from being true at the same time. But the relations between propositions A and E are so different from those between A and O, and E and I, that the term "contradictory" has been chosen to indicate that mutual inconsistency between A and O, and E and I, by which only one of them can be true, and only one of them false, at the same time. But A and E are called *Contraries*, because although the truth of A implies the falsity of E, and, *vice versa*, the truth of E implies the falsity of A, yet the falsity of A does not imply the truth of E, nor the falsity of E the truth of A. The mutual inconsistency existing between the universals and their opposite particulars does not exist between universals. Hence they are called contraries to distinguish them from contradictories.

OPPOSITION

It remains to determine the relations between A and I, E and O, and I and O. If it be true that "All men are mortal," it must be true that "Some men are mortal." So, if it be true that "No men are trees," it must be true that "Some men are not trees." This we express by saying that if A be true, I is true, and if E be true, O is true, because the part must be included in the whole. But if it be true that "Some men are wise," it does not follow that "All men are wise;" and if it be true that "Some men are not wise," it does not follow that "No men are wise." This we express by saying that if I be true, A is indeterminate, and if O be true, E is indeterminate. This is because we can affirm nothing of the whole when we affirm something only of the part. If we were to take cases supposing the falsity of A, we should find I indeterminate, or the falsity of E, we should find O indeterminate. But the falsity of I does not leave A indeterminate, nor does the falsity of O leave E indeterminate. This variable relation is expressed by calling A and I, or E and O, *subalterns* of each other. But I and O are called *subalternates*, and A and E are each called a *subalternans*.

When we compare I and O we find that they represent propositions of opposite quality; that is, one is affirmative and the other negative, and in that respect they are the opposite of each other. But the relation between them is the reverse of that between A and E. If it be true that "Some metals are elements," the law of contradiction already established between I and E will make the proposition "No metals are elements" false, and by subalternation O, "Some metals are not elements," will be indeterminate. That is, nothing follows about O from the truth of I, and also nothing about I from the truth of O. But if it be false that "Some men are trees," it then follows by contradiction that the proposition "No men are trees," is true, and by subalternation, "Some men are not trees," is true also. This we express by saying that if I be false, O is true, and if O be false, I is true. But both cannot be false at the same time, and both may be true. This relation is expressed by calling them *subcontraries*, as A and E are called *contraries* because they cannot both be true at the same time, but both may be false.

These various relations of the propositions A, E, I, and O are represented by a diagram which has been but slightly modified since Aristotle. It is called the Square of Opposition.

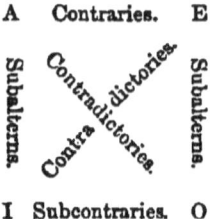

The relations, as we have developed them, can easily be applied to this scheme. They are embodied in the following rules, which it is important to keep in mind:

1. Of contradictory propositions, one must be true and the other false.
2. Of contrary propositions, both cannot be true at the same time, and both may be false.
3. Of subcontrary propositions, one only can be false, and both may be true at the same time.
4. Of subalterns, both may be true or both may be false at the same time. But if the subalternans be true, the subalternate is true, and if the subalternate be false the subalternans is false; if the subalternans be false, the subalternate is indeterminate, and if the subalternate be true, the subalternans is indeterminate.

2d. Application of Opposition and its Principles.—Nothing can be determined *formally* about the relation of opposition between the propositions A, E, I, and O, unless we assume identity of *matter*. A difference of matter simply isolates the two propositions and throws them out of all relation to each other in the scheme of opposition. The question is therefore suggested, What laws determine the nature of these relations of contradiction, contrariety, and subalternation, and what must be taken into account when applying them to actual discourse? Keynes furnishes a complete answer to the first of these questions.

"The inferences," he says, "based on the square of opposition, may be considered to depend exclusively on the three fundamental Laws of Thought, namely, the Law of Identity—A is A; the Law of Contradiction—A is not A; and the Law of Excluded Middle—A is either B or not B." For example, from the truth that "All men are mortal" I may infer by the Law of Identity that "Some men are mortal," and by the Law of Contradiction the falsity of the proposition that "Some men are not mortal." By the Law of Excluded Middle we can infer from the falsity of the proposition "All men are mortal," the truth of the proposition that, "Some men are not mortal." The Law of Identity means that a thing can be affirmed of itself, or conceptions which agree with each other can be affirmed in that sense. The Law of Contradiction means that two conflicting or contradictory conceptions cannot be affirmed of a thing at the same time. Thus I cannot affirm that a man is both "mortal" and "not mortal" at the same time. The Law of Excluded Middle means that of two contradictories one must be true. But a fuller discussion of these laws must be postponed to a later chapter. This brief account will suffice for the applications with which we have to deal at present.

In discourse and controversy we have to be careful about the real nature of our conceptions and propositions. We are liable to mistake, at times, a contrary for a contradictory judgment, or an indefinite for a definite judgment, or subcontraries for contraries. This will particularly be the case when propositions are one thing in form and another thing in matter. Thus singular propositions are treated as universal in form and are therefore contraries. But in matter, subject and predicate being the same and their quality the opposite of each other, they are contradictories; as, for example, "Socrates is a man," the only possible negative of which is "Socrates is not a man." Pure universals have two opposites, the contrary and the contradictory, but singulars have only one, which is in reality

10

the contradictory, as will be seen in the case given. Thus if it be true that "Socrates is a man," the negative, "Socrates is not a man," is false, and *vice versa*. So far it seems like a case of contraries. But if "Socrates is a man" be false, it is true that "Socrates is not a man," and *vice versa*. This makes it a case of contradiction, because if they were contraries they might both be false at the same time. This would mean that the assumed or proved falsity of the proposition "Socrates is a man," would leave the proposition "Socrates is not a man" indeterminate. But we observe that it cannot be so, and hence singular judgments *in respect to quality* have to be treated as contradictories, but *in respect to quantity* as universals. This will determine the relation between the two propositions "Socrates is a man" and "Socrates is a horse." They must be regarded as contraries, not as contradictories. They contain different matter in their predicates, but the same matter in their subjects; so that although the predicates are both *positive* concepts they are mutually exclusive as species, and so relatively negative in comparison with each other. The two propositions are related as contraries because the truth of either denies the other, while both may be false. If it be true that "Socrates is a man," it cannot be true that "Socrates is a horse," and so if he were a horse he could not be a man. But if it be false that "Socrates is a man," it does not follow that he is a horse, because he might be anything else except a man. Hence terms representing co-ordinate species will be contraries, not contradictories, in the scheme of opposition.

But what will be made of the propositions "Socrates is a man" and "Socrates is a Greek?" Of course, *formally* neither these nor the previous propositions can be treated of under the principles of opposition, and I am not designing so to treat them. I am endeavoring to give the purely formal rules some modifications to suit their *material* application. The last two propositions, although singular, as the two previous ones, are somewhat different because of the relation between the two predicates. These, instead of being co-ordinate species, are genus and species. Hence if the first be true it does not follow that the second is either true or false ; but if it be false the second is false. It is interesting to note from this that they are subalterns, the proposition "Socrates is a man" being the subalternate, and "Socrates is a Greek" being the subalternans. We may generalize, therefore, in such cases, by saying that in singular judgments with the same subject, when the predicate is a genus in one and a species in the other, the genus is the mark of the subalternate and the species of the subalternans.

When the predicate is the same the subjects can never be a genus and species, and the propositions remain singular at the same time. Hence no relation of opposition in such cases is determinable. Thus, in the propositions, "Socrates is a Greek," and "Plato is a Greek," nothing can be said about the truth or falsity of one when the other is either true or false.

Hence, when the predicates are the same in both propositions, and the subject of one a genus, and of the other a species, both propositions formally are universals, but materially they are A and I, or universal and particular, and so are subalterns again. But in this case the genus marks the subalternans and the species the subalternate. Thus "All Greeks are men," and "Socrates is a man." The truth of the first implies that of the second, and the falsity of the first leaves the second indeterminate (except on a condition to be discussed again), while the truth of the second leaves the first indeterminate, and the falsity of the second implies the falsity of the first. We must keep in mind, however, that we assume all along that Socrates is a Greek, and hence an individual or a species of the genus.

The matter is still more complicated when we come to consider universal propositions or judgments. Suppose we take the first example with the same subject and with co-ordinate species for predicates. Thus "All men are bipeds," and "All men are rational." It is apparent in such cases that neither agreement nor conflict between them can be inferred from either the truth or the falsity of one of them. Hence a relation of opposition is not determinate here, any more than in the singular propositions "Socrates is a man," and "Socrates is blind." If the predicates be such inconsistent concepts as "quadrupeds" and "newspapers," the propositions might be treated as contraries, but there is no criterion for determining this opposition. It requires to be a uniform concomitant of some other characteristic. But if the predicates be genus and species, and the subject identical, as "All men are vertebrates," and "All men are bipeds," the problem is not different from the previous one, because genus and species, or conferentia and differentia, always agree and never conflict, but never imply each other. No relation of opposition therefore can be established between such propositions.

If the predicate remain the same, and the subjects are genus and species, as "All Europeans are Caucasians," and "All Frenchmen are Caucasians," the propositions are evidently subalterns, the genus marking the subalternans, and the species the subalternate, as already shown. But as between "All Europeans are Caucasians," and "Some Frenchmen are not Caucasians," we evidently have complex cases, and so also with "All Frenchmen are not Caucasians," because the relation between genus and species determines the particularity of the last proposition in comparison with the first. Of course such propositions as "All men are bipeds," and "All men are quadrupeds," or "All wise men are good," and "All wise men are bad," will be contraries, owing to the nature of the concepts biped and quadruped, and good and bad. But the form of the terms cannot determine this fact, and hence no rule for estimating them can be established. It will be important to observe, however, that *good* and *bad*, *wise* and *ignorant*, *rational* and *irrational*, *beautiful* and *ugly*, etc., are often used as contradictory conceptions, and sometimes only as contraries. The latter is the true conception of them, as there is a third alternative between them, owing to the fact that they are both positive terms. A true contradiction can exist only between positive and negative terms, never between two positives, or between a positive and a nego-positive. The latter are only contraries, and so establish contrariety between propositions having them as predicates, but with the same subject.

It is very important to keep these facts and distinctions in mind in order to understand the mental processes involved in one man's asserting that "Washington was a good man," and another controverting it by the counter-assertion that "Washington stole a horse,".or one man's affirming that "Americans are a mercantile people," and another's asserting that "A. B. C. of the Americans are great scientists." In the first two the propositions are evidently contraries. In the second instances, everything depends upon whether the indefinite proposition, "Americans are a mercantile people," be regarded as a universal or a particular, A or I. If the former, they are contraries, supposing that commerce and science are incompatible callings, as in practice, generally at least, they seem to be, although perhaps not ideally or theoretically so. But if the proposition be I, they are subcontraries, under the same assumption as before. But the cases only show how we are to analyze the conceptions when a controversy is involved.

In the relation between subcontrary judgments we must be careful not to admit the ambiguous use of the word *some*, as this would at once imply the truth of the opposite proposition. But in the true conception of opposition I never implies O, and O never implies I. Hence the term "some" must always mean *a part, and it may or may not be all*, as has already been shown.

Errors in argument, so far as an assumption of the relation between propositions may be concerned, are occasioned in two ways; first, in assuming an agreement between them when they are inconsistent, and second, in assuming a conflict when they are consistent with each other. The latter is always called the fallacy of *Ignoratio Elenchi*, which we shall have to explain again. But it will be illustrated in the course of the present discussion. We are to examine proof and disproof of judgments so far as that can be accomplished under the scheme of opposition.

By the principles of opposition we can only prove I and O as subalternates by assuming or proving A and E, and disprove A and E by assuming or proving I and O as their contradictories. Thus we prove I if we admit A, and we prove O if we admit E. But we disprove A if we admit O, and disprove E if we admit I. In such formal propositions as "Some men are mortal" and "All men are mortal," this is evident from the rules. But in such propositions as "Europeans are white," and "Germans are white," it is not evident until we notice the logical relation existing between the conceptions "Europeans" and "Germans." I prove the truth of the latter when I admit the former, assuming that "Germans" are a species of the genus "Europeans," because they are related as subalterns, the former being the subalternans, and the latter the subalternate, the proposition to be proved. I may prove that "Scientists are learned," by assuming or proving that "Educated men are wise," if "learned" and "wise" are identical, and "educated men" be the genus of which "scientists" are the species. But I cannot prove that "Germans are good" by assuming or proving that "Europeans are not bad," without making the second assumption that "good" and "not bad" are identical.

In disproof the matter is a more important and interesting one. Here we have to do with A and E as conclusions, and I and O as conditions of their validity or invalidity. "In order to prove the falsity of A, it is sufficient to establish the truth of O, and it is superfluous, even if possible, to prove E; similarly E is disproved by proving I, and it is superfluous to prove A. Any person who asserts a universal proposition, either A or E, lays himself under the necessity of explaining away or disproving every single exception brought against it. An opponent may always restrict himself to the much easier task of finding instances which apparently or truly contradict the universality of the statement, but if he takes upon himself to affirm the direct contrary, he is himself open to easy attack. "Were it to be asserted, for instance, that 'All Christians are more moral than Pagans,' it would be easy to adduce some examples showing that 'Some Christians are not more moral than Pagans,' but it would be absurd to suppose that it would be necessary to go to the contrary extreme, and show that 'No Christians are more moral than Pagans.' In short, A is sufficiently and best disproved by O, and E by I. It will be easily apparent that, *vice versa*, O is disproved by A, and I by E; nor is there, indeed, any other mode of disproving these particular propositions." The error in disproof, however, may lie in certain assumptions about the relations between the two propositions after the proper one has been proved. Thus it may be no disproof of the assertion that "John Smith is good," to prove that he is not civilized, because the conceptions " good " and " not civilized " are not necessarily contradictory, nor even contrary. I simply evade the issue. Again, it is no disproof of the assertion that "Cromwell was a usurper," to say that " Foreign nations acknowledged his authority," any more than it would be a proof of his legitimacy to make the same statement. Likewise it is no disproof of a man's badness to say that he is religious, any more than we should prove he is white by showing that he is not black. If I assert that " Government is a necessary institution," it is no disproof of it to show that some governments are bad. Many arguments, however, are conducted upon just such logic, where agreement or conflict are simply assumed by both parties to exist between conceptions which may be inconsistent in certain accidental relations, but not necessarily so. Thus, if I assert that free-trade is the right policy for a country, it is no disproof of it to show that protection helps manufacturers; nor is it a disproof of it to show that I have myself instituted a policy of protection. The untrained mind, however, is likely to suppose the argument valid in both cases. In purely formal Logic this is all very easy, but in applied Logic, or actual discourse, we require dexterity in discovering the relations between conceptions in order to test the connection of general principles with them. To illustrate, take the conceptions just mentioned in the last example, protection and free-trade. Without some knowledge of economics, perhaps, we should not suspect or detect any relation of either agreement or conflict. "Protection," as a formal and logical conception, might suggest nothing more than the idea of "self-defence" or "self-preservation," and so carry everything before it by this association, while no conceivable meaning might be attached to "free-trade," except the vague sense of contradiction or opposition from the mere fact that some one so considered it. But an examination of their real meaning from a knowledge of economics shows them to be contradictories in reality, and not merely contraries, although they are both positive terms, in perhaps every other connection than this one. This is made clear by the fact that "protection" is a restriction upon free commerce, "free-trade" is the absence of that restriction. We see in this distinction the presence and the absence of certain qualities which defines respectively a positive and negative term.

In many cases the error can be tested by taking into account the assumptions made in any particular assertion. But this introduces the syllogism into the problem, while it may be possible to decide the matter upon the principles of opposition. This is all that we have here considered, and the student should practise the application of these principles on every possible occasion, not so much for the sake of familiarity with the laws of formal Logic, as for a better understanding of the means for dealing with the subject-matter of discourse and controversy. As an illustration of what a mental process is, and what law of opposition is involved or implied, I quote at random a statement from an article in a monthly periodical: "Those who oppose nationalism on the ground that the present social condition is, by reason of its privations, a blessing, ought to understand that the exact opposite of a false proposition is by no means certain to be the true one—though it is a favorite argumentative short-cut to assume this to be the case." Now whether he knew it or not, the writer was using the law regulating the relation between contraries in asserting that "the exact opposite of a false proposition was by no means certain to be the true one." But the assertion charges a fallacy upon his opponent which would have been technically called an *ignoratio elenchi*. The case could, perhaps, have been made stronger by connecting the error directly with the violation of a law of Logic. Illustrations of the same kind are plentiful, and the student can profitably spend his time in looking for them.*

* Special references on the subject of Opposition need not be indicated, as the general agreement upon it leaves no points of importance in doubt. The works previously mentioned deal with the subject very briefly.

CHAPTER X.

IMMEDIATE INFERENCE

1st. Definition and Divisions.—An inference of any kind is merely the explicit statement of what is implicit in a previous assertion or thought. If this account of it is not clear, we may define it as drawing one conception or truth from another, which implies the one drawn, or it is the carrying out into a following proposition a thought which was virtually contained in an antecedent judgment. This last definition is Hamilton's, slightly modified. It can be illustrated in several ways. I can infer that "Europeans are white," if I know that Caucasians are white, and that Europeans are Caucasians. Or I may infer that the weather will be clear, when I see the clouds breaking. The first form of inference is said to be deductive, and the second inductive. But for the present we need only to know that a certain fact or facts may suggest to consciousness other facts not explicitly stated or conceived in the first case, and which are said to follow from it. An inference, then, is what follows from another thing, in so far as the latter is a conceived truth. A case quite different from the previous illustrations is the succeeding one. From the proposition "The sciences are useful studies," I can infer that "Some of the useful studies are sciences," or from "All negroes are black," that "Those who are not black are not negroes." It is evident that these forms of inference are quite different from the previous instances. Hence we proceed to the two kinds of inference.

There is, first, the inference known as *Mediate Inference*, or Reasoning, which is illustrated by the first instances. It denotes *reasoning by the agency of a middle term.* Thus in all mediate inference we have first to compare two terms with a third, called the middle term, and then infer that these two terms can be compared with each other. But it is quite different with the second class of inferences. This is called *Immediate Inference,* or Reasoning, and denotes *reasoning without the use of a middle term.* Mediate inference requires two, immediate inference but one proposition as a basis. This is exemplified in the illustrations. Immediate inference is usually divided into Conversion, Obversion, Contraposition, Added Determinants, and Complex Conceptions. But I shall slightly modify this division so as to read Conversion, Obversion, Contraversion, Inversion, and Contribution, the last of which comprehends the last two in the previous classification. Opposition is not treated as a mode of inference. But it would not be far amiss to so treat it, or at least to consider it as involving processes of immediate inference. For we pass directly to certain conclusions from certain other conditions or premises. But it differs from other modes of inference in that it requires the supposition of certain material truths in order to render it possible, while in other forms of immediate inference we require only a formal relation between subject and predicate. Perhaps, however, if we regard the conditions of opposition as formal, it may be possible to give the transition from A to O, or from E to I, by contradiction, or from A to E, by contraries, as sufficiently formal and inferential to speak of the process as involving a kind of immediate inference. Although I believe this to be the case I have remained by traditional usage and discussed it by itself. Under immediate inference, therefore, I begin with Conversion.

2d. Conversion.—Conversion is the transposition of subject and predicate, or the process of immediate inference by which we can infer from a given proposition another having the predicate of the original for its subject, and the subject of the original for its predicate. But there are certain limitations under which the transposition can take place. For instance, from the proposition that " All horses are animals," I cannot infer that "All animals are horses;" nor from "All metals are elements" that "Some elements are not metals," although this may actually be the case. The rules, therefore, which limit the process of conversion are two.

(a) The quality of the converse must be the same as that of the convertend.

(b) No term must be distributed in the converse which is not distributed in the convertend.

These rules may be abbreviated so as to read: *Do not change the quality of the proposition*, and *Do not distribute an undistributed term*. We may *indistribute* a term which is distributed, but *not vice versa*. The Convertend is the proposition to be converted: the Converse is the converted proposition.

The forms of Conversion are two, according as the *quantity* of the proposition is changed or remains the same. If the quantity of the converse remains the same as that of the convertend, the conversion is called *Conversio simplex*, or Simple Conversion; if the quantity is changed, it is called *Conversio per accidens*, or Limited Conversion, usually Conversion by Limitation. We have only to illustrate them and to ascertain the extent of their application to the several propositions A, E, I, and O.

Take a proposition in A, "All apples are fruit." In this proposition, as already shown, the predicate is not distributed. This means, as illustrated in Fig. 5 (p. 130), that other things also may be contained in the class "fruit," so far as can be determined by the assertion given. Hence, if in transposing subject and predicate we say, "All fruits are apples," we shall be asserting more than the original proposition will permit. In the original we have said nothing about the whole of the term "fruit," and so cannot be permitted to do so in the converse. Hence we can assert something only of a part of it, if we can assert anything at all. That we may assert something is evident from the fact that some degree of identity or connection exists between the subject and predicate in the convertend, and this same relation can be asserted in the converse. By limiting our statement, therefore, to the part of the predicate of which we actually affirm something, we are able to infer from the original proposition that "Some fruits are apples." This is evidently true, if the original be true. Here the quantity of the proposition is changed, but its quality remains the same; that is, the quantity of the convertend is universal and its quality affirmative, while the quantity of the converse is particular and its quality affirmative. We have, therefore, converted A into I. To convert "All apples are fruit" into "All fruits are apples" would be to violate the second rule. Hence A cannot be converted into A. To change the quality of the proposition A in conversion, that is, into either E or O, would be to violate the first rule, and it is apparent that we cannot infer an exclusion between a subject and predicate from an affirmed connection or identity between them. Hence A cannot be converted into either E or O, and we have found that A cannot be converted into A, but that it can be converted into I, which is to say that propositions in A cannot be converted *simply*, but only *by limitation*.

There is one exception in the conversion of A. This is the case of *definitions*, and of singular propositions with a singular term for predicate. Definitions are universal affirmatives, and are nevertheless convertible by simple conversion. This is a case of A being converted into A. But this is only because the nature of the case enables us to know the extension or distribution of the predicate. Formally the proposition is like all others and not convertible simply. But materially we know from the fact that the proposition is a definition, that the extension of the subject and predicate is the same, and hence convertible from A into A. Without assuming its material nature, however, we could know nothing of this distribution, and hence formally it would have to be treated as all other propositions in A, which, being neither formally nor materially distributed in the predicate, can only be converted by limitation. Nevertheless it is important to observe that in some of our material reasoning the mind may be correct in its processes on the ground that its datum is a definition; that is, subject and predicate are conceived as identical, while formally the reasoning may seem to be fallacious.

The exclusive proposition, although it appears to be a universal, may or may not be so. The "only" means *some, and it may and it may not be all, but certainly nothing else.* When converted, it becomes a universal, but not before this.

Proposition I, "Some men are vertebrates," can only be converted simply, or by Simple Conversion. We cannot infer from it that "All vertebrates are men," for the same reason that we could not convert A into A. It is because the predicate is undistributed in the convertend, and must not be distributed in the converse. It would violate rule second to convert it into E and O, and for the same reason that A could not be converted into E or O. Hence I must be converted into I, "Some men are vertebrates" into "Some vertebrates are men." This is *Conversio simplex*, or Simple Conversion, because the form of the converse is the same as that of the convertend.

Proposition E, "No books are pens," can be converted either simply or by limitation. In this the predicate is distributed, and this fact will permit of its distribution in the converse. Besides, since something is said excluding the whole of the predicate from the subject, we can assert this in the converse. Hence we can infer that "No pens are books." By subalternation, from this we can infer "Some pens are not books." The first is the *simple*, and the second the *limited* converse of the original. Hence E is convertible into either E or O. O might be called a *weakened* converse in this case, because E might as well be inferred.

In regard to propositions in O, as, "Some men are not Caucasians," a peculiar difficulty exists. First, the converse must be of the same quality as the convertend, according to the first rule. It must, therefore, be negative. But second, negative propositions distribute the predicate. Hence, whether we convert "Some men are not Caucasians," by simple conversion or by limitation, the predicate in the converse, which is undistributed in the convertend, will be distributed, and hence violates the second rule. O cannot, therefore, be converted by the ordinary method.

It has been usual, however, to apply an indirect method, called *Conversion by Negation*. Take, for example, "Some realities are not material objects." If we infer that "Some or All material objects are not realities," we commit a fallacy, because the predicate of the converse is distributed, while it is not distributed as subject in the convertend, because the proposition is particular. But if we attach the negation to the predicate in the original we have "Some realities are not-material objects," or "Some realities are immaterial objects." The proposition thus becomes I, which we can convert simply into "Some immaterial objects are realities." This proposition, then, can be inferred from the original, and the process of reaching it is called Conversion by Negation. The same process is applicable to similar propositions. Thus "Some pleasant acts are not just," would become "Some pleasant acts are not-just, or unjust," and then by conversion, "Some unjust acts are pleasant." "Some men are not agreeable" would be converted into "Some disagreeable persons are men."

But it must be observed that the quality of this so-called converse is affirmative while the convertend is negative, and hence the process of conversion by negation is a violation of the first rule. Besides, we have been led by it to affirm something positive about non-material or immaterial objects in the assumed converse, when the convertend merely denies something about *material* objects. While this may be allowable by some other process, it is not permissible by conversion. The violation of the first rule decides that matter. Hence we conclude that proposition O is really not convertible at all. This is the general opinion of logicians. In his smaller work Jevons assumes the validity of the process, but in a later treatise he concedes his error, and comes over to the general view. The proposition which is supposed to be the converse in this case is really the result of a double process, obversion and conversion, and hence is the converse of the obverse, or the contrapositive, called also the contraverse.

The forms of Conversion may be summarized in the following table, showing also the impossible forms:

Convertend.		Converse.		Impossible Forms.
All S is P.	A	Some P is S.	I	A, E, and O.
Some S is P.	I	Some P is S.	I	A, E, and O.
No S is P.	E	No P is S.	E	A and I.
Some S is not P.	O	Some P is not S. (None.)	O	A, E, I, and O.

3d. Obversion.—This is sometimes called "Immediate Inference by Privative Conception." This will serve as a good name when the propositions are affirmative, and when a privative term can be found for the purpose. But when the proposition is negative, and when a privative term is not accessible, the shorter name *Obversion* is much to be preferred. *The process consists in negating the copula and the predicate without conversion.* That is, the quality of the obvertend must be changed, and new matter introduced into the obverse, which shall be the negative concept of the predicate in the obvertend. Thus the proposition "All men are mortal" must be obverted by changing the quality of the proposition and making the predicate of the obverse the negative of "mortal." Hence the obverse of it will be, "All men are not not-mortal," or "All men are not immortal," or "No men are immortal." Of the proposition "All oaks are trees," it will be "No oaks are not trees." It will be apparent that this is, after all, identical in meaning with the original proposition, although the meaning *is stated in a negative way* for the sake of emphasis. The two negatives together make an affirmative, and we can have a negative proposition of the obverse only on the supposition that one of the negative particles qualifies the copula, and the other the predicate, so as to make it a negative conception. In some cases this negative particle can be joined to the predicate, in others not.

In regard to the negative proposition the obversion is accomplished simply by connecting the negative particle with the predicate, which both changes the quality of the proposition and the character of the predicate. Thus if I say "All men are not angels," the obverse will be "All men are not-angels." A better illustration, because of the negative term "unpleasant," is, "All pain is not pleasant," the obverse of which is "All pain is unpleasant," assuming for the present that "not pleasant" and "unpleasant" are the same. This result can be brought about by the more complicated process described in the definition, which is really implied in the case just mentioned. Thus, if I say "No pain is pleasant," I can obvert, as required by the definition, in negating the copula and predicate, as follows, "No pain is not not-pleasant," or "No pain is not unpleasant." But this is very awkward, and as two negatives make an affirmative by cancelling each other, we assume this effect and have either the original proposition or its obverse, as, "All pain is unpleasant." A negative proposition is therefore most conveniently obverted by transferring the negative particle to the predicate, which changes its quality. In regard to obversion, therefore, it is noticeable that it can be applied to all four forms of propositions—A, E, I, and O.

4th. Contraversion or Contraposition.—This process is usually called by the latter name, but a few logicians have used the former term, which seems decidedly preferable to the latter, because it indicates to some extent by the very name the nature of the process involved. Jevons, using the term Contraposition, calls it a form of conversion, but this is only partly true, because the conversion takes place after the process of obversion has been performed, as will be shown after defining what the process is.

Contraversion, or Contraposition, *consists in the negation of the copula and of the predicate with conversion.* That is, we negate the copula or proposition and the predicate, and then convert. It amounts to the same thing to take the negative of the predicate for the subject of the contraverse, and deny the connection between it and the subject of the contravertend if the contravertend be affirmative, and affirm the connection if the contravertend be negative. This can be better explained by an example. Take the proposition "All men are mortal." By the very terms of this judgment the class "men" is wholly included in the class "mortal," as in Fig. 5 or Fig. 14. Hence it is necessarily excluded from everything outside the circle "mortal." I can therefore affirm that "All men are not in the class of those who are not mortal;" or more briefly, that "No men are immortal." By simple conversion of E we get "No immortals are men." But noticing that the inclusion of "men" among the "mortal" excludes those who are not mortal from the class "men," I may as well affirm that fact directly, and hence from the original I can infer at once, from "All men are mortal," that "All not-mortals are not men." The result is the same by either process, and hence we have only to apply the principle to any similar proposition. Thus "All oaks are trees," when contraverted will be "All not-trees are not oaks," etc. We seem in all such cases to reach the result without any roundabout process. But nevertheless that of obversion is actually involved, and it was the failure to remark this fact which has led Jevons to deny contraversion or contraposition of I and O, and admit it only in an indirect way for E.

But if we observe that obversion is implied in the process, to be followed by the conversion of the obverse, we shall find that contraversion is applicable to E and O as well as to A, and in the same manner, but is not applicable to I because its obverse, O, cannot be converted. Thus take a case in E, "No Caucasians are negroes," or "All Caucasians are not negroes." Obverting this according to rule we have, "All Caucasians are not-negroes," and then converting we have, "Some not-negroes are Caucasians." Or, "No men are perfect." Obverted it is "All men are not-perfect," and converted, "Some not-perfect (imperfect) beings are men." Then again take O, "Some men are not Americans." Obverting we have, "Some men are not-Americans," and converting we have, "Some not-Americans are men." It will be noticed in this that we have a result which has been called Conversion by Negation, as already explained. But having shown that O cannot in reality be converted, it remains to observe that those who have assumed that it could be converted in this indirect manner have confused the process with contraversion or contraposition. O is, therefore, contravertible. I remains to be considered.

Take the proposition "Some men are wise." Obvert it and we have "Some men are not not-wise (foolish)." Thus we have proposition O, "Some men are not foolish," and as we have seen that O cannot be converted, we can proceed no further with the case. I, therefore, cannot be contraverted. A, however, is contraverted into E ; E into I; O into I, and I not at all. It might be added, also, that A can be contraverted into O as well as E, because O is the subalternate of E, and so is equally true with it.*

* The first thing to be remarked about Contraversion is, that, unlike Conversion, the process involves a change in the quality of the judgment, a change from the affirmative in the contravertend to the negative in the contraverse, and from the negative in the contravertend to the affirmative in the contraverse. The accompaniment of this fact and the cause of it, perhaps, is the second incident, which is that there is a change of matter as well as form in the proposition. The subject in the contraverse is new matter, and we may well question the validity of the whole process, since immediate inference aims to deduce from a content already given nothing more than it contains ; that is, immediate inference purports to give the same kind of matter, not always the same quantity, in the conclusion that is given in the premise. That new matter appears in contraversion or contraposition is evident from the fact that in the original proposition nothing is said about the negative of the predicate, and it may seem gratuitous to introduce it into the contraverse. Thus, "All men are mortal" asserts nothing about the "not-mortal," or the "immortal." Why, then, assert anything about them in the contraverse ? Again, a more important objection is that the inference can be at best only *formal*, because we do not know from "All men are mortal" that there are any "not-mortals." Undoubtedly we can assert that "All not-mortals are not men," but if we suppose that the contravertend expresses a real and material fact, does it follow that the contraverse also expresses a real fact ? If so, it would imply that there are *real* "not-mortals," about which, as a fact, we know nothing. Of course, if there are any beings other than "mortal," the contraverse will be materially true, but this existence of them cannot be inferred from the original proposition. In immediate inference the conclusion should be the same really and materially as the premise ; that is, it should be materially as well as formally true if the premise is

Conversion and Contraversion may be compared in the following table, and it will be seen at once how they contrast with each other:

Original Proposition.	Converse.	Impossible.	Contraverse.	Impossible.
All S is P . A.	Some P is S . I.	A. E. O.	All not-P is not S . E. Some not-P is not S . O. (None.)	A, I.
Some S is P . I.	Some P is S . I.	A. E. O.		A, E, I, O.
No S is P . E.	No P. is S . E.	A. I.		
Some S is not P . O.	Some P is not S . O. (None.)	A, E, I, O.	Some not-P is S . I. Some not-P is S .-I.	A, E, O. A, E, O.

First, it is noticeable that there is a change of quality in Contraversion. Then where we convert A into I only, we both. Of course, if the premise is only formal, or consists of pure conceptions, the conclusion cannot assert more. But this only shows that we cannot go beyond the matter of the original proposition. What, then, becomes of the validity of contraversion?

The only possible reply to this question is, that we are always assured by the law of contradiction between the predicate and its negative that they are mutually exclusive, and that what is affirmed of one can be denied of the other, and *vice versa;* and second, that *practically* there are real existences, in the case of the largest number of propositions, which belong to a class that is the negative of the predicate. But this reply is not satisfactory, because the law of contradiction, as mentioned, can only give a formal inference and not a material one, and because the principle does not justify a material inference where other objects than those in the predicate are not known to exist; and their existence is not implied by the material truth of the original. However, it is to be remarked that, *if* any objects exist which are excluded from the predicate of the original proposition, the inference will then be materially true, and that this inference is *contingent* upon that condition, including, on the one hand, the existence of certain objects other than the predicate, and, on the other, the exclusion or contradiction between them and the given predicate. We, therefore, conclude from the existence of this condition that contraversion is, at least *materially* considered, a conditional inference ; not conditioned upon the material truth of the contravertend, but upon the material existence of data that are excluded from the predicate of the original, or contravertend. In conversion and obversion the converse and obverse are always materially the same in kind as the propositions from which they are drawn, so that, if the originals are materially true, the derivatives are also. But in contraversion, as we have seen, the material truth of the contraverse is not deducible from anything act-can contravert it into both E and O, as we can convert E into both E and O, and contravert it into I only. Again, as I can be converted into I, and O cannot be converted at all, so I cannot be contraverted, and O can be into I. The impossible cases denote that the forms included under that head cannot be derived from the original proposition by the process indicated.

In the practical application of Contraversion we must be careful not to confuse nego-positive with *infinitated* negative conceptions. Thus if we say that "All just acts are expedient," the contraverse or contrapositive is, "All not-expedient acts are not just." But we are not entitled to say "All not-expedient acts are unjust," because all that we know from the original assertion is that they are excluded from the "just," and the negative of this, which is "not-just," includes the "unjust," and those which are neither just nor unjust, namely, indifferent acts, such as physical actions, and we do not know from the original whether the "not-expedient" acts are excluded from those which are neither just nor unjust, or included in the "unjust." We only know that they are excluded from the "just." Besides, to say that "All not-expedient acts are unjust" is to produce a proposition in A, when the contraverse should be E. The conception "unjust" is not necessarily co-extensive with the negative "not-just," which is infinitated. If it were, there would be no objection to their identification, and then the original contraverse, "All not-expedient acts are

ually affirmed in the contravertend, but is conditioned solely upon the existence of objects other than the predicate of the latter. It may then be regarded as *formally* true, but *materially* indeterminate. The same will be true of Inversion, of which we have yet to speak, and would be true of Obversion except for the fact that it expresses merely in a negative way the same as the original proposition. In Conversion, however, the converse is both formally and materially true; that is, materially true when the convertend is so. The following is a classification of immediate inferences based upon the above considerations:

Immediate inference
- Formally and materially true
  - Conversion.
  - Obversion.
- Formally true, but materially indeterminate
  - Contraversion.
  - Inversion.

not just," could be *obverted* into "All not-expedient acts are unjust " (not-just). But only on that condition can it be done. The same remarks apply to the substitution of *inexpedient* for "not-expedient." A better and more self-evident illustration of what is here maintained can be seen in the proposition "All human actions are free," the contraverse of which must be, "All not-free actions are not human," not "All not-free actions are inhuman." Or again, "Human kindness is a virtue," the contraverse of which is "All that is not virtue is not human kindness," not "All that is not-virtue is inhuman kindness." The absurdity of the last is palpable, but it is precisely what we have to guard against.

5th. Inversion.—Inversion is not often treated by logicians, and it is not a process of any practical importance. Besides, it has first to be obtained through such combinations of the three previous processes as not to be apparent at first sight. But the fact that it is possible requires mention in a complete treatment of immediate inferences. Inversion is the process of inferring from one proposition another which shall contain for its subject the negative of the subject in the original, and for its predicate the predicate of the original. The result can be obtained in more than one way. We may alternate conversion and obversion, or use contraversion with either or both the other processes. As contraversion, however, is merely a combination of conversion and obversion, we may use only the former method. We may begin with either conversion or obversion.

Take the proposition "All horses are animals." Convert and we have "Some animals are horses." Then obvert this result, which gives "Some animals are not not-horses." Here comes the turn for conversion, but as the proposition is O we can proceed no further. But if we begin with obversion the first transformation gives us "No horses are not-animals," then convert and we have "All not-animals are not horses." Obvert again and we have "All not-animals are not-horses " (remembering that obversion is only a change of quality in the proposition), and then converting this obverse we have "Some not-horses are not-animals." Obverting this again we have "Some not-horses are not animals," which is the required proposition, having the negative of the original subject for its subject, and the original predicate for its predicate.

If we take proposition E, "No men are quadrupeds," and begin with conversion, we have "No quadrupeds are men." Obvert this and we get "All quadrupeds are not-men," and applying conversion we have "Some not-men are quadrupeds," the required inverse. If we begin with obversion we shall find that the result cannot be obtained.

Without going through the process with I and O, which the student can do for himself, we shall content ourselves with the assertion that they cannot be inverted. A and E are the only two invertible, A into O, and E into I. It must be remarked, however, that the material truth of the conclusion in either case is dependent upon the same conditions as in contraversion. Take the case of the inversion of A into the form "Some not-S is not P," from "All S is P," and it is clear from "All S is P" that, so far as we know, all things outside of S are included *in* P, and if so, we could hardly conclude that some of them were not so included, and hence "not P." But if there are any objects outside of P, then it is clear from the subalternate of the contraverse of "All S is P," that "Some not-S is not P." But it can be materially true only on that condition. Hence I have placed inversion in the class of inferences which are materially indeterminate. It has no practical importance, however, and is applicable only to A and E, and not to I and O. It should be observed that it involves a change of quality in the deduced proposition. All four processes can be compared in the following table:

Original.	Converse.	Contraverse.	Inverse.
All S is P . A.	Some P is S . I.	All not P is not S . E.	
		Some not P is not S . O.	Some not-S is not P . O.
No S is P . E.	No P is S . E.		
	Some P is not S . O.	Some not P is S . I.	Some not-S is P . I.
Some S is P . I.	Some P is S . I.	(None.)	(None.)
Some S is not P . O.	(None.)	Some not P is S . I.	(None.)

6th. Contribution.—I have called two processes of immediate inference by this name because the same general principle is involved in both. They are called *Immediate Inference by added Determinants*, and *Immediate Inference by Complex Conceptions*. The essential characteristic in common between them is that *what is affixed to the subject as a modifier may also be affixed to the predicate in the same sense*. What is contributed to one may be contributed to the other. Its simplest illustration is in mathematics. Thus if $x = a$, $x + 1 = a + 1$.

(*a*) *Inference by Added Determinants.*—This consists in merely adding some adjective or similar term to both subject and predicate. Thus if "A brick house is a dwelling," "A good brick house is a good dwelling." There is one important limitation to this process, however, and it is that the addition must represent the same quality and quantity to both subject and predicate. Thus, "Dogs are quadrupeds" may be modified by adding "useful" to both terms, but not by saying "The largest dogs are large quadrupeds." Nor can we infer from "Dwarfs are men," or "A cottage is a building," that "Large dwarfs are large men," or "A large cottage is a large building." The reason for the error in these cases is that the addition of quantity in each case is not the same. When the superlative degree is added it implies a comparison between the special individuals mentioned or described and the whole class which is thus limited, so that the same addition is not made to subject and predicate. It is the same with such terms as "large," "long," "small," "short," etc., in certain cases, namely, when an unequal comparison is suggested. Terms, therefore, *expressing quantity* must be used carefully in this form of inference. Terms *expressing quality*, however, can be used with perfect freedom, provided they are not used equivocally.

(*b*) *Inference by Complex Conception.*—This consists in the addition of words, phrases, or clauses that make the subject and predicate complex in their nature. Thus from "Apples are fruit" I may infer that "A box of apples is a box of fruit," or from "Pigeons are birds," "A flock of white pigeons is a flock of white birds," etc. But here again we have to be warned against the addition of expressions which indicate the addition of a greater quantity to one side than to the other. Thus we cannot infer from "Voters are men," that "The majority of voters is the majority of men." The same principle holds here as in inference by added determinants.

Inferences by contribution have little importance, as they are seldom employed, and if fallacies occur in this use they are so easily detected and corrected that little provision requires to be made against their occurrence.

7th. Antithesis.—There is a very common error in dealing with propositions, which it is necessary to remark. We have seen that an exclusive proposition implies a complementary opposite, and it frequently occurs that we infer from an ordinary universal, A or E, proposition one which is its opposite both in regard to the subject and predicate. Thus, if we assert that "All white objects are visible," many persons think themselves entitled to infer that "All not-white objects are not visible;" or from "All good men are wise," that "All bad men are not wise." But such an interpretation and inference is a case of illicit distribution of the predicate. If the proposition were either exclusive or a definition, the inference might be drawn. Formally, however, no one ever has the right so to treat it. In ordinary antithesis this relation is assumed, as is so frequent in the Book of Proverbs; for example, "In the multitude of words there wanteth not transgression; but he that refraineth his lips doeth wisely." And in all ordinary cases where subject and predicate are coterminous, the same can be done. In such instances we are entitled to infer the complementary opposite. But we should not generalize a privilege of this kind and apply it to all universal propositions. In the universal propositions as defined, we have no data for asserting anything outside the subject, unless we assume something to exist outside the predicate, which we have no right to assume from the given proposition. But if it be known or assumed that something exists outside the predicate, we may then infer a contrapositive or the contraverse, and the limited converse of this which has been shown to be the *inverse* of the original proposition. Any other inference than those based upon this supposition, is an illicit assumption of distribution of the predicate when the original proposition is affirmative, and an illicit distribution of the negative of the subject when the proposition is negative. Thus, in the first case, if we infer from "All men are mortal" that "All not men are not mortal," we assume a distribution of the predicate which is not stated in the original. In the second case, if we infer from "No men are quadrupeds" that "All not-men are quadrupeds," we assume a distribution of the negative of the subject "men." This cannot be true, because although we know by the infinitation of the concept "not-men" that some things excluded from "men" are included in "quadrupeds," there is nothing in this idea to justify the assertion that *all* "not-men" are so included. The subalternans cannot be inferred from the truth of the subalternate. This error may be called illicit contrast or illicit antithesis.*

* References on Immediate Inference are the following: Keynes: Formal Logic, Part II., Chaps. III., IV. and V.; Jevons: Studies in Deductive Logic, Chap. V.

CHAPTER XI.

PRINCIPLES OF MEDIATE REASONING

1st. Definition.—Mediate inference, as already defined, is reasoning by means of a middle term. A middle term is one which is compared with two others, and on the ground of which a connection or relation can be established between the two terms thus compared with it. Suppose I wish to prove that "Machines are useful." This is the conclusion, and its proof will depend upon two other propositions containing the two terms in the conclusion and comparing them with a middle term. If, therefore, I can obtain assent, first, to the proposition that "Mechanical arrangements for the application of power are useful," and second, to the proposition that "Machines are mechanical arrangements for the application of power," the conclusion will follow as a matter of course. The terms "machines" and "useful" are compared with the single term "mechanical arrangements," etc., and found to agree with it, and so it must be inferred that they agree with each other. The conclusion is thus *mediated* by a middle term. The reasoning involved in this process is called the *Syllogism*, or *Syllogistic Reasoning*. It differs from immediate inference in requiring more than one proposition, and more than two terms to start from. The two rules which determine or condition the character of the syllogism are :

(a) *Every syllogism should have three, and only three terms.*

(b) *Every syllogism should have three, and only three propositions.*

The three terms are called the *Major*, the *Middle*, and the *Minor* terms. Of the three propositions two are called the *Premises*, and one the *Conclusion*. The Major term is the *predicate* of the conclusion ; the Minor term is the *subject* of the conclusion. The Middle term is found only in the premises, and may be either the subject or the predicate as the case requires. The Major Premise contains the Major and Middle terms, and the Minor Premise the Minor and Middle terms. Without the conclusion being expressed there is no absolute rule for determining which of the premises is the major and which the minor. We are at liberty to choose either of them and to try the consequences by the laws of the syllogism. The usual method of the logician is to state the major premise first, the minor premise second, and the conclusion last. But in common discourse and reasoning the minor premise may come first, and the major second; or the conclusion may even come first, and the two premises after it, as reasons for its truth. As an illustration of a syllogism having the minor premise first, the following is an instance:

The earth revolves around the sun.
Bodies revolving around the sun are planets.
Therefore the earth is a planet.

It is evident from the conclusion, and from what has been said about the terms which it contains, that the minor premise is the first and the major premise the second. As a case of the conclusion being stated first, the following is an instance:

"Comets consist of matter, for they obey the law of gravitation, and whatever obeys the law of gravitation is matter."

The symbols which we shall employ for the several terms of the syllogism are the letters S, M, and P, with a slight modification of their previous usage, but yet essentially the same. S shall stand for the *minor term*, which is the subject of the conclusion, and P for the *major term*, which is the predicate of the conclusion. They thus stand respectively for subject and predicate, as heretofore, *but only in the conclusion*, since S or the minor term is not always the subject in the minor premise, and P, the major term, is not always the predicate of the major premise. M will stand for the *middle term*. The combination of these terms will represent the premises and conclusion.

Terms.	Propositions.
M = Middle term.	M is P = Major Premise.
S = Minor term = Subject of Conclusion.	S is M = Minor Premise.
P = Major term = Predicate of Conclusion.	S is P = Conclusion.

PRINCIPLES OF MEDIATE REASONING

2d. Rules of the Syllogism.—The rules of the syllogism are classified in slightly different forms, but substantially in the same manner. It is usual to include among them the two which I have already mentioned, and to include in one of them a statement about the *ambiguous middle*. But I prefer to specify two classes of rules affecting different aspects of the syllogism. The first two already mentioned, and with them a precaution against an ambiguous middle, relate to the logical matter of the syllogism. The remaining six relate to the quantity and quality of propositions, and the distribution of terms, as effecting the conclusion. We shall classify them according to these divisions.

(*A*) Rules affecting the logical matter of the syllogism.
    1. Every syllogism must have three terms, and only three.
    2. Every syllogism must have three propositions, and only three.
    3. No term shall be used ambiguously.

The violation of the first of these rules gives rise to the fallacy known as *Quaternio Terminorum*, or the Fallacy of Four Terms. The third rule is a corollary of the first, because an ambiguous term is a word with two meanings, and hence is equivalent to two terms, so that it is only a form of introducing *four* terms into the syllogism, producing in effect a fallacy of *Quaternio Terminorum*.

(*B*) Rules affecting the quantity and quality of propositions and the distribution of terms.
    4. The middle term must be distributed at least once in the premises.
    5. No term must be distributed in the conclusion which was not also distributed in one of the premises.
    6. No conclusion can be inferred when both premises are negative.
    7. If one premise be negative the conclusion must be negative and, *vice versa*, in order to prove a negative conclusion one of the premises must be negative.

8. No conclusion can be drawn when both premises are particular.

9. If one of the premises be particular the conclusion must be particular.

The last two are frequently considered as corollaries of the preceding. All of them are so important that they should be thoroughly committed to memory by the student for constant reference.

3d. Symbolization of the Syllogism and Representation of Formal Fallacies.—The syllogism can be symbolized in the same manner as propositions, except that it requires a larger number of circles for each syllogism, because there are three terms and three propositions. The representation depends upon the quantitative relation between subject and predicate, and can be used to indicate whether any two terms agree or exclude each other. When the relation between subject and predicate is properly observed, and when the proper distribution of terms is made, the conclusion will be *valid*. But if the middle term be not distributed in one of the premises, or if the major or minor terms be distributed in the conclusion when they are not distributed in the premises, a fallacy occurs, which is called a *Formal Fallacy*, because it is a violation of the formal principles of the syllogism. Hence such reasoning will be formally *invalid*. The conclusion may happen to be *materially* true, but under the circumstances it has not been correctly obtained or deduced from the premises, as a conclusion. These formal fallacies are called *Illicit Processes*, respectively, of the major, middle, and minor terms. If the major term be distributed in the conclusion and is not distributed in the premise, the inference is an *Illicit Process of the Major Term*. If the middle term be not distributed at least once in the premises, the inference is an *Illicit Process of the Middle Term*. If the minor term be distributed in the conclusion and is not distributed in the premise, the inference is an *Illicit Process of the Minor Term*.

There are two ways in which we can represent symbolically the valid and the illicit processes of formal inference. We shall employ both of them. We must remember the distinction, however, between the formal truth or falsehood of the inference and the material truth or falsehood of the propositions. The two may have no relation to each other. Thus we may reason correctly from false premises, or falsely from correct premises. Hence, as we are dealing here only with the manner of getting the conclusion, we may set aside all questions about the nature of the premises materially considered. The student must always be prepared to detect the difference between the truth of the conclusion and the validity or invalidity of the process by which we obtain it. It is a common fault of our minds to accept the reasoning if we accept the conclusion, and to impeach the reasoning if we do not accept the conclusion. But this is a fallacy of itself, and we must dispel such assumptions from our minds, or we shall not be able in all cases to detect the true or false character of the reasoning. On the other hand, we must not be induced by the purely formal correctness of the reasoning to accept the material truth of the conclusion on that account. The most perfect reasoning is that in which the premises are materially correct and the reasoning formally correct. This will give a conclusion which is both formally and materially valid. Now to illustrate only the formal process, and to symbolize it accordingly, we shall take first a valid case, the ordinary syllogism: thus "Metals are elements," "Iron is a metal," "Therefore iron is an element," will be represented by the circles in Fig. 18, and the distribution of terms in the accompanying diagram.

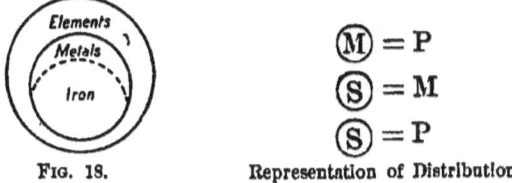

FIG. 18. Representation of Distribution.

In Fig. 18, "metals" are represented as included in the class "elements," and "iron" in the class "metals." It must, therefore, follow that "iron" is included in the class "elements," as the lesser is included in the greater, or the part in the whole. The diagram shows that the middle term has been distributed at least once, and that no term has been distributed in the conclusion which was not distributed in the premises. The propositions are, of course, affirmative. But we may take a case where one of them is negative, and so have an instance of exclusion. Thus, "Iron is a metal," and "Wood is not a metal." Therefore "Wood is not iron." This syllogism will be represented also by three figures, but one of them is excluded from the other two, as required by the negative proposition.

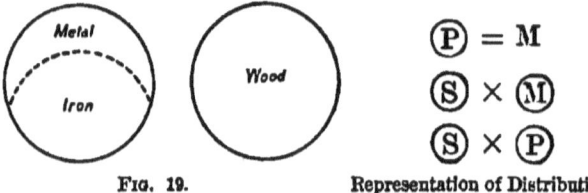

FIG. 19.   Representation of Distribution.

Fig. 19 represents "iron" as included in class "metal," and "wood" as excluded from the same class, so that it must be excluded from all contained within the class "metal." The diagram shows the observation of the rules in another form.

Instead of a dotted line we might use a smaller circle in either case, as is usual, and perhaps better, in most instances. When the predicate is not exhausted in the subject the complete smaller circle is much more convenient, and hence in the future it will be employed, especially in testing the forms of illicit process.

Had it been used in the previous symbols Fig. 18 would stand as Fig. 20, and Fig. 19 as Fig. 21.

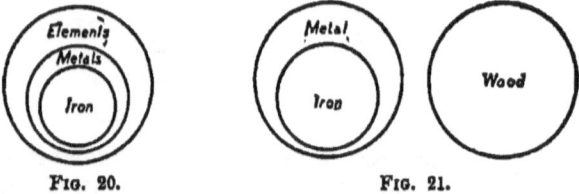

FIG. 20.   FIG. 21.

The student or the teacher may vary the forms of figures to suit other valid modes of reasoning. We turn next to test

PRINCIPLES OF MEDIATE REASONING 177

and represent the forms of illicit process. Take first the case of illicit middle. It occurs in the following syllogism, "Men are mortal." "Horses are mortal" "Therefore horses are men." It is represented in Fig 22.

FIG. 22.   Representation of Distribution.

As both propositions of the premises are affirmative, the predicate, which in this case is the middle term, is not distributed, and hence an attempt to draw the conclusion, "Horses are men," or S is P, is illicit or invalid, because a violation of Rule 4. It is represented by Fig. 22, in that both "horses" and "men" are included in the class mortal, but we have no right to infer therefrom that "horses" are included in the class "men." The fact that we know it is not the case helps us to perceive the impossibility of it. But even in cases where the conclusion would be true as a fact, we should have no right to draw it in this manner. Thus we may say, "All Americans are men," and "All Virginians are men," but we should have no right to infer that "All Virginians are Americans," although it is true as a matter of fact. The fallacy here is that of an *Illicit middle term*.

Illicit process of the major may occur as follows, and is likely to be a very frequent fallacy. If from "Men are mortal," and "Horses are not men," we infer that "Horses are not mortal," we have more in our conclusion than is given in the premises. Fig. 22 will also represent this form. But the diagram for distribution of terms will be different. It will be as follows, and represents the major term as distributed in the conclusion when it is not distributed in the premises. The major premise is affirmative, and hence the predicate, which in this case is the major term, is not dis-

12

tributed. But the minor premise being negative, the conclusion, according to Rule 7, must be negative. We have seen that all negative propositions distribute the predicate, and as this is the major term in the conclusion, it must be distributed, if drawn at all, although it is not distributed in the major premise. The error is, therefore, an *illicit process of the major term*. Fig. 22 represents it by showing that the class "horses" is excluded from the class "men," as asserted by the minor premise, and yet is not excluded from the class "mortal," as the conclusion would indicate. The reasoning is therefore a violation of Rule 5.

Illicit process of the minor term is quite as easily represented. It occurs in the following syllogism: "All horses are quadrupeds," and "All quadrupeds are animals." "Therefore all animals are horses." It is represented in Fig. 23.

$$\text{(P)} = M$$
$$\text{(M)} = S$$
$$\text{(S)} = P$$

FIG. 23.   Representation of Distribution.

In the minor premise the predicate S, or "animals," is not distributed, because the proposition is affirmative, but in the conclusion it is distributed, because the proposition is universal, and thus to distribute it is violation of Rule 5. The impossibility of the conclusion is evident in the representation, on the ground that the larger class cannot be included in the smaller, or the whole in the part, as it is represented in the conclusion actually drawn. Had we said "Some animals are horses," the conclusion would have been valid as conforming to all the rules. But the minor term is distributed in the conclusion first drawn, and is not distributed in the premise.

The violation of Rule 6 occurs in the following syllogism, and is represented by either of Figs. 24 and 25, "No solids are liquids," and "No metals are liquids." "Therefore no metals are solids," or "Metals are solids."

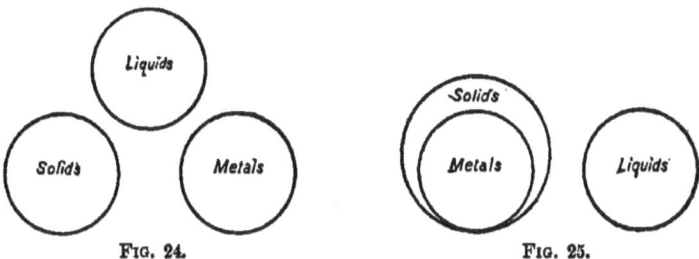

Fig. 24.   Fig. 25.

Fig. 24 represents the premises as negative propositions, in which the subjects "solids" and "metals" are both excluded from "liquids," and excluded from each other. But their common exclusion from the middle term does not necessarily imply their exclusion from each other, as is apparent in Fig. 25, where they are both excluded from "liquids," and yet one of them is included in the other. All that is necessary to make the premises negative is to have the major and minor terms excluded from the same middle term, and this leaves the connection between the major and minor terms entirely indeterminate. We can, therefore, draw no conclusion at all, either affirmative or negative, because there is nothing said or denied in the premises to imply one or the other.

Diagram of Distribution.

There is an apparent exception to Rule 6 in such cases as the following, which require analysis.

That which is not wise cannot be useful.
Intemperance is not wise.
∴ Intemperance cannot be useful.

Here it seems very apparent that we have negative premises, and yet the mind does not wince at the conclusion, nor feel that anything wrong has occurred in drawing it. The reasoning resists all attempts to invalidate it. But the reason for this is, that although the propositions *seem* to be EEE, they are in reality EAE of the first figure, and therefore valid. The first proposition is made negative by the "cannot," but in the second the nature of the middle term is such that in order to have it the same in both premises the predicate of the minor premise must be negative, and the "not" in its real import must go with the "wise." The logical subject in the major premise is the whole thought expressed by the middle term, which is "the not-wise acts," and in the minor premise this conception can be brought out only by construing the predicate in the same way, and so regarding it as equivalent to "that which is not wise," or "one of those things which is not wise," and therefore "a not-wise act," and we find that the proposition becomes A with a negative conception for the predicate. That is, we simply obvert the minor premise, so that it becomes in reality affirmative. Whatever we may think of the form of the proposition, therefore, it becomes materially an affirmative one, or is thought as such, and the reasoning is determined accordingly.

The fact calls attention to the double use of the negative particle in propositions which must be reckoned with in dealing with them. It may be attached to the copula, denying the connection between subject and predicate, and making the quality of the judgment negative, or it may be attached to the predicate, making the quality of the judgment affirmative and the predicate negative. This is but a case of obversion, but it shows that we may often mistake the mere form of a proposition for its matter. We must, therefore, be on our guard against mistaking what is really a negative attribute for a negative assertion.

Rule 7 is proved and illustrated by Figs. 19 and 21. A violation of it would appear in the attempt to draw the affirmative conclusion, "Wood is iron," because in the premises "wood" is excluded from "metals," and so must be excluded from whatever is contained in that class.

Rules 8 and 9 can be best tried and illustrated when we come to consider the Moods and Figures of the syllogism in the next chapter. The circles would take up more space than is necessary, while practically few fallacies are incident to the 8th Rule.

CHAPTER XII.

MOODS AND FIGURES OF THE SYLLOGISM

1st. Moods.—The Mood of a syllogism is that characteristic of it which is determined solely by the *quantity and quality* of its propositions. The Mood can never be separated from the Figure, but it is not determined by the same properties. It is necessary to consider both of them in order to ascertain the valid and invalid forms of reasoning, but this result can be most easily and most briefly attained by first considering the Moods. Every syllogism, as we have seen, must contain three propositions, and only three. But there are four forms of propositions, namely, A, E, I, and O, from which to choose, or which may be combined in various ways to produce the requisite number for a syllogism. Thus it is possible to take all three propositions in A. In such a case the major and minor premises, and the conclusion, would each and all be propositions in A. Or, we might have one in A, one in E, and one in O. It is conceivable that all three propositions should be E, or I, or O. We happen to know, however, from Rules 6 and 8, that such forms are not valid; but apart from those rules we could not so decide the matter. Hence we must represent at present all the conceivable moods, and test them afterward. Apart from these rules, therefore, any combination of three propositions is possible, as representing the conceivable ways in which a syllogism might be formed. If the three propositions were in A, the syllogism for AAA would represent a mood in which the major premise was A, the minor premise A, and the conclusion A. This order represents the order of the premises and the conclusion. The mood AEO would therefore mean that the major premise was A, the minor E, and the conclusion O. Now as the conceivable moods may represent all possible combinations, either of the same kind or of different kinds of propositions, AEO being one mood an EAO being another, there will be a large number of possible moods. When completed we find the possible number to be 64. They appear as follows:

AAA	AEA	AIA	AOA	EAA	EEA	EIA	EOA
AAE	AEE	AIE	AOE	EAE	EEE	EIE	EOE
AAI	AEI	AII	AOI	EAI	EEI	EII	EOI
AAO	AEO	AIO	AOO	EAO	EEO	EIO	EOO
IAA	IEA	IIA	IOA	OAA	OEA	OIA	OOA
IAE	IEE	IIE	IOE	OAE	OEE	OIE	OOE
IAI	IEI	III	IOI	OAI	OEI	OII	OOI
IAO	IEO	IIO	IOO	OAO	OEO	OIO	OOO

But these are not all valid. Some of them violate one rule, and some another. For example, EEA violates Rule 6; EAI violates Rule 7; IOA violates Rules 7 and 8. By thus applying the rules of the syllogism to each one we find that a large number of them are to be rejected as *pseudo-moods*, or false moods, because no conclusion can be valid in them. In this way 52 are rejected because they violate some one or two of the Rules 6, 7, 8, and 9. Some of them also violate Rule 5, but this law is not applied until we come to test the moods in the Figures of the syllogism. There remain, therefore, 12 *possibly* valid moods; we say possibly, because it is not proved that they are valid when they are not found to conflict with any of the above four rules. They must first be tested in the Figures, and it will then be found that some of them are valid and some are invalid in one or more of the Figures, and one mood, IEO, proves to be invalid in all of them, on the ground of an illicit process of the major term. The following are the 12 moods, with IEO bracketed for the reason just specified:

AAA	EAE	IAI	OAO
AAI	EAO	[IEO]	
AII	EIO		
AEE			
AEO			
AOO			

MOODS AND FIGURES OF THE SYLLOGISM

These remain to be tested by the Figures, and as there are four Figures there will be 48 forms which have yet to be considered, four of which have to be rejected as involving IEO, and leaving 44 cases in which the reasoning might be valid. But as a large number are found to violate Rule 5, in one or more of the Figures, they have to be rejected. What remains will appear in a moment.

2d. Figures.—The Figure of a syllogism is that characteristic of it which is determined by *the position of the middle term*. As there are two propositions, each with a subject and predicate, in the premises of every syllogism, there are *four* possible positions for the middle term. It may be the subject of both, the predicate of both, the subject of the major, and the predicate of the minor, or the predicate of the major and subject of the minor premise. These positions are represented in the following diagrams:

1st Fig.	2d Fig.	3d Fig.	4th Fig.
M = P	P = M	M = P	P = M
S = M	S = M	M = S	M = S
S = P	S = P	S = P	S = P

Each of the 12 moods can be tested in the four Figures, making, as already said, 48 possible forms in all, which might be valid, were not some of them violations of Rule 5 in one or more of the Figures. Thus in mood AAA we may have four different positions of the middle term, which is represented in the following diagrams, showing those in which it is valid and those in which it is invalid:

1st Fig.	2d Fig.	3d Fig.	4th Fig.
A (M) = P	A (P) = M	A (M) = P	A (P) = M
A (S) = M	A (S) = M	A (M) = S	A (M) = S
A (S) = P	A (S) = P	A (S) = P	A (S) = P
Valid.	Invalid.	Invalid.	Invalid.

In the second Figure, as shown, AAA gives rise to the fallacy of undistributed or Illicit Middle; in the third Figure, to the fallacy of Illicit Minor; and also Illicit Minor in the fourth Figure. It is therefore valid formally only in the first Figure. Take another example, the mood AEE:

1st Fig.	2d Fig.	3d Fig.	4th Fig.
A Ⓜ = P	A Ⓟ = M	A Ⓜ = P	A Ⓟ = M
E Ⓢ × Ⓜ	E Ⓢ × Ⓜ	E Ⓜ × Ⓢ	E Ⓜ × Ⓢ
E Ⓢ × Ⓟ	E Ⓢ × Ⓟ	E Ⓢ × Ⓟ	E Ⓢ × Ⓟ
Invalid.	Valid.	Invalid.	Valid.

We find AEE valid only in the second and fourth Figures. In the first and third Figures it gives rise to the fallacy of Illicit Major. By testing each mood in this way we reject a large number of them as invalid in some Figures, although valid in others. Only one is not valid in any of them, and this, as remarked, is IEO. When all the valid moods, therefore, are selected out of the 44, we find there are 24 of them, which are given below. Of this number 5 are called *Weakened Conclusions*, because, although valid, they are of no practical use, because a *particular* conclusion is drawn when a *universal* one might as well have been drawn. They are placed in brackets.

1st Fig.	2d Fig.	3d Fig.	4th Fig.
AAA	AEE	AAI	AAI
AII	AOO	AII	AEE
EAE	EAE	IAI	IAI
[EAO]	[EAO]	EAO	EAO
EIO	EIO	EIO	EIO
[AAI]	[AEO]	OAO	[AEO]

Several other lists with a slightly different order might be given, but the present one is probably convenient enough for the memory. The arrangement has been made as nearly as possible to indicate in the same line, and with as little space, those moods which are valid in more than one Figure, and at the same time to collect allied moods together, so far as permissible. But since it may appear a difficult task to memorize the whole 24 forms, and to locate them in their right Figures, I have adopted a mnemonic system which may be convenient for many students, although I do not find it necessary to use it for myself. The meaning of the system can be easily explained. The vowels indicate the *Moods;* the consonants B, C, D, and M represent, respectively, the first, second, third, and fourth Figures, M being chosen instead of F, solely for the sake of euphony. L and N are simply connecting consonants where we cannot use B, C, D, or M. There are only eleven words in all in the system, corresponding to the eleven valid moods. Each word indicates the mood and Figures in which a given form is valid. They are as follows : *

Balana   Caleme   Dilami
Badami   Calemo   Dolano
Badini   Calono
Becane
Becadom
Becidom

The shortest possible list of the valid moods is the following, and it will not appear so formidable to the memory as the previous instances. It ignores the weakened conclusions because it is apparent that I can be drawn wherever A is drawn, an O wherever E is drawn. The numbers following each mood signify the Figures in which it is valid. These moods can probably be best remembered without resort to a mnemonic system. AAI, 3. 4 and EAO, 3. 4 are omitted.

AAA, 1.   AEE, 2. 4.   EAE, 1. 2.   IAI, 3. 4.
AII, 1. 3.   AOO, 2.   EIO, 1. 2. 3. 4.   OAO, 3.

In examining the list of valid moods and Figures, we observe that the first Figure is the only one which can have A for a conclusion, and is the only Figure in which all of the four propositions A, E, I, O can be proved. In the second Figure we observe only negative conclusions. No affirmative conclu-

* This is not intended to take the entire place of the usual mnemonic verses to be considered under the reduction of the Moods and Figures, but only to aid in the detection of those which are valid in more than one Figure. Those who adopt the common system need not adopt the one I have given.

sion is possible in this Figure because the premises would have to be affirmative in order to give it. But since the middle term is the predicate in both premises of this Figure, it is undistributed whenever the proposition is affirmative, and hence an inference would be a case of illicit middle. The third Figure gives only particular conclusions. If the major premise be negative, a universal conclusion would be an illicit minor; if the minor premise be negative, any conclusion would be an illicit major. Hence only particular conclusions can be drawn, and that only when both premises are affirmative, or the major premise negative, except in the Mood OAO. The fourth Figure is peculiar in being easily reducible into the first Figure, and hence it is of little practical use. We have only to change the position of the premises, or *mutate* them, as it is called, in order to convert the fourth Figure into the first. It requires mention, however, because certain moods are valid in it which are not valid in the first.

In regard to the practical importance and usefulness of the Figures there is some difference between them. We have already remarked that the first Figure is the only one to give conclusions in all four propositions, A, E, I, and O; the second will give them in E and O; the third in I and O; and the fourth in E, I, and O. The first Figure, therefore, is the only one to give universal affirmatives, or A. This fact makes it the most important and useful of the forms of reasoning, because the demands of human belief and conduct require universal truths of some kind which do not give exceptions such a place and influence as to impair the value of general principles. Thus if the needs of society require that "All men should do right," and this proposition was not accepted on its bare statement, we should require to prove it, and upon premises of which it was a part or an equal. But if we could prove from those premises only that "Some men should do right," the truth would be so indefinite that there would be no telling whether the obligation was incumbent upon a sufficient number of men to make the principle worth asserting. If the "some" means two or three, or a hundred, the immunities which the remaining majority *may* have, so far as premises and conclusion are concerned, make the principle perfectly ineffective, especially as the particularity and indefiniteness of the conclusion is such as to make it, *practically*, a universal negative; that is, if we only prove that "Some men should do right," and are not able to specify what "some" or portion of them are under that obligation, the indefiniteness of the assertion prevents us from asserting the obligation of anybody in particular. But if we can prove the universal, the conclusion applies definitely to everybody. Hence the first Figure has an importance corresponding to its efficiency for establishing such conclusions.

There is another point of interest in this Figure to which Keynes calls attention. It is that only in it have we both the subject of the conclusion as the subject in the premises, and the predicate of the conclusion as the predicate in the premises. This makes it unnecessary for the mind to change its conception from one relation to the other in coming to its conclusion; it can still think of its terms in the conclusion as having the same functions and logical relations as in the premises. It, therefore, accounts for the fact that reasoning so generally seems more natural in the first Figure than in any other. We resort to it for greater clearness and effectiveness.

The second Figure is especially adapted to disproof, or the proof of universal negatives, or propositions in E. It is so adapted, partly because one of the premises must be negative, and partly because of the kind of comparison which can be established by it between subjects and predicate. It is clear that if two things cannot agree in their predicate they cannot agree with each other. Hence to disprove an assertion we have only to prove that one instance, or more, included in the general statement does not agree with the predicate and we have the contradictory established. Thus, if it be asserted that "All forms of government are beneficial," we may disprove this by showing that "An arbitrary despotism is not beneficial," which would involve in a comparison of the two statements a syllogism of the second Figure, whose conclusions would be, "Arbitrary despotisms are not governments," which is the opposite of what is implied in the nature of the case, and hence the opponent will have the option of giving up the universality of his assertion that "All governments are beneficial," or of maintaining that despotisms are not forms of government.

This seems, however, like a mere proof of an exception, and not of the *total* falsity of the universal. A better instance of the case is the following. Suppose it be asserted that "Strikes are justifiable." The most complete disproof of this would be the truth of the proposition that "No strikes are justifiable." Now, if we can assert that "All interferences with the rights of property and capital are not justifiable," and can have it taken for granted that strikes are such interferences, we have proved the direct opposite of the original. The syllogism would stand thus with the original proposition as the minor premise, instead of the major, as in the first case.

Interferences with the rights of property and capital are not justifiable.
Strikes are justifiable.
Strikes are not justifiable.

Now as the minor premise is asserted by one person and the major by another, with the supposition that it will be admitted by the first person, the conclusion intended to be enforced is, "Strikes are not interferences with the rights of property and capital," while it is assumed that it is clear they are such interferences. Hence if they are such, and yet the major premise is admitted, it will follow that "All strikes are not justifiable," the direct opposite of the original proposition asserted. The second Figure is, therefore, well adapted to the method of disproof.

The third Figure is adapted to the proof of exceptions to universals, or disproves universals by their contradictories, as the first Figure disproves them by their opposites. Or, when it does not prove exceptions, it proves certain particular truths which will be of considerable value against the assertion of

MOODS AND FIGURES OF THE SYLLOGISM 189

this universal contradictory. As an illustration of this latter fact, as well as the former, take the assertion that "No philosophers are wise." Now, if we wish to disprove this, and at the same time prove that "Some philosophers are wise," we have only to prove the following:

>Plato, etc., were wise.
>Plato, etc., were philosophers.
>∴ Some philosophers were wise.

Here we both prove that "Some philosophers are wise," and disprove the universal negative which had been asserted against this possible fact. A particular exception is proved. Therein lies the value of the third Figure, when it is found difficult or impossible to prove the universal opposite of a given assertion.

The fourth Figure is not regarded by logicians as having any great practical value. It is not so frequently used as the others, and can so often be changed into the first Figure by mutating the premises, that its practical importance does not require special notice.

CHAPTER XIII.

REDUCTION OF MOODS AND FIGURES

1st. Direct Reduction.—The number of valid Moods and Figures is so great, and the apparently irregular character of them so perplexing that the mind often finds it difficult to remember them correctly. The old logicians, therefore, invented a mnemonic device in aid of the memory, and which at the same time would contain some indications of the various processes required for what is called their *reduction*. The mnemonic verses consist of barbarous Latin terms added to a few genuine ones, indicating all the Moods and Figures that are valid, except the instances of *weakened conclusion*. The verses are given below, with the indications of the accent to be employed in reading them according to the laws for scanning Latin poetry:

Bărbără, Cĕlārēnt, Dărĭĭ, Fĕrĭōquĕ prĭōris :
Cĕsărĕ, Cāmēstrēs, Fēstĭnŏ, Bărŏkŏ, sĕcūndæ:
Tĕrtĭă, Dārāptī, Dĭsămĭs, Dătīsĭ, Fĕlāpton.
Bōkārdō, Fĕrīson, hăbĕt : Quārta insŭpĕr āddit
Brāmāntĭp, Cămĕnēs, Dĭmărīs, Fĕsāpŏ, Frĕsīson.

The first line indicates the moods of the first Figure, the second the moods of the second Figure ; the third, and the first two words of the fourth line, the moods of the third Figure, and the last line those of the fourth Figure. The first Figure is called by logicians the *perfect*, and the remaining three the *imperfect* Figures, because they are to be reduced by certain processes to the first. The letters indicating how this is to be done require to be explained.

The capital letters B, C, D, F, in the last four lines, indicate the mood in the first Figure to which the mood in the other Figures beginning with that letter is to be reduced. Thus *Camestres* is to be reduced to *Celarent*, etc.

The vowels *a, e, i, o* indicate the mood of the syllogism. Thus in *Ferison* the mood is EIO etc.

p indicates that the preceding proposition is to be converted *per accidens* or by limitation.

s at the end of a word indicates that the conclusion of the new syllogism has to be converted by simple conversion in order to obtain the given conclusion : in the middle of a word it denotes that the preceding proposition is to be converted simply in the process of reduction. This difference between the use of *s* at the end, and its use in the middle of a word is remarked by Keynes. But I am inclined to see little importance in the distinction, because *s* in both cases denotes that the preceding proposition is to be converted simply, so that the converted conclusion of the reduced mood becomes the proper conclusion of the corresponding "perfect" Figure and Mood. Thus in *Camestres* we are to reduce the mood, as already said, to Celarent, or AEE of the second Figure to EAE of the first. The *m* denoting that the premises are to be transposed, we have to convert the minor premise and make it the major premise of the new syllogism, and make the major of *Camestres* the minor of the new. Then, if we convert the conclusion of *Camestres* by simple conversion, we shall have the proper conclusion of the new.

<table>
<tr><td>*Camestres.*</td><td></td><td>*Celarent.*</td></tr>
<tr><td>A is C</td><td rowspan="3">Reduced to</td><td>C is not B</td></tr>
<tr><td>B is not C</td><td>A is C</td></tr>
<tr><td>∴ B is not A</td><td>∴ A is not B.</td></tr>
</table>

Keynes also makes the same distinction between the use of *p* at the end, and its use in the middle of a word. But as in the case of *s*, I do not find it important. In all cases it denotes that the preceding proposition is to be converted *by limitation*, and so to appear in the new syllogism. Thus in *Felapton* A is to be converted *per accidens* or by limitation. In

192   ELEMENTS OF LOGIC

Bramantip, however, the subalternans of the converse is taken after the conversion.

m denotes that the *premises* of the "imperfect" moods are to be transposed or mutated in order to form the premises of the new and "perfect" mood. Thus in *Camenes* A and E must be transposed so as to become E and A of *Celarent*.

k signifies that the mood is to be reduced *indirectly*; and the position of the letter is, as affirmed by Keynes, to indicate that in the process of indirect reduction the first step is to omit the premise preceding it; that is, we take instead of it the contradictory of the conclusion as our premise. *Baroko* and *Bokardo* are the moods to which indirect reduction is usually applied, although the process can be applied to others. The present mnemonic lines, however, do not designate how this can be done in the case of the other moods. They can be more easily reduced directly, and it is superfluous to apply the indirect process.

One or two more examples of direct reduction will suffice for illustration, and we can pass to consider the indirect process. Take the cases of *Festino* and *Bramantip*, which are to be reduced respectively to *Ferio* and *Barbara*. If we perform the processes as designated by the several letters they will stand as follows:

Festino.
A is not C
Some B is C
∴ Some B is not A

Reduced to

Ferio.
C is not A
Some B is C
∴ Some B is not A.

Bramantip.
A is C
C is B
Some B is A

Reduced to

Barbara.
C is B
A is C
∴ Some A is B,
or All A is B.   Subalternans.

In the case of Bramantip there is apparent a defect in the mnemonic verse, as the term ought to indicate the process of changing the converted I into its subalternans A. One or two concrete examples will be helpful to a better understand-

REDUCTION OF MOODS AND FIGURES 193

ing of the process. We may take *Cesare* and *Disamis*, to be reduced to *Celarent* and *Darii*. Following the directions, we have in *Cesare* and *Celarent* the two succeeding syllogisms:

Cesare.
Men are not quadrupeds.
Horses are quadrupeds.
∴ Horses are not men.

Reduced to

Celarent.
Quadrupeds are not men.
Horses are quadrupeds.
∴ Horses are not men.

Disamis.
Some men are negroes.
All men are vertebrates.
∴ Some vertebrates are negroes.

Reduced to

Darii.
All men are vertebrates.
Some negroes are men.
∴ Some negroes are vertebrates.

The process can easily be applied to the other moods without further illustration. It will be good practice for the student to choose his own examples.

2d. Indirect Reduction.—Some difficulty attends the reduction of *Baroko* and *Bokardo*, because there are no moods in the first Figure representing AOO and OAO, the former giving an illicit major and the latter an illicit middle. Besides, no conversion of O is possible, and hence if the reduction of these two moods be possible it must be in some indirect way.

There are two ways in which it is done, and there is a difference of opinion as to whether the first form is indirect or not. Jevons speaks of it as indirect; Keynes as direct. But inasmuch as the major premise in both moods is contraverted, and the minor premise of *Baroko* obverted, it will hardly be amiss to follow Jevons's opinion, because neither of these processes are provided for in the mnemonic system we have adopted. If the matter of reduction were of great practical importance, it might be incumbent upon us to discuss this question. But the chief matter of interest to the scientific student of Logic is the fact and the mode of reduction, and we content ourselves with stating them.

We take up the first method of treating *Baroko* and *Bokardo*, illustrating them as before:

Baroko.
A is C
Some B is not C
∴ Some B is not A

Reduced to

Ferio.
All not-C is not A
Some B is not-C
∴ Some B is not A.

13

It should first be remarked in this case that the mnemonic *Baroko* is named so and introduced with the capital B to suit its indirect reduction to *Barbara*. But in reducing it to *Ferio*, it should be called *Faroko*. The obversion of the minor premise is accomplished as usual by connecting the negative particle with the predicate, making the proposition affirmative. We give a concrete illustration of the same:

Baroko.
Metals are elements.
Some solids are not elements.
∴ Some solids are not metals.

Reduced to

Ferio.
All not-elements are not metals.
Some solids are not-elements.
∴ Some solids are not metals.

The reduction of *Bokardo* by the same method is as follows, except that we have to transpose the premises:

Bokardo.
Some A is not C
All A is B
∴ Some B is not C

Reduced to

Darii.
All A is B
Some not-C is A
∴ Some not-C is B.

A similar remark should be made about *Bokardo* as was made about *Baroko*. To suit the reduction we have given, it should be called *Dokardo*. We also give a concrete illustration of the process:

Bokardo.
Some trees are not oaks.
All trees are organisms.
∴ Some organisms are not oaks.

Reduced to

Darii.
All trees are organisms.
Some not-oaks are trees.
∴ Some not-oaks are organisms.

The conclusion in the form to which *Bokardo* is reduced is also contraverted, as can be observed by the student.

The second method of reducing the same moods is undoubtedly indirect, as, instead of depending upon the forms of conversion, obversion, or contraversion, it resorts to the square of opposition for a contradictory of a premise for the conclusion, and of the conclusion for a premise in *Barbara*, as the illustrations below will distinctly show:

Baroko.
A is C
Some B is not C
Some B is not A

Reduced to

Barbara.
A is C
B is A
∴ B is C.

REDUCTION OF MOODS AND FIGURES   195

In this process the major premise remains the same. But instead of taking any form of the minor premise, converted or otherwise, the contradictory of the conclusion is assumed. In *Baroko* the conclusion is O; its contradictory is A. This is done on the supposition that if O, "Some B is not A," is not true, then "B is A" is true, and if this be assumed hypothetically in the premises we shall have a syllogism in *Barbara* with a conclusion which is the contradictory of the omitted minor premise, "Some B is not C." This contradictory is "B is C," and we must either admit one of our premises to be false or allow that our original conclusion is true. But as we assume that our original premises and conclusion are true, the impossibility of their contradictories being true is an indirect proof of them, as it is always regarded an indirect proof of a proposition to prove the impossibility of its opposite or contradictory, and reduction is proving one mood and Figure by another. *Bokardo* may be treated in a manner similar to *Baroko*.

Bokardo.   *Barbara.*

Some A is not C ⎫   ⎧ All B is C
All A is B ⎬ Reduced to ⎨ All A is B
∴ Some B is not C ⎭   ⎩ ∴ All A is C.

In this instance the major premise is omitted, and in its place stands hypothetically the contradiction of the conclusion in *Bokardo* for the major premise of *Barbara*. We, therefore, have a conclusion which is the contradictory of "Some A is not C," the major premise of *Bokardo*. If this premise be assumed to be true, the impossibility of the truth of its contradictory in the reduced form is an indirect proof of it, and we have the indirect reduction as before. The method is sometimes called the *reductio per impossibile*, or the proof of a thing by showing its contradictory to be impossible.

Little or no practical importance can be attached to any forms of reduction, except when we find it necessary to test a conclusion which does not seem so clear to us in one Figure or form of reasoning as in another. They may sometimes be helpful in completing certain enthymemes which can be formed in either the first or the second Figures, and be valid in the first, but invalid in the second. But as reduction does not apply to invalid moods it is only the principle of it that can be applicable in such cases. Hence we do not generally find the process to have more than a purely scientific interest and importance.

CHAPTER XIV.

FORMS OF SYLLOGISTIC REASONING

THE syllogism, as it has already been explained, appears to have a very simple form, and to consist of three propositions arranged in a particular way. While this is the regular form of reasoning, and expresses the real nature of the mental process involved in all instances of ratiocination, arguments are not always formulated in the regular way. Some propositions may be omitted in expression, although included in the thought of the reasoner, and in other cases, two or more syllogisms may be involved in the argument. These two conditions give rise to two divisions of the syllogism, besides the simple form already considered. They are the *incomplete* and the *complex* forms of it. We shall classify them before discussing them in detail. The following is an outline of them:

Syllogisms
- Complete
  - Simple = The ordinary form already discussed.
  - Complex = Prosyllogism and Episyllogism.
- Incomplete
  - Simple = Enthymeme
    - 1st order.
    - 2d order.
    - 3d order.
  - Complex =
    - Epicheirema.
    - Sorites.

The complete forms of the syllogism are, of course, an explicit statement of all that is involved in the mental process. They very seldom appear in ordinary discourse, and then only to give it more cogency. The usual mode of stating an argument is either to state the facts and to allow the inference to be drawn by others, or to abbreviate the process by assuming the most apparent of the premises. But when we wish to indicate with perfect clearness all that the mind takes into account we formulate our thoughts into complete syllogisms. The first of these is the simple syllogism of three propositions, and which has already been discussed. We then take up the second to consider it briefly.

1. PROSYLLOGISM AND EPISYLLOGISM.—This consists of two syllogisms, the conclusion of one, the *prosyllogism*, being a premise in the other, the *episyllogism*. The following are illustrations:

A is B	Men are vertebrates.
C is A	Europeans are men.
∴ C is B	∴ Europeans are vertebrates.
D is C	Italians are Europeans.
∴ D is B	∴ Italians are vertebrates.

In the formal illustration here given the prosyllogism ends with the conclusion "C is B," and the episyllogism begins with the same proposition as a premise. In the material instance the proposition, "Europeans are vertebrates," is the conclusion of the prosyllogism and a premise of the episyllogism.

2. ENTHYMEME.—An enthymeme is an incomplete syllogism in which either one of the premises, or the conclusion, may be omitted. If the major premise be the one which is omitted the enthymeme is said to be of the *first order;* if the minor premise be omitted, it is of the *second order*, and if the conclusion be omitted, it is of the *third order*. The signs of the enthymeme are such words as indicate a reason for the truth of a proposition. They are *for, because, since, inasmuch as*, and in the conclusion, *therefore, consequently*, etc. The last two words denote that an inference is drawn from some preceding fact or statement, and are signs of an enthymeme only when a single premise is given. Those which give a reason for a truth are usually necessary when the syllogism is not complete, or when the conclusion is stated first and requires to have its dependence upon the premises indicated. As an illustration of an enthymeme we have the proposition "Atmospheric air must have weight because it is a material substance." The conclusion in this example is, "Atmospheric air must have weight." If we were stating a mere fact, it would not require

proof, but we often desire to support such assertions by reasons that will show their truth apart from the acceptance of them on authority, and so we give a reason for them. In this case the reason for the truth or assertion that "Atmospheric air must have weight" is, that "it is a material substance." In this reason it is tacitly assumed that "all material substances have weight." It is the omission of this premise that makes the reasoning an enthymeme. Whenever such propositions are taken for granted it is because they either require no proof or they are sufficiently evident to justify their omission, and the process of thought or reasoning can be abbreviated or economized. It is the usual form in ordinary speech. When resolved into the form represented by their proper logical order and dependence, the propositions take the order laid down for the common syllogism. The above instance can easily be resolved by the student into a simple syllogism.

There are forms of the enthymeme in which the signs are not expressed, but which have to be determined by the evident relation of the thoughts stated. Every sentence, however, which is definitely introduced by a particle denoting a reason for something else is an express statement of reasoning. But such forms of proof become stilted and inelegant, and hence discourse may sometimes be best conducted by affirming facts or truths which of themselves imply the dependence of something else upon them. We require to be on the alert in such cases in order either to detect the existence of reasoning at all, or to discover the form in which it is implied. This may also be true of statements involving complete syllogisms. An interesting instance of reasoning in the form of an enthymeme without the usual signs is the following:

"The high prices caused by the new tariff law have severely taxed the mind of the *Chronicle-Telegraph*, but it has at last evolved this curious and interesting explanation: 'For a long time past everything, or nearly everything, entering into the daily consumption of the people has been unusually cheap. But close observers of the trend of events believe we are approaching an era of higher prices, and that it is near at hand, and that it is to be looked upon as a natural reaction from the period of constant decline in the value of merchandise which has lasted for several years.'" The reasoning involved in this case is apparent when we enunciate that the present high prices are a reaction from previous low prices, and that such reactions are natural, and not caused by tariff laws. The passage beginning with the words, "For a long time past," is intended to express a fact, from which it follows that high prices must occur without reference to tariff influences, and hence we have in thought a process of reasoning without the use of any of the signs. The premise which is omitted is, that "All periods of extremely low prices are followed by a natural reaction." If the other premise be formulated we should all see how the conclusion would follow, which is intended to be the contrary of the first proposition in the passage. Most arguments in ordinary discourse are of this general kind, and the student would do well to formulate them for the sake of discovering where the reasoning is.

3. EPICHEIREMA.—An epicheirema is a syllogism in which one or both of the premises is supported by a reason which implies an imperfectly expressed syllogism: in other words, it is a syllogism in which one or both of the premises is an enthymeme either of the first or of the second order. The epicheirema may be *single* or *double*. It is single when only one of the premises is an enthymeme, and double when both are enthymemes. The following is an illustration of the single epicheirema:

>A is B, for it is P
>C is A
>∴ C is B.

A material form of the same is the following, but the enthymeme in this case is in the minor premise:

>Vice is odious.
>Avarice is a vice, because it depraves.
>Therefore avarice is odious.

The double epicheirema takes the following form :

A is B, because it is P
C is A, because it is Q
∴ C is B.

Or, Man has a mind, because he is rational.
Europeans are men, because they are civilized.
Therefore Europeans have minds.

The single epicheirema when resolved or completed becomes a prosyllogism and an episyllogism, or two syllogisms. The double epicheirema when completed becomes three syllogisms, representing two prosyllogisms and two episyllogisms, one of the three being both a prosyllogism and an episyllogism, the former in relation to the following, and the latter in relation to the preceding syllogism. The student may practise their resolution until familiar with the processes involved in them. The resolution of the double epicheirema above is as follows :

Man has a mind : ⎧ Whatever is rational has a mind.
for he is rational = ⎨ Man is rational.
⎩ Man has a mind.

Europeans are men : ⎧ Whatever is civilized is man.
for they are civilized = ⎨ Europeans are civilized.
⎩ Europeans are men.
Therefore Europeans have minds.

The third syllogism appears in combining the conclusions of the first two to form its premises, from which we obtain the conclusion that "Europeans have minds."

4. SORITES.—A sorites is so called because the propositions form what is regarded as a " chain," or a continuous series, of premises from which a conclusion is drawn at the end of the series. It consists of enthymemes of the *third* order, as the epicheirema consists of enthymemes of the first and second orders. When completed, therefore, it forms a prosyllogism and episyllogism. To complete it we have only to supply the intermediate conclusions implied by the proper prem-

ises. The form of the sorites is twofold. It may be *progressive* or *regressive*. The following are illustrations:

Progressive series.

<table>
<tr><td>A is B</td><td></td><td>Bucephalus is a horse.</td></tr>
<tr><td>B is C</td><td></td><td>A horse is a quadruped.</td></tr>
<tr><td>C is D</td><td>or,</td><td>A quadruped is an animal.</td></tr>
<tr><td>D is E</td><td></td><td>An animal is an organism.</td></tr>
<tr><td>∴ A is E</td><td></td><td>∴ Bucephalus is an organism.</td></tr>
</table>

Regressive series.

<table>
<tr><td>A is B</td><td></td><td>An animal is an organism.</td></tr>
<tr><td>C is A</td><td></td><td>A quadruped is an animal.</td></tr>
<tr><td>D is C</td><td>or,</td><td>A horse is a quadruped.</td></tr>
<tr><td>E is D</td><td></td><td>Bucephalus is a horse.</td></tr>
<tr><td>∴ E is B</td><td></td><td>∴ Bucephalus is an organism.</td></tr>
</table>

The difference between the progressive and regressive sorites is merely in the appearance of the *Figure* of the syllogism. In the progressive series it seems to be of the fourth Figure, and the propositions are expressed in that order, but the actual reasoning is performed by a virtual transmutation of the premises, when we choose the minor term from the first proposition and the major term from the last, and hence is of the first Figure. In the regressive series both the form of expression and the reasoning are in the first Figure, and represents the major premise as the first in order.

The completion of a sorites may be illustrated as follows: The incomplete form is sometimes called the *occult*, and the complete the *manifest* form.

<table>
<tr><td>A is B</td><td></td><td>Bucephalus is a horse.</td></tr>
<tr><td>B is C</td><td></td><td>A horse is a quadruped.</td></tr>
<tr><td>∴ A is C</td><td>or,</td><td>∴ Bucephalus is a quadruped.</td></tr>
<tr><td>C is D</td><td></td><td>A quadruped is an animal.</td></tr>
<tr><td>∴ A is D</td><td></td><td>∴ Bucephalus is an animal.</td></tr>
</table>

The sorites may be *constructive* or *destructive*. It is constructive when the premises and conclusion are affirmative, and hence must be represented by the moods AAA, AAI, and AII, of the first Figure. But it is destructive when the conclusion is negative, and therefore when the major premise is negative. The destructive sorites must then be represented by the moods EAE, EAO, and EIO, of the first Figure.

There is a dispute among logicians as to whether a sorites can exist in the second and third Figures. I agree with Mill and Keynes that it is possible, but its importance is so slight, and its occurrence in practice so infrequent, that I do not think it deserves any special attention.

The rules for the valid forms of the sorites should be stated, because the number of cases in which this mode of reasoning is valid is exceedingly limited. Any number of premises may be used, but the quantity and quality of the propositions constituting them are under strict limitations. The rules for the process, therefore, are two, and are as follows :

1. Only one premise can be negative, and this must be the prime major.
2. Only one premise can be particular, and this must be the final minor.

Since the sorites is an incomplete prosyllogism and episyllogism, there are intermediate major and minor premises. The prime major, therefore, will denote the last premise in the progressive, and the first premise in the regressive, series. The final minor will denote the first premise in the progressive, and the last in the regressive, series. If any other premise than the prime major be negative, there will be a fallacy of illicit major, and if any other than the final minor be particular, there will be an illicit middle.

CHAPTER XV.

HYPOTHETICAL REASONING

THE forms of reasoning may be divided in the same manner as propositions. In fact they are determined by the nature of the propositions which constitute them. We have already observed that propositions may be divided into Categorical, Hypothetical, and Disjunctive. Arguments or forms of reasoning are divided according as the premise or premises are categorical, hypothetical, or disjunctive. A categorical syllogism is one in which all the propositions are categorical. Of this kind we have already treated, and it remains to consider the hypothetical form.

1. HYPOTHETICAL SYLLOGISMS.—A hypothetical syllogism, or form of reasoning, is one in which one or more of the propositions is hypothetical or conditional. Most frequently it is the major premise alone that is conditional, while the minor premise and conclusion are categorical. The major premise consists of two propositions, one of which is dependent upon the other. The proposition expressing the condition is called the *antecedent*, and is introduced by some such words as *if, suppose, allow, granted that, provided that*, etc., all of which indicate that a statement is made under conditions restricting its application. The proposition depending upon this condition is called the *consequent*, and is either a categorical or a disjunctive proposition. An illustration of an antecedent and consequent so defined is: "If the sea is rough, it is dangerous." "If the sea is rough" is the antecedent, and "it is dangerous" is the consequent. The rules regulating the validity of hypothetical inferences depend upon understanding the use of these terms.

The general form of the hypothetical syllogism is that of an antecedent and consequent, or a conditional and a categorical proposition for the major premise, a categorical proposition for the minor premise, and a categorical proposition for the conclusion. But as there are two propositions in the major premise, the minor premise may affirm or deny either the antecedent or the consequent. This gives *four* forms to be considered, two of which are valid and two of which are invalid modes of reasoning. The two valid moods are called, respectively, the *modus ponens* and the *modus tollens*, or the constructive and the destructive hypothetical syllogism. The *modus ponens* means that if the antecedent be affirmed the consequent will follow. The major premise only asserts this relation conditionally. But if the antecedent be affirmed categorically in the minor premise the conclusion will follow categorically. The only effect of a hypothetical minor premise is to make the conclusion hypothetical. The *modus tollens* means that if the consequent be denied, the antecedent must be denied. The two forms can be illustrated as follows:

Modus ponens.
If A is B, C is D
A is B
∴ C is D

Constructive hypothetical syllogism.
If iron is impure it is brittle.
or,   Iron is impure.
∴ Iron is brittle.

Modus tollens.
If A is B, C is D
C is not D
∴ A is not B

Destructive hypothetical syllogism.
If the sun shines it is light.
or,   It is not light.
∴ The sun does not shine.

The rule, therefore, for testing the validity of hypothetical reasoning is, that *either the antecedent must be affirmed or the consequent denied.* If this rule be violated in either of its conditions a fallacy occurs. This is illustrated in the two remaining, but invalid, moods of the hypothetical syllogism:

If A is B, C is D
C is D
∴ A is B

If it rains it is cloudy.
or,   It is cloudy.
∴ It is raining.

This is a case of the *fallacy of affirming the consequent*. The fact that "A is B," or that "it is raining," is not the only condition that C should be D, or that it should be cloudy. Some other condition may be true, so that if "A is B" follows from the affirmation of the consequent, the other condition would follow also. But there is nothing to determine that, and hence the known condition must remain indefinite so far as drawing it from the truth of the consequent is concerned. If any other condition, however, may exist as a determinant of the consequent, that condition may be the very opposite of the one specified. Thus, C may be D, if A is not B, or it may be cloudy if it is not raining, and in this case it will be apparent that we cannot equally draw opposite conclusions from the truth of the consequent. Hence, we have no right to draw any inference whatever. A concrete illustration will make this still clearer. "Thus, if a man's character be avaricious, he will refuse to give money for useful purposes; but it does not follow that every person who refuses to give money for such purposes is avaricious. There may be many proper reasons or motives leading him to refuse; he may have no money, or he may consider the purpose not a useful one, or he may have more useful purposes in view." No inference, therefore, can be drawn from an affirmation of the consequent.

A second fallacy comes from *denying the antecedent*, as represented in the following form:

If A is B, C is D   If gold were cheap it would be useful.
A is not B   or, Gold is not cheap.
∴ C is not D   ∴ Gold is not useful.

The fallacy is due to the same causes as before. The antecedent is not the only possible condition of the consequent, and hence no conclusion denying the existence or truth of the consequent can be drawn from the denial of the antecedent. It is apparent in the concrete case that other qualities besides cheapness might make gold useful, and therefore the absence of this quality would not remove the usefulness of the metal.

Difficulties appear in some cases in which it would seem that a negative conclusion does follow the denial of the antecedent. To take a case we have :

> If men are white they are Caucasians.
> Men are not white.
> ∴ They are not Caucasians.

In supposing whiteness as the only condition of being a Caucasian, as we generally do, the absence of it would imply that the person was not a Caucasian. And so perhaps with the following instance :

> If fire is hot it will burn.
> Fire is not hot.
> ∴ It will not burn.

The usual assumption is that fire and the power to burn are the same, so that to deny the heat of the former is to deny the capacity of it to burn. In such cases a denial of the antecedent seems to involve an inference to the negation of the consequent.

But we must not be deluded by such instances. They are in effect cases in which the major premise is an *exclusive* hypothetical proposition. When converted it becomes equivalent to a *modus tollens*, in which the consequent instead of the antecedent is denied. This explains how the reasoning is valid. Thus if whiteness is the only consideration of being Caucasian, the hypothetical proposition would be "If only men are white they are Caucasian," which is equivalent to saying, "If men are Caucasians they are white." Hence when we affirm that "they are not white," in the minor premise, we should be denying the consequent and not the antecedent. *Materially*, therefore, the reasoning is correct, and it would always be so when the antecedent expresses the *only* condition of the consequent. But *formally* we cannot consider it so, as there is nothing in the form of the proposition to indicate whether the antecedent expresses the only condition or not. Since we are treating only of formal reasoning, we have to regard all propositions alike which are only formal in their character. If the *exclusive* nature of the hypothetical proposition were expressed, we might formulate a rule for it, but when it is not so expressed, we must consider it formally as under the general law. Nevertheless in practical reasoning we should always be prepared to distinguish when the mind tacitly supposes the material conditions which might make the material reasoning correct while it is formally wrong. In this way we could admit the truth of the conclusion and yet show that it has not been obtained in accordance with the formal law of hypothetical reasoning, or that we are liable to frequent fallacies, if we allow the correctness of this material reasoning to lead us into the indiscriminate use of its privileges in hypothetical syllogisms at large.

There are three forms of hypothetical propositions and syllogisms which require notice because of the misunderstanding to which they may give rise. They consist of negative propositions, while those we have illustrated consist of affirmative propositions. We must show that the case is not altered by the use of negative propositions, but that the whole matter turns upon the connection between the antecedent and the consequent. There are, therefore, three more forms in which the major premise of the hypothetical syllogism may be expressed. They are :

(a) If A is B, C is not D
(b) If A is not B, C is D
(c) If A is not B, C is not D.

The peculiar characteristic to be remarked about these propositions is their quality and the mode of affirming and denying the antecedent or the consequent. In proposition (a) the antecedent will be treated in the same manner as in previous instances, but the consequent will be *affirmed* in the minor premise by saying "C is not D," in which case the fallacy of affirming the consequent is committed. But the consequent would be denied by saying "C is D," and then we should be obliged to draw the conclusion that "A is not B." In proposition (*b*) the consequent being an affirmative proposition would be treated as before, and the antecedent would be affirmed by making the minor premise to be "A is not B," when we should have a *modus ponens*. But it would be denied by the form "A is B," and the usual fallacy would be committed. In proposition (*c*) both the antecedent and the consequent must be treated as we have treated the consequent in proposition (*a*) and the antecedent in proposition (*b*).

2. REDUCTION OF HYPOTHETICAL SYLLOGISMS.—It does not appear from the manner in which hypothetical syllogisms have been discussed that the process of reasoning involved can be reduced to the forms of categorical syllogisms. This, however, is the fact, and in order to understand how the ordinary laws of reasoning are applicable it is necessary to reduce them to the categorical form. They are convenient, often, for the purpose of emphasizing the conditional character of the major premise, and insuring the acceptance of the conclusion on those conditions when the minor premise is accepted. The usual object, however, is to have the major premise accepted formally, or the connection between antecedent and consequent, and then to show that the antecedent is true or the consequent false, in order to obtain a conclusion which is not clear or admitted at the outset of the argument.

Nevertheless, in spite of the superior convenience, at times, of the hypothetical forms of reasoning, they can all be reduced to the categorical form. In all cases we may regard *the antecedent of the hypothetical major premise as the subject of the categorical proposition, and the consequent of the hypothetical proposition as the predicate of the categorical.* In some instances this change is a very simple one; in others it can be effected only by a circumlocution. It can be done simply when the terms of the antecedent and consequent can be made to form a phrase representing a noun and its modifiers. Thus we have the examples :

If iron is impure it is brittle. } { Impure iron is brittle.
It is impure. } { Iron is impure.
∴ It is brittle. } { ∴ Iron is brittle.

If the weather is stormy sea travel will be dangerous.
The weather is stormy.
∴ Sea travel will be dangerous.

This may be reduced to the following:

Stormy weather is a cause of dangerous sea travel.
The present weather is stormy.
∴ The present weather is a cause of dangerous sea travel.

In all such instances we practically supply a new *minor term* in order to complete the categorical form, but it is only a particular case under the general in the major premise.

But all instances of the hypothetical syllogism are not so easily reduced. In many of them we have to resort to a circumlocution in such phrases as "*the case of*," "*the circumstances that*," etc. Thus in the hypothetical syllogism below we must use this means of its conversion:

If Aristotle is right, slavery is a proper form of society.
But slavery is not a proper form of society.
∴ Aristotle is not right.

By using the phrase "*the case of*," this becomes in the categorical form:

The case of Aristotle being right is the case of slavery being a proper form of society.
But slavery is not a proper form of society.
∴ Aristotle is not right.

This is clearly a syllogism in the second Figure of the mood AEE. It is evident, therefore, that we may easily determine the valid and invalid forms of hypothetical reasoning in terms of the categorical syllogism. We shall illustrate all four forms in order to make this clear. First, the *modus ponens*:

If water is pure it is good.   Pure water is good.
It is pure.   This water is pure.
∴ It is good.   ∴ This water is good.

In this we have a syllogism in Barbara, or AAA of the first Figure, and therefore valid.

The *modus tollens* will appear as follows:

If water is pure it is good. } { Pure water is good.
It is not good.   } { This water is not good.
∴ It is not pure.   } { ∴ This water is not pure.

Here we have again a case of Camestres, or AEE in the second Figure, and valid.

The case of denying the antecedent is as follows:

If water is pure it is good. } { Pure water is good.
It is not pure.   } { This water is not pure.
∴ It is not good.   } { ∴ This water is not good.

In this instance we have a case of AEE in the first Figure, which is invalid. The fallacy is that of *illicit process of the major term*. The major term is not distributed in the major premise, but is distributed in the conclusion. Hence, when we attempt to draw a conclusion after denying the antecedent the fallacy is that of an *illicit major term*.

The next is an illustration of affirming the consequent:

If water is pure it is good. } { Pure water is good.
It is good.   } { This water is good.
∴ It is pure.   } { ∴ This water is pure.

We have in this illustration a case of AAA in the second Figure, and invalid because the middle term is undistributed. Hence all cases of affirming the consequent are instances in which we commit the fallacy of *illicit process of the middle term*, or undistributed middle. The valid forms, therefore, are AAA of the first, and AEE of the second Figure. The invalid forms are AEE of the first, and AAA of the second Figure.

When the hypothetical propositions are negative, we may either obvert them into their corresponding affirmatives, or consider the invalid forms as due to attempts to reason with negative premises.

CHAPTER XVI.

DISJUNCTIVE SYLLOGISMS

A DISJUNCTIVE syllogism is one which is determined by the presence of a disjunctive proposition in one of its premises, and sometimes in the conclusion also. A disjunctive proposition we have already learned to be one which contains alternative or mutually exclusive conceptions between which the choice of the mind is to be made, and which are accompanied by the disjunctives *either* and *or*. Wherever these terms are found in such propositions they are meant to imply that only one of the two things can be affirmed, and the other denied. Thus when I say that "The weather is either clear or cloudy," I mean that it cannot be both at once, but that if it is one it cannot be the other. It is this fact which determines the right to draw the inference in the disjunctive syllogism.

"It is a disputed question whether in a disjunctive proposition the alternatives should be regarded as in all cases mutually exclusive; whether, for example, in the proposition 'A is either B or C,' it is necessarily implied that A cannot be both B and C. There are really involved here two questions which should be distinguished.

"(1) In ordinary speech do we intend that the alternatives in a disjunctive proposition should be necessarily understood as excluding one another? A very few instances, I think, will enable us to decide in the negative. Take, for example, the proposition, 'He has either used bad text-books, or he has been badly taught:' would any one understand this to exclude the possibility of his having been badly taught and having used bad text-books as well? Or, suppose it laid down as a condition of eligibility for some appointment that every candidate must be a member either of the University of Oxford, or of the University of Cambridge, or of the University of London. Would any one regard this as implying the ineligibility of persons who happened to be members of more than one of these Universities? Jevons instances the following proposition, 'A peer is either a duke, or a marquis, or an earl, or a viscount, or a baron.' We do not consider this statement incorrect because many peers, as a matter of fact, possess two or more titles.

"(2) Still this does not definitely settle the question. Granted that in common speech the alternatives of a disjunction may or may not be mutually exclusive, it may nevertheless be maintained that this is only because common speech is elliptical, that in Logic we should be more precise, and that the statement 'A is either B or C' (where it may be both) should therefore be written, 'A is either B and not C, or C and not B, or both B and C.'

"This is a question of interpretation or method, and I do not apprehend that any burning principle is involved in the answer that we may give. For my own part I do not find any sufficient reason for diverging from the usage of every-day language. On the other hand, I think that if Logic is to be of practical utility, the less logical forms diverge from those of ordinary speech the better. And further, condensed forms of expression do not conduce to clearness, or even ultimately to conciseness. For where our information is meagre, a condensed form is likely to express more than we intend, and in order to keep within the mark we must indicate additional alternatives." *

The purport of these remarks is that the disjunction "*either — or*," is capable of a double import. The first is, that the terms may denote alternatives, either one of which may be sufficient to satisfy the terms of the proposition, although both may exist in the same connection. Thus in the case of a candidate's eligibility depending upon membership in either the University of Oxford, or the University of Cambridge, etc., we mean that membership in any one of them is sufficient, and

* Keynes's Formal Logic, Part II., Chap. IX., p. 167.

that non-membership in the others will not be an obstacle in that case. The second meaning is that the two alternatives shall be mutually exclusive. This is the form which is necessary for correct formal disjunctive reasoning, and as it frequently occurs, we have to take it into account in a complete exposition of the syllogism. The former case, when "either — or" does not express mutual exclusion between the alternatives, gives rise to what is called an *incomplete disjunction*. The fallacy incident to this fact will be noticed again. At present we have only to consider those cases where "either — or" expresses mutually exclusive alternatives. Upon this assumption definite rules can be established for disjunctive reasoning. In the meantime we shall use the terms either — or as the only accessible symbols for a formal disjunction.

Before enunciating these laws and illustrating them we shall classify the forms of the disjunctive syllogism. There are two general divisions, the *categorical* and the *dilemmatic*, or the definite disjunctive syllogism and the dilemma. The former consists of a disjunctive proposition in the major premise, and a categorical in the minor premise, giving a categorical conclusion. The dilemma consists of a hypothetical proposition in the major premise, and a disjunctive in the minor premise. The subdivisions of these two general forms is illustrated in the following table:

Disjunctive syllogisms
- Categorical
  - Modus ponendo tollens.
  - Modus tollendo ponens.
- Dilemmatic
  - Constructive
    - Simple.
    - Complex.
  - Destructive
    - (Simple.)
    - Complex.

The first form of the categorical disjunctive syllogism is called the *modus ponendo tollens*, because it means that by *affirming one of the alternatives we must deny the other*. This is the meaning of the Latin phrase denominating it. It is illustrated as follows:

A is either B or C
But A is B
∴ A is not C

Oak trees are either tall or short.
They are tall.
∴ They are not short.

The *modus tollendo ponens*, which is the second form, is so named because by *denying one of the alternatives we must affirm the other*. It is illustrated thus:

 A is either B or C   The air is either cool or warm.
 A is not B   It is not cool.
∴ A is C   ∴ It is warm.

In these cases we assume that the alternatives are mutually exclusive, and that the subject cannot be both at once, or that there can be no other alternative. If this assumption were not made the conclusion would be invalid, as a case of *incomplete disjunction*. This is illustrated in the following instance:

 Macaulay either had great talents or he was very studious.
 He had great talents.
∴ He was not very studious.

This conclusion does not necessarily follow because the alternatives are not necessarily exclusive of each other. A man may be both talented and studious. Hence when the disjunction is incomplete in the major premise, it gives rise to a fallacy which is a *petitio principii*, and which will be explained again. This fallacy, however, will not occur, if we assume the disjunction to be complete. If we really assume that Macaulay was either one or the other of the two alternatives, and not possibly both of them, or anything else, the conclusion is valid. Very frequently in such cases we mean that the disjunction shall be perfect, and hence the reasoning cannot be criticised. Thus, if I say "All birds are either white or black," and then, after affirming that "they are not white," infer that "they are black," I would be wrong only because I was wrong in the major premise. If I really meant that these were the only two alternatives, the conclusion would be true. We see, therefore, that there is no *formal* fallacy in disjunctive reasoning, but that it occurs in the *matter* of the assumption in the major premise.

The laws of disjunctive syllogisms seem to be quite different from the categorical and the hypothetical. We seem to infer a negative conclusion from affirmative premises, and an affirmative conclusion when one of the premises is negative; a negative conclusion in the *modus ponendo tollens*, and affirmative in the *modus tollendo ponens*. But it can easily be shown that this is not exceptional. This can be done in two ways. First, the major premise, which is a disjunctive proposition, contains both an affirmative and a negative assertion, with the implication that one is true and the other false. The reasoning is, therefore, based upon the law of contradiction in the square of opposition, so that a negative conclusion is involved in the negation expressed, or implied in the major premise, and an affirmative conclusion when the minor premise is negative. But the clearer exposition of the case is the second. As we have already remarked a disjunctive proposition is one which is categorical in its form and conditional or hypothetical in its matter. Its meaning, therefore, must be determined by reducing it to its equivalent, and we shall see that the disjunctive syllogism can be resolved into the hypothetical, and this hypothetical into the categorical, so that, after all, the regular laws of reasoning apply to the disjunctive syllogism, although in a modified and less apparent form.

When we say that "A is either B or C," and imply that there are no other alternatives, we mean that *if A is B it is not C*. This, we see, is a hypothetical proposition with a negative consequent. Or we may mean that *if A is not B it is C*, in which case we have a negative antecedent. We have then only to state the minor premise, as in the disjunctive syllogism, and the reasoning becomes hypothetical. This can be illustrated in the following manner:

A is either B or C   · If A is B, it is not C
A is B   A is B
∴ A is not C   ∴ A is not C.

In the hypothetical form we have, therefore, a case of *modus ponens* which is valid. After the reduction of the disjunctive form the *modus tollens* appears thus:

$$\left.\begin{array}{l} \text{A is either B or C} \\ \text{A is C} \\ \therefore \text{A is not B} \end{array}\right\} \left\{\begin{array}{l} \text{If A is B, it is not C} \\ \text{A is C} \\ \therefore \text{A is not B.} \end{array}\right.$$

But as the nature of the disjunctive syllogism is such that we can always make it a *modus ponens* in the hypothetical, and as it is always formally valid, we do not require to test its laws by either the *modus tollens* or the invalid forms of hypothetical reasoning. It therefore suffices to convert it always into the one form for the purpose of discovering the law underlying its logical process, and this is the law of the hypothetical syllogism, which we have already ascertained to be the same as that of categorical syllogism.

The dilemma, or dilemmatic disjunctive syllogism, is subject to the laws of hypothetical reasoning, because its major premise is hypothetical. The first form is that of the *simple constructive dilemma*.

If A is B, C is D ; and if E is F, C is D
But either A is B, or E is F
∴ C is D.

We observe in this and all cases of the simple constructive dilemma that the consequent is the same for both antecedents. This gives as its distinctive mark a *categorical conclusion*. A concrete illustration is the following:

"If a science furnishes useful facts, it is worthy of being cultivated ; and if the study of it exercises the reasoning powers, it is worthy of being cultivated ; but a science either furnishes useful facts, or its study exercises the reasoning powers ; therefore it is worthy of being cultivated."

The second form of the dilemma is the *complex constructive dilemma*.

If A is B, C is D ; and if E is F, G is H
But either A is B, or E is F
∴ Either C is D, or G is H.

This is different from the simple constructive dilemma in that the consequents are different, and this fact gives as its distinctive mark a *disjunctive conclusion*. As an instance of it we have the following argument:

"If a statesman who sees his former opinions to be wrong does not alter his course he is guilty of deceit ; and if he does alter his course he is open to the charge of inconsistency ; but either he does not alter his course or he does ; therefore he is either guilty of deceit or he is open to the charge of inconsistency."

The *destructive dilemma* is supposed always to be *complex*, because it can otherwise be resolved into two distinct hypothetical syllogisms, and because no disjunctive proposition occurs in it. If a simple destructive dilemma occurred, it would be in the following form:

If A is B, C is D ; and if E is F, C is D
But C is not D
∴ Neither A is B, nor E is F.

This form of reasoning is possible, and it is only a question of definition as to whether we shall call it disjunctive and dilemmatic. But we should have to change the conception of "disjunction" in order to include it in that form. The complex dilemma, therefore, is the only one that complies with the conditions. It is as follows:

If A is B, C is D ; and if E is F, G is H.
But either C is not D, or G is not H.
∴ Either A is not B, or E is not F.

A concrete illustration is found in the following argument: "If this man were wise, he would not speak irreverently of Scripture in jest ; and if he were good, he would not do so in earnest ; but he does it either in jest or in earnest, therefore he is either not wise or not good."

The fallacy incident to the dilemma is the same as in hypothetical reasoning, and does not require special discussion.

CHAPTER XVII.

CLASSIFICATION OF FALLACIES

1st. Definition.—The term "fallacy" is from the Latin *fallo*, denoting deception, illusion, error. In Logic it must be distinguished from such words as illusion. An illusion is a misinterpretation of the data of sense perception : a fallacy is an error in reasoning. The term, however, is often applied to those errors which are liable to occur in the interpretation of ambiguous propositions, made so by the displacement of a word or a phrase. But in the true logical sense these errors are not fallacies. They may give rise to fallacies in reasoning by rendering the data uncertain and ambiguous, but they are not errors in reasoning itself, they are only errors in interpretation. As logicians, however, have uniformly included them in their treatment and classification of fallacies, we shall continue this practice for the sake of the practical convenience they possess in ordinary reasoning, although the proper place to deal with them is in Rhetoric.

In discussing the laws of the syllogism we have been trying to ascertain the rules or laws which regulate *right* reasoning. We have now to examine the illegitimate modes of inference, or the mental processes which result in fallacies, or erroneous reasoning. We require some means of knowing when the rational faculty is liable to go astray, as well as when it has conformed to the true principles of reasoning. In order to do this we must classify and explain the various forms of fallacy.

2d. Divisions.—As already indicated, the term fallacy is used in a broad sense to cover both errors of *interpretation*, or of grammatical and rhetorical form, and errors of *inference*, or logical reasoning. This gives rise to a twofold division of fallacies, into *Hermeneutic* and *Logical* fallacies. I employ the term "hermeneutic" to denote that they are errors of interpretation, and hence of the perception of the meaning of a proposition. The error is intellectual, but not ratiocinative. This class of error or fallacy will require very brief consideration. I recognize but two forms of it. First, the so-called *Fallacy of Amphibology*, and second, the *Fallacy of Accent*.

We quote the language of Jevons upon each of these forms of error : "The Fallacy of Amphibology consists in an ambiguous grammatical structure of a sentence, which produces misconception. A celebrated instance occurs in the prophecy of the Spirit in Shakespeare's *Henry VI.*: 'The Duke yet lives that Henry shall depose,' which leaves it wholly doubtful whether the Duke shall depose Henry, or Henry the Duke. This prophecy is doubtless an imitation of those which the ancient oracle of Delphi is reported to have uttered ; and it seems that this fallacy was a great resource to the oracles who were not confident in their own powers of foresight. The Latin language gives great scope to misconstructions, because it does not require any fixed order for the words of a sentence, and when there are two accusative cases with an infinitive verb, it may be difficult to tell, except from the context, which comes in regard to sense before the verb. The double meaning which may be given to 'twice two and three' arises from amphibology ; it may be 7 or 10, according as we add the 3 after or before multiplying. In the careless construction of sentences it is often impossible to tell to what part any adverb or qualifying clause refers. Thus, if a person says, 'I accomplished my business and returned the day after,' it may be that the business was accomplished on the day after as well as the return ; but it may equally have been finished on the previous day. Any ambiguity of this kind may generally be avoided by a simple change in the order of the words ; as, for instance, 'I accomplished my business, and on the day after returned.' Amphibology may sometimes arise from confusing the subjects and predicates in a compound sentence, as if in the sentence, 'Platinum and iron are very rare and useful metals,' I were to apply the predicate useful to platinum and rare to iron, which is not intended. The word 'respectively' is often used to show that the reader is not at liberty to apply each predicate to each subject."

"The Fallacy of Accent consists in any ambiguity arising from a misplaced accent or emphasis thrown upon some word of a sentence. A ludicrous instance is liable to occur in reading Chapter XIII. of the First Book of Kings, verse 27, where it is said of the prophet, 'And he spoke to his sons, saying, Saddle me the ass, and they saddled *him*.' The italics indicate that the word *him* was supplied by the translators of the authorized version, but it may suggest a very different meaning. The Commandment, 'Thou shalt not bear false witness against thy neighbor,' may be made by a slight emphasis of the voice on the last word to imply that we are at liberty to bear false witness against other persons. Mr. De Morgan, who remarks this, also points out that the erroneous quoting of an author, by unfairly separating a word from its context, or italicising words which were not intended to be italicised, gives rise to cases of this fallacy.

"It is curious to observe how many and various may be the meanings attributable to the same sentence according as emphasis is thrown upon one word or another. Thus the sentence, 'The study of Logic is not supposed to communicate a knowledge of many useful facts,' may be made to imply that the study of Logic *does* communicate such a knowledge although it is not supposed to do so; or that it communicates a knowledge of a *few* useful facts; or that it communicates a knowledge of many *useless* facts. This ambiguity may be explained by considering that if you deny a thing to have the group of qualities A, B, C, D, the truth of your statement will be satisfied by any one quality being absent, and an accented pronunciation will often be used to indicate that which the speaker believes to be absent. If you deny that a particular fruit is ripe and sweet and well-flavored, it may be unripe and sweet and well-flavored; or ripe and sour and well-flavored; or ripe and sweet and ill-flavored; or any two or even all three qualities may be absent. But if you deny it to be ripe and sweet and *well-flavored*, the denial would be understood to refer to the last quality. Jeremy Bentham was so much afraid of being misled by this fallacy of accent that he employed a person to read to him, as I have heard, who had a peculiarly monotonous manner of reading."

As already remarked, although these errors in the interpretation of propositions are not strictly fallacies, according to the present usual acceptation of that term, it may be well to have given them this consideration, because they are often the source of logical fallacies in giving wrong assumptions to start from. But they are not the result of violating any logical laws such as have been laid down. It is these violations with which the proper discussion of fallacies is concerned. Hence we turn to the second class, which we have denominated Logical Fallacies.

Logical Fallacies are errors in reasoning or inference, and not of interpretation. An error of interpretation is an error of intellectual perception ; an error of reasoning is an error of judgment in the passage from one proposition or conception to another assumed to be contained in the former. The data or premises may be correctly interpreted and yet the inference be a wrong one. As an illustration take the simple conversion of propositions in A. We may be correct in our conception of the proposition "All nations are aggregates of men," but wrong in the inference from it, by simple conversion, that "All aggregates of men are nations." And so on with all other forms of inference where the conclusion is not legitimately deduced from the premises.

Logical fallacies are divided into *formal* and *material*, according as the error is in the form of the reasoning or in the subject-matter of reasoning. A formal fallacy is an error which arises from a violation of the formal laws of inference. It is incident to the mere form of statement, or, as it is often said, is a fallacy *in dictione* or *in voce*. It requires only a knowledge of what the formal laws of reasoning are to detect such fallacies. On the other hand, a material fallacy is one which is due to some peculiarity in the matter of the reasoning, and hence arises independently of the form of statement, and so is said to be *extra dictionem*. The formal laws may be conformed to, but owing to some ambiguity of meaning or assumption of facts which are not true the conclusion may be materially vitiated in spite of the correctness of the formal reasoning. The material fallacy can be detected only by those who are familiar with the subject-matter of the discourse or argument. In Political Economy, for instance, any one familiar with the laws of reasoning might be able to detect formal errors in reasoning, but in order to discover the fallacies due to material considerations, that is, to the matter of the subject, the student must understand Political Economy. It is the same with all other subjects when the question regards material fallacies.

The further classification or subdivisions of formal and material fallacies must be considered in separate paragraphs. We take up briefly the formal fallacies.

1. FORMAL FALLACIES.—These have been sufficiently defined as mere violations of the principles of the syllogism which we have previously enunciated. They are determined by the number of terms, the distribution of terms, and the nature of the premises in the syllogism. Each species may be considered briefly.

(a) *Fallacy of Four Terms, or Quaternio Terminorum.*— One rule of the syllogism is that it shall not contain more than three terms : the presence of a fourth term vitiates the conclusion, because it prevents that comparison with a middle term which is necessary to reasoning. A simple illustration of the *Quaternio Terminorum* is the following :

> Men are mortal.
> Socrates is a Greek.
> ∴ Socrates is mortal.

The impossibility of drawing an inference in such cases is so apparent, and the temptation to do it is so unlikely that errors of this kind scarcely deserve notice. They are not common enough to require any special warning against them. It is only in the modified form of Equivocation that they are frequent. This occurs when the form and matter of a term are different, that is, when the same term has different meanings. There are, of course, cases where the terms are not grammatically the same, but which are logically identical in meaning. These would only apparently be cases of four terms. The one circumstance which determines a case of four terms is a distinction of material import that is not likely to be confused with any form of a concept. This will distinguish such instances from Equivocation, which is a modified form of Quaternio Terminorum.

There is, perhaps, a sense in which the fallacy of four terms is a material fallacy, in that new matter is introduced into the syllogism besides what is necessary to give it legitimacy. But this aspect of it is hardly worth serious consideration, although it may deserve mention for the purpose of recognizing the possibility.

(b) *Illicit Process of the Middle Term.*—This fallacy has already been explained and illustrated. It is due to a failure to distribute the middle term at least once in the premises. It may occur in several ways. One illustration of it will suffice :

$$\begin{array}{lll} M = P & \text{Some Pennsylvanians are Americans.} & I \\ \circledS = M & \text{All Philadelphians are Pennsylvanians.} & A \\ \therefore \circledS = P & \text{All Philadelphians are Americans.} & A \end{array} \right\} \text{Fig. I.}$$

(c) *Illicit Process of the Major Term.*—This is due to the distribution of the major term in the conclusion when it is not distributed in the premises. This also may occur in several ways :

$$\begin{array}{lll} \circledM = P & \text{All men are mortal.} & A \\ S \times \circledM & \text{Some animals are not men.} & O \\ \therefore S \times \circledP & \therefore \text{Some animals are not mortal.} & O \end{array} \right\} \text{Fig. I.}$$

(d) *Illicit Process of the Minor Term.*—This fallacy is due to the distribution of the minor term in the conclusion when it is not distributed in the premises. One of the many ways in which this occurs is the following:

$$\begin{aligned}\text{(M)} &= P \\ \text{(M)} &= S \\ \therefore \text{(S)} &= P\end{aligned} \quad \begin{aligned}&\text{All Germans are Caucasians.} \\ &\text{All Germans are men.} \\ &\therefore \text{All men are Caucasians.}\end{aligned} \quad \left.\begin{aligned}A \\ A \\ A\end{aligned}\right\} \text{Fig. III.}$$

(e) *Fallacy of Negative Premises.*—This is due to an attempt to draw a conclusion when both premises are negative and requires no illustration.

(f) *Fallacy of Particular Premises.*—This is due to an attempt to reason with particular premises. This case when tested turns out to be a fallacy due to illicit distribution of terms, either illicit middle or illicit major.

We might also include in the formal fallacies breaches of Rules 7 and 9 (p. 173), or attempts to draw an affirmative conclusion when one of the premises is negative, and to draw a universal conclusion when one of the premises is particular. These errors, however, have not received any special name.

2. MATERIAL FALLACIES.—Material fallacies, as defined, are due to something in the matter of reasoning. They are, excepting one instance, the Petitio Principii, cases of introducing *new* matter into the syllogism, while the form remains the same, and so are modifications of the Quaternio Terminorum. This introduction of new matter may be either in the *premises* or in the *conclusion*. If it be in the premises it must be in connection with the middle term, which will give some form of Equivocation as the first material fallacy. If it be in the conclusion, it must be in connection with the major or minor terms, which will give some form of Inconsequence. Those of Equivocation correspond to the formal fallacies of Undistributed Middle and Quaternio Terminorum. Those of Inconsequence correspond to the formal fallacies of Illicit Major and Illicit Minor. The exceptional case of Petitio Principii, which we have mentioned, is an instance of assuming matter which is not admitted or not proved. There seems, therefore, to be three forms of material fallacy. But when we consider that

fallacies of Inconsequence are assumptions of matter not in the premises, as the Petitio Principii is an assumption of unproved premises, we may reduce the last two to what may be called Fallacies of Presumption, and take as the first class those of Equivocation. Material fallacies we therefore divide into two classes, those of Equivocation and those of Presumption. These two classes require some further explanation.

The fallacies of Equivocation are all due to the equivocal or ambiguous use of terms. They most frequently occur in connection with the middle term of the syllogism, although some logicians consider them possible in connection with the major and minor terms. I prefer to limit them to the middle term for the sake of convenience, although we might admit Equivocation in the major and minor terms as at the same time a fallacy of Inconsequence. The whole matter, and our reasons for the classification, will not appear clear until the material fallacies have been explained in detail. At present it must suffice to obtain their classification.

The fallacy of Equivocation with logicians generally is limited to what is called *Ambiguous Middle*, and is not identified with the two fallacies of Accident, and the two of Composition and Division. But since they all turn upon an equivocal use of the middle term, and since, in my own experience with students of Logic, these various fallacies are constantly confounded with each other, I am convinced that they should be classed together under a common principle. I, therefore, use the term Equivocation in a much more comprehensive sense than is usual with writers on Logic, and so to include the fallacies of Accident, Simple and Converse, of Composition and Division, and the ordinary case of Ambiguous Middle, which I shall call Specific Accident. In this way the student has only to determine, first, whether the fallacy turns upon the use of an equivocal term, or upon the presumption of matter in the premises or the conclusion, and then he can proceed to determine the special form of Equivocation. The peculiarities of these forms will be examined in the next chapter.

CLASSIFICATION OF FALLACIES

In the so-called fallacies of Presumption, as already indicated, we may assume the truth of the premises when they should be proved, or we may assume new matter in the conclusion not contained in the premises, which may be admitted. This distinction gives rise to two general divisions of the fallacies of Presumption, namely, the *Petitio Principii*, or Begging the Question, and the *Non Sequitur*, or False Consequent. They will be discussed in the next chapter. The following table summarizes our classification:

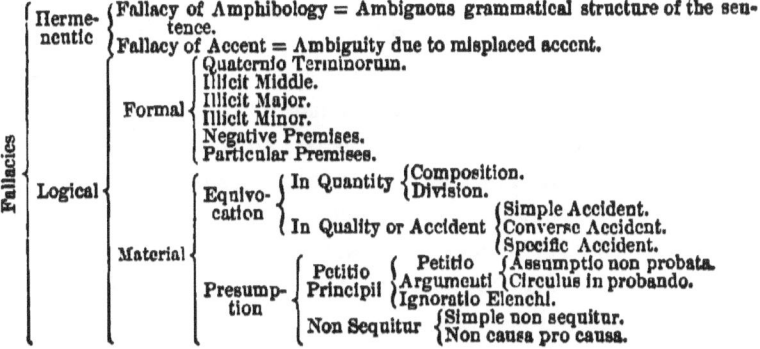

CHAPTER XVIII.

MATERIAL FALLACIES

1st. Fallacies of Equivocation.—These have been defined as caused by the equivocal use of terms. I have divided them into two classes, those of *quality*, or Accident, and those of *quantity*. Those of quality or accident are so called because the fallacy arises from some confusion due to differences of meaning in regard to the *attributes* denoted by a term in a proposition. Thus, if I say "Iron is a metal," I affirm "metal" of it in its proper form, as an aggregate of certain qualities or attributes. Now, if I also say "Rust is iron," I use the term "iron" in a slightly different sense, affirming that the substance, or *generic*, not the *specific*, qualities of it are identical with "rust;" that is to say "rust" is "iron" only in its substance not in its form. This fact prevents me from drawing the conclusion that "Rust is a metal." The fallacies of quantity are so called because they are due to the different senses in which a merely numerical aggregate of *individuals* can be taken. Thus, "All the trees" may be taken collectively or distributively, and so give rise, as we shall see, to an equivocation. We consider this form of fallacy first in order, and it is perhaps the easier to detect. It is that of Composition and Division.

1. FALLACY OF COMPOSITION AND DIVISION.—Both fallacies arise from the confusion of a *collective* and a *distributive* term, but one of them is the converse of the other. The mode of determining them can be expressed in the following formula:

Composition { In the major premise the middle term is used *distributively*.
{ In the minor premise the middle term is used *collectively*.

Division { In the major premise the middle term is used *collectively*.
{ In the minor premise the middle term is used *distributively*.

In the fallacy of Composition it will thus be seen that we argue from a distributive to a collective use of the term; and, *vice versa*, in Division we argue from the collective to the distributive use of a term. Probably a simpler means of determining the matter in each case would be to observe whether, as a whole, the proposition was used distributively or collectively, and not to make the decision more difficult by looking for this distinction in the middle term of the premises.

One of the best illustrations of a fallacy of Composition is the following:

All the angles of a triangle are less than two right angles.
A, B, C are the angles of a triangle.
Therefore A, B, C are less than two right angles.

In the major premise the proposition is true, if we suppose that the expression "all the angles of a triangle" is taken distributively; that each angle taken alone is less than two right angles; for taken together they are equal to two right angles. The conclusion, therefore, cannot be true, unless A, B, C are taken distributively. For if we mean in the major premise that *each* angle is less than two right angles, and in the conclusion that *all together* are less than two right angles, we infer what we have no right to infer; that is, we argue from what is distributively true of A, B, C, to what is supposed wrongly to be collectively true of them. A similar case often occurs in arguments like the following:

Thirteen and seventeen are prime numbers.
Thirty is thirteen and seventeen.
Therefore thirty is a prime number.

In the major premise "thirteen and seventeen" are used distributively, and in the minor premise collectively. Thirty not being identical with "thirteen and seventeen," considered distributively, cannot be identical with that which is identical with them in this sense, and hence the fallacy of composition.

In the first illustration the fallacy grows out of the ambiguous use of the word *all*, which in such cases may have either a collective or a distributive signification. Thus if we were to argue that because "All the peers derive their titles from the crown," and "The House of Parliament consisted of all the peers," therefore "The House of Parliament derived its title from the crown," we should be committing again the fallacy of Composition. "We must not argue that because every member of a jury is very likely to judge erroneously, the jury as a whole are very likely to judge erroneously; nor that because each of the witnesses in a law case is liable to give false or mistaken evidence, no confidence can be reposed in the concurrent testimony of a number of witnesses." And we may add that we cannot argue from the truth of all the incidents in a story to the truth of the story as a whole. A novel may interweave a large number of true facts and incidents and yet not be true or historical in its totality. "It is by a fallacy of Composition that protective duties are still sometimes upheld. Because any one or any few trades which enjoy protective duties are benefited thereby, it is supposed that all trades at once might be benefited similarly; but this is impossible, because the protection of one trade by raising prices injures all others."

The best illustration of the fallacy of Division is the converse of the one given for Composition. It is as follows:

All the angles of a triangle are equal to two right angles.
A is an angle of a triangle.
Therefore A is equal to two right angles.

In the major premise the middle term is used collectively, as in no other way could we say that "*all* the angles of a triangle are equal to two right angles." Hence we mean, not that each individual angle is so, but only that *all together* are. In the minor premise the middle term is distributive, and hence in the conclusion we show that we have argued from what is true collectively, or of an aggregate, to what is true only in a distributive sense. The fallacy is, therefore, one of

Division. If I were to argue from the fact that Congress or Parliament had voted a subsidy that Mr. A. or Lord B. had voted for the same, I should be committing the same fallacy. So also would be the argument that because houses make a city a given mansion would make a city because it is a house. We commit this fallacy when we imagine that because the aggregrate of expense is large, the number of individual items of expense will be large.

2. FALLACIES OF ACCIDENT—It is important to keep these distinct from the fallacies of Composition and Division. The latter have to do with *numerical* or *mathematical* aggregates and individuals, the former with *logical* or *metaphysical* wholes which represent totals of attributes. Unless we keep this in view we are liable to confuse them. But if we remember that Composition and Division turn upon the collective and distributive use of terms, and the fallacies of Accident upon the confusion of *essentia* and *accidentia*, or genus and species (conferentia and differentia), or of the abstract and concrete, we shall have no difficulty in the judgment of particular cases. We divide the fallacies of Accident or Quality into three kinds, namely,

(a) *Simple Accident*, or argument from the essence or conferentia to the accident or differentia. Its dictum in old Latin is, *a dicto simpliciter ad dictum secundum quid*, meaning "from an absolute or unconditioned statement to one which is conditioned or accidental."

(b) *Converse Accident*, or argument from an accident or differentia to the essence or conferentia. Its dictum in old Latin is, *a dicto secundum quid ad dictum simpliciter*, meaning the reverse of that for Simple Accident.

(c) *Differential* or *Specific Accident*, an argument from accident to accident, or from differentia to differentia. Its dictum would be, *a dicto secundum quid ad dictum secundum quid*, meaning "from a conditioned to a conditioned assertion."*

* This classification of Ambiguous Middle with the fallacies of Accident is entirely new, so far as I know, but I think the exposition of it will quite justify the innovation.

Jevons defines the fallacy of Simple Accident to be an argument "*from a general rule to a special case,*" and the fallacy of Converse Accident, "*from a special case to a general one.*" There is considerable ambiguity in this account of the case, because the expression "general rule" is equivocal. It may denote what is numerically or mathematically "general," or what is logically "general." In the former instance it simply denotes what is true of a large number or the majority of a class, but in the latter it denotes what is true of the genus, essentia, or conferentia. If we accept the former meaning we are liable to confuse the formal fallacy of an undistributed middle with the material fallacy of accident. Thus if we were to argue that because "men have the right to vote," and "criminals are men ;" "therefore criminals have the right to vote," our reasoning would be perfectly correct as long as the major premise was regarded as a universal proposition. But as it stands it is what is called a "general" or indefinite proposition, and may simply denote that men as a "general rule" have a right to vote. In such a case the proposition is a particular one and the middle term is undistributed. But we should be arguing from a "general" to a special case, and yet the fallacy is not one of Accident. We prefer, therefore, to define the fallacies of Accident more accurately by indicating that the "general rule" or case must mean the genus, essence, or conferentia, and the "special case" must mean the species, accident, or differentia. This fallacy, therefore, is an argument from one of these properties or group of properties to the other ; that of Simple Accident from genus to species, from conferentia to differentia, from essence to accident, from abstract to concrete, etc.; that of Converse Accident from species to genus, etc., and that of Differential Accident from species to species, or differentia to differentia, etc. There can be no fallacy in arguing from genus to genus, or essence to essence, because these always represent the same or identical properties.

One of the oldest examples of Simple Accident is the following :

What you bought yesterday you eat to-day.
You bought raw meat yesterday.
Therefore you eat raw meat to-day.

De Morgan humorously remarks of this ancient illustration: "This piece of meat has remained uncooked, as fresh as ever, a prodigious time. It was raw when Reisch mentioned it in the 'Margarita Philosophica' in 1496; and Dr. Whately found it in just the same state in 1826." It is not so accurate an illustration as is desirable according to the definition, because the subject of the major premise is so indefinite, and is hardly a genus. But in the conclusion the predicate is asserted of the subject, with the accidental quality of rawness added, while in the major premise that predicate is asserted only of the substance or essence of what was bought, and hence we mistakenly argue from meat in general, and without qualification to meat in a particular form. Another and perhaps better illustration is the following:

Pine wood is good for lumber.
Matches are pine wood.
Therefore matches are good for lumber.

Here the predicate of the major premise is asserted of the substance or essence of "pine wood," not of all forms of it, while matches are pine wood not only in essence, but in a particular form or accident. We cannot affirm of this differential accident what is true only of the essence or conferentia. So also we cannot argue from the fact that oxygen and hydrogen will burn, that water will burn because it is oxygen and hydrogen. "It would be a case of the simple fallacy of Accident to argue that a magistrate is justified in using his power to forward his own religious views, because every man has a right to inculcate his own opinions. Evidently a magistrate as a man has the rights of other men, but in his capacity of a magistrate he is distinguished from other men, and he must not infer of his special powers in this respect what is true only of his rights as a man." All fallacies which attempt the substitution of a particular thing for the generic form belong to this head.

An illustration of the fallacy of Converse Accident is the following:

>Intoxicating liquors act as a poison.
>Wine is an intoxicating liquor.
>Therefore wine acts as a poison.

In this case we are arguing from the excessive use to all uses of wine, an inference that is fallacious. The major premise is true only of a particular mode of using liquors, or of the excessive use of them, while the conclusion, unless interpreted with a similar qualification, asserts the same thing of *all* forms of using them. "It is undoubtedly true that to give to beggars promotes mendicancy and causes evil; but if we interpret this to mean that assistance is never to be given to those who solicit it, we fall into the converse fallacy of Accident, inferring of all who solicit alms what is true only of those who solicit alms as a profession." Another formulated instance appears in the following illustration:

>Loyalty to the government is the duty of all citizens.
>Loyalty to Charles I. was loyalty to the government.
>Therefore loyalty to Charles I. was the duty of all citizens.

We may look at this instance in more than one way. In the first place, the major premise means that loyalty is a duty to legitimate governments or to such as execute the law, while the minor premise asserts the fact that loyalty to Charles I. was loyalty to the government whatever its nature was, and hence the conclusion asserts loyalty to Charles I. to be a duty without qualification, and without distinguishing between him as a magistrate and as a man. In the second place, loyalty to Charles I. may have been loyalty to him as a private person, say by his servants, while all citizens could not be loyal to him in this capacity, and so it is an error to argue from this particular kind of loyalty to every form of it including civil allegiance.

MATERIAL FALLACIES

When these fallacies of Accident occur we may formulate the following means of determining one form from the other, the simple from the converse, and *vice versa*. The principle is the same as in the fallacies of Quantity, Composition, and Division.

Simple Accident
{ In the major premise the middle term must be a *genus*, *essentia*, or *conferentia*, the predicate affirmed of these making an abstract proposition.
In the minor premise the middle term must be a *species*, *accidentia*, or *differentia*, the predicate so affirmed of the subject making a concrete proposition. }

Converse Accident
{ In the major premise the middle term must be a *species*, *accidentia*, or *differentia*, the predicate affirmed of these making a concrete proposition.
In the minor premise the middle term must be a *genus*, *essentia*, or *conferentia*, the predicate so affirmed of the subject making an abstract proposition. }

These rules are worded for the first Figure of the syllogism, and are designed to indicate a clear means of deciding the case when it cannot be done without such help. But the syllogistic form of inference is not the only, and probably not the most frequent, form of committing this fallacy, although the process can no doubt be thrown into the form of the syllogism. The fallacy may often occur in the immediate inference by subalternation and by added determinants and complex conceptions, or if not by these, by a process which very much resembles them. If we infer from the contemptible character of one "reformer" in a particular cause, that all "reformers" are bad, we are committing the fallacy of converse Accident. On the other hand, if we infer from the exchangeable value of money, that the old Confederate currency has exchangeable value, because it is money, we commit the fallacy of Simple Accident.

Without reference to the distinction between the two forms of these fallacies, it may be said that wherever we attempt to make an interchange of essence and accidence, or abstract and concrete, under the same term, a fallacy of Accident is committed. An assertion sometimes seems to be made of the whole of a concrete subject, when in fact it is made only of its essential or of its accidental forms. Thus, when we say that sulphuric acid is poisonous, we can assert this predicate, not of its essence, but of a particular accidental quantity of it, because it can be taken with impunity in certain forms or amounts. But we can neither argue from its poisonous character in large amounts to its injury in small amounts, nor to its harmfulness in small quantities from its dangers in large quantities. So of the assertions that "Governments are useful," "Truth is sublime," "Charity is a virtue," etc. This will be the case in many, if not nearly all abstract ideas and propositions. Indeed we might say that we are extremely liable to commit fallacies of Accident in arguing from the concrete to the abstract and from the abstract to the concrete. We are certain to do so when the concrete and abstract are viewed *logically* and not *mathematically*, as previously explained. When we say "Governments are useful," we really affirm the predicate of governments in the abstract, or perhaps *generally* of actual governments. But we mean usually to speak of certain ideal forms of social organization and not necessarily of any or all particular concrete forms of them. Hence we speak of them in their essence or conferentia, and so are not allowed to infer by subalternation that what is true of them in this sense is true of them in the concrete. On the other hand, if we asserted the predicate of them in the concrete as bad, we could not immediately infer that the same was true of them in the ideal or abstract sense. Although I have spoken of these inferences as apparently immediate, they may be converted into mediate arguments by supplying a suppressed and perhaps implied premise. Their immediacy appears in the assumption that what can be affirmed of government in general can be affirmed of governments in particular included in the class.

But the error lies precisely in this assumption, which does not allow for the two senses in which the conceptions "government," "truth," "charity," etc., can be used. It is here that we can put to practical use the distinction between *logical* and *mathematical* generals, or between the logical and the mathematical genus ; that is, between the genus as the *sum* of the species, and the genus as a name for the conferentia. The latter is the logical, as the genus, taken as the sum of the species, is the mathematical conception of a class. If we say that " governments are useful," in the mathematical sense, we mean all individual and particular governments in the class, and so no fallacy will be committed in an inference to the species. But if we say it in the logical and abstract sense, we are using the term in a sense which may not be true of any individual case whatever, and hence the fallacy.

Another illustration will perhaps make the matter still clearer. When we say "Men are mortal," we have made an assertion which applies to *all individual* men. The predicate is asserted mathematically of men. It is not asserted of man in the abstract, but of all men in the concrete, all forms and conditions of men. Hence when we form a minor premise involving any number or species of men, the conclusion follows necessarily because they are included in the same sense in the major premise. The middle term has only a mathematical signification and so admits of no fallacy. But suppose we affirm "Meat is healthy food," here is a statement which may be taken either mathematically to denote all specific kinds of meat, or in an abstract logical sense to denote that the substance so called is healthy food, and so it would be spoken of in its essence, essential qualities, or conferentia, while we would not intend to include the same matter in its raw state, its stale or decayed condition, or in unlimited quantities. Hence we could not argue from the universal truth in the first case to the particular case in the second. This has already been illustrated in the first syllogism representing the fallacy of Simple Accident. In fact such statements are meant to affirm the predicate of certain well-known, perhaps usual and normal forms of the subject, and so exclude the cases involved in the conclusion when the fallacy of accident is committed. But it is not apparent from the form of statement, because the mathematical conception of universal propositions is the usual and the most natural one. But it is just such substitutions that the student must be on guard against, as liable, in more serious situations than we have illustrated, to lead him astray.

The technicalities of law offer a very rich field for the fallacies of Accident. It is the difference between one case and another that occupies the barrister in his attempts to show that they are not included in the general rule. In philosophy the Cartesians committed the fallacy by denying that hardness, weight, etc., were essential qualities of matter, and then inferring that a cubic foot of iron had no more matter in it than a cubic foot of air, because space or extension was regarded as the essence of matter. De Morgan quotes an amusing story from Boccaccio which illustrates the fallacy, but in too obtrusive a form to deceive any one, and yet it illustrates the whole case:

"A servant who was roasting a stork for his master was prevailed upon by his sweetheart to cut off a leg for her to eat. When the bird came upon the table the master desired to know what had become of the other leg. The man answered that storks never had more than one leg. The master, very angry, but determined to strike his servant dumb before he punished him, took him next day into the fields, where they saw storks, standing each on one leg, as storks do. The servant turned triumphantly to his master; on which the latter shouted, and the birds put down their other legs and flew away. 'Ah, sir,' said the servant, 'you did not shout to the stork at dinner yesterday; if you had done so, he would have shown his other leg too.'"

Not all fallacies of Accident are so easily detected as this, but they illustrate the same principle and the same logical characteristics. They are perhaps as frequent as any other form of logical error, and in fact the inclination to make those substitutions of two different things under the same name, and separated only as essence and accident, is so common that it has been well worth the pains to dwell upon the subject at great length.

It remains to show that the fallacy of Ambiguous Middle, or what I have called Differential Accident, is rightly included under the general head of Accident. We have intimated that all are forms of equivocation, that is, substitutions of one meaning of a term for another, and now we have to show that a closer relation than is usually recognized by logical writers exists between the fallacies of Accident and the ordinary Ambiguous Middle. An illustration will be the best means of proving the case.

The end of life is its perfection.
Death is the end of life.
Therefore death is the perfection of life.

The ambiguity of the word "end" is perfectly apparent, and we might be content with calling the fallacy merely one of equivocation. But if we observe closely, although the word end denotes two different things, there is a common idea at the basis of them which makes the equivocation possible. This common characteristic constitutes the generic or conferential idea of the word. But it is the differential quality which in each case determines the nature of the assertion. In the major premise "end" means the *object* or *purpose* of life of which perfection is asserted. In the minor premise it means the *termination* of life, which is made identical with death. Now, the common idea or conception which enables us to apply the word "end" in both cases is the notion of *limit*, or the point of interruption in a line, beyond which we need not go for a given purpose. Hence the notion of *object* is the differentia of one use, and *termination* that of the other, so that the attempt to argue from one to the other, on the ground of a common medium, is an attempt to pass from one accident to another. It will be the same in all equivocal terms where the confusion is not due to a mistaking of the genus or conferentia for the species or differentia, and *vice versa*. Thus, again to use Jevons's example,

All criminal actions ought to be punished by law.
Prosecutions for theft are criminal actions.
Therefore prosecutions for theft ought to be punished by law.

Both the terms "criminal" and "action" are used in a double sense. In the major premise "criminal" denotes what is *immoral*, and "action" a form of conduct, as an act of the will. In the minor premise "criminal" denotes merely *pertaining to a crime*, without implying any judgment upon its character, and "action" denotes, after its old Latin use, a *suit* at law. These are simply differential or specific meanings of the term, which has no generic application apart from such as are given, and so the argument is from one of these to the other through the common conception implied in the terms. The fallacy is a modified form of Quaternio Terminorum. But we should call it the fallacy of Differential or Specific Accident in order to classify it correctly and in order to understand its characteristics. The expression *Ambiguous Middle* should be reserved for a more comprehensive use, as equivalent to equivocation, the two terms to be used interchangeably.

2d. Fallacies of Presumption.—According to our previous explanation of these fallacies, something is presumed or assumed which we have no right to take for granted in the terms of the syllogism. They are presumptions in regard to the matter or contents of the reasoning. The presumption may be regarding the material truth of the premises, or it may be regarding the introduction of new matter into the conclusion when the premises are admitted. However correct the formal reasoning may be, the conclusion may be vitiated materially, either by assuming the premises when they should be proved, or by introducing a fourth term into the conclusion. We have then, as indicated in the classification of fallacies, two kinds of materially false inferences of Presumption, the *Petitio Principii* and the *Fallacia Consequentis*, or *Non Sequitur*.

1. FALLACY OF PETITIO PRINCIPII.—This is ordinarily called *Begging the Question*, and means the assumption of a fact or a premise without proof, or, as in the argument called reasoning in a circle, is an attempt to prove a proposition by itself. This is a form of assuming it when it should be proved by some more general and accepted truth. The Petitio Principii

we divide into two distinct forms, the *Petitio Argumenti*, which is committed in the presentation of an argument or when attempting the proof of a proposition, and the *Ignoratio Elenchi*, which is committed in the refutation or the attempt to disprove a proposition; it is simply a little more complicated *petitio principii*. The Petitio Argumenti we again subdivide into two forms, the *assumptio non probata*, or assumption of unproved premises, which may be different from the conclusion, and the *circulus in probando*, or reasoning in a circle, the assumption of premises which are the same as the conclusion.

The *assumptio non probata* can be illustrated by any syllogism whatever. Thus if we were trying to prove that "All men were mortal," and assumed that "All organic beings are mortal," with the minor premise that "All men are organic beings," we could be charged with begging the question by one who did not admit the proposition "All organic beings are mortal." He might admit that the formal reasoning was perfectly correct, and that the conclusion would be true if the premises were; but he would insist that the material inference was false because the premise was not admitted or not proved. It does not matter which premise is disputed, the effect is the same. We can charge a *petitio principii* upon a man when we dispute the major and admit the minor premise, and *vice versa*, or when we dispute both premises. It is sufficient to question one of the conditions of the conclusion.

It is not merely the failure to prove one's premises that constitutes the fallacy of begging the question. This failure must be one which occurs when proof is needed or demanded. It is, perhaps, most frequent when trying to convince some one else of a given truth, although it may occur whenever we are trying to prove to our own minds a conclusion without assuring ourselves sufficiently of the stability of the premises upon which the conclusion rests. But it is most frequent in arguments with others, because the one condition of proof in such cases is that an opponent or reader admits the principles upon which the conclusion is to be established. We cannot prove to him a truth with premises he does not admit. If we assume these without his acceptance, our reasoning has no cogency, and he is at liberty to say that we are begging the question, and this without disputing either the formal accuracy of our process or the truth of our proposition. He merely claims that the case is not proved. A proposition in a conclusion may be true, although it has not been proved in the premises. The advantage of proving it lies in making it a special case, included under a general law or class, so that when a person has admitted the larger he must perforce admit the smaller. But there are instances in which we may dispute the *universality* of a principle or premises either to show that the conclusion *may*, so far as we know, be an exception, or to assert that it is not proved by such a case, however true it may be in reality. Suppose we wish to prove that "All cattle have cloven feet." If, in order to do so, we assert that "All ruminants are cloven-footed," and "All cattle are ruminants," the conclusion will follow, provided the premises are accepted. But we can be charged with begging the question if the major premise, "All ruminants are cloven-footed," is not true, although it may be true that "All cattle are ruminants," and also that they are all cloven-footed. But the proposition is not proved except by the universality of the major premise in this case. It is one thing to perceive the truth of a proposition as a matter of fact, and it is another to prove it by means of a higher condition. The charge of *petitio principii*, then, must not be construed as properly meaning that the conclusion is denied, but only that it is not proved. We should be committing a counter-fallacy if we supposed that this error was a disproof of the proposition in question.

We may too hastily impute the fallacy of begging the question. This is virtually done when we demand proof for a premise merely because we see that the conclusion must be accepted if the premise is admitted. It is often employed in order to evade the issue and escape conviction. It may be permissible sometimes to carry the demand for proof back through several steps, but the danger is that it will most frequently be dishonestly done, or be the mark of a weak cause.

De Morgan describes the case in the following language: "There is an opponent fallacy to the *petitio principii* which, I suspect, is of more frequent occurrence; it is the habit of many to treat an advanced proposition as a begging of the question the moment they see that, if established, it would establish the question. Before the advancer has more than stated his thesis, and before he has had time to add that he proposes to prove it, he is treated as a sophist, on his opponent's perception of the relevancy of his first step." In such emergencies the person presenting the argument must ascertain whether his opponent admits in each case that the conclusion will follow if the premises are true, and by continuing this process he will either expose the motive of his opponent or morally weaken his demand for proof.

This fallacy is very likely to occur in the disjunctive syllogism, and especially in the dilemma ; for we may assume the disjunction to be complete when it is not. There may be more than the two alternatives usually assumed in the case. An instance of this occurs in the sophism which was used by early Greek philosophers to prove the impossibility of motion. It was said that a thing must either move where it was, or where it was not. It was absurd to suppose that it could move where it was not, and if it moved it could not be in the place where it was, and therefore it was inferred that its motion was impossible. But this conclusion, or the premises rather, lost sight of the third alternative, namely, that a body might move from the place where it was to a place where it was not, or had not been the moment before. The omission of this alternative in the premise made the argument a *petitio principii*. There is a traditional answer to this argument which we shall notice under the Ignoratio Elenchi.

"Jeremy Bentham pointed out that the use even of a single name may imply a *petitio principii*. Thus in a church assembly or synod, where a discussion is taking place as to whether a certain doctrine should be condemned, it would be a *petitio principii* to argue that the doctrine is *heresy*, and therefore it ought to be condemned. To assert that it is heresy is to beg the question, because every one understands by heresy a doctrine which is to be condemned. Similarly, in Parliament, a bill is often opposed on the ground that it is unconstitutional and therefore ought to be rejected; but as no precise definition can be given of what is or is not constitutional, it means little more than that the measure is distasteful to the opponent. Names which were used in this fallacious manner were aptly called by Bentham *Question-begging Epithets.*"

The *circulus in probando* is a species of *petitio principii*, which consists in "arguing in a circle," or in assuming as proof of a proposition some assertion which is identical with it in its import, or in trying to prove a proposition by itself. Thus to say that "Man is wise because he is rational," is to argue in a circle, because "rational" is substantially identical in meaning with "wise." So also would it be to argue that "The weather is warm because it is summer, and it is summer because it is warm," and "Men never practise excess because they are not immoderate in their habits." Jevons's illustration is the following: "Consciousness must be immediate cognition of an object; for I cannot be said really to know a thing unless my mind has been affected by the thing itself." Here "to know" and "immediate cognition," are identical in import and cannot be used to prove each other.

It is mostly in long arguments that this fallacy can be committed without ready detection. When we argue that a person should submit himself to the guidance of his party, or his government, because they maintain what is right, and then proceed to prove this by asserting they are right because they ought to be submitted to; or if we argued that lead had more matter in it than a given amount of wood, because it was heavier, and that it was heavier because it had more matter in it, the circle would be so narrow that it would be easy of detection. But when the circular *petitio principii* occurs at the end of a long discourse, as it often may do, it may be committed without easy discovery. Only the closest observation can secure us against it. It is likely to be committed by the use of synonyms which are taken to express more than the conception involved. Jevons and Whately have remarked that the English language, being composed of two or more languages, is liable to this fallacy, because it frequently has several synonymous terms for the same conception.

The *Ignoratio Elenchi* is the second general class of fallacies which we have included under the head of Petitio Principii. It may not seem clear why we have chosen to consider it a species of begging the question. The reason for so doing cannot be fully appreciated until the fallacy has been defined. It has been called by Whately and others the fallacy of *Irrelevant Conclusion*. This is a true enough description of it, except that the definition does not exclude the *fallacia consequentis* or *non sequitur*. We prefer, therefore, to define it as *Ignorance of the Issue or Argument*, and hence it consists in arguing to the wrong point, or in proving one thing in a way that seems to prove something else; or proving something which is not the contradictory of the thing asserted. It will be apparent from the account of it that it occurs in the process of refutation, or in proving something which is supposed to be the opposite of what is believed or affirmed by an opponent. We commit the fallacy by assuming the conclusion we reach to be in contradiction with that against which we are arguing, when it is not a contradictory. In refutation it is our business to prove a contradictory of a given assertion, but if we prove something which is not denied by our opponent, we are evading the issue, and proving something that is irrelevant. Thus, in assuming the contradiction which is not a contradiction, or something to be denied by our opponent which is not denied by him, we indirectly beg the question. This fact is, our reason for classing the Ignoratio Elenchi with the Petitio Principii. It is much more complicated than the simple case, but when we consider that it is merely the counter-petition of one who is adducing an argument in refutation to that of the person producing proof of a proposition, we shall perceive the right to regard it as we have done. It is the assumption of what is not a fact, or of what is not admitted to be a fact by an opponent, and such an assumption is of the nature of a *petitio principii*. But the fact is concealed and complicated by the circumstance that two syllogisms and the law of contradiction are involved in the explanation of the case. That the issue is evaded can generally be determined without resolving the fallacy into a petitio principii. But it is this kind of fallacy nevertheless.

A good illustration of the Ignoratio Elenchi is the following: Suppose a man is accused of being a thief, and I prove that he is not a thief. Now the proper disproof of this assertion is to prove the contradictory, namely, that he is a thief. But if my opponent instead of proving this, proves that *the man is a rogue*, he commits the *ignoratio elenchi*, because I have not denied the latter proposition, or asserted the contradictory of it. He virtually begs the question by assuming that to prove him a rogue is to prove him a thief, and that he has proved the contradictory of my assertion when his proposition is not denied. Omitting the premises which might be involved in establishing either side, and those involved in proving the proportion assumed to be disproof, the whole relation may be represented as follows:

Proof.	*Disproof.*	*Ignoratio Elenchi.*
∴ A is not a thief.	∴ A is a thief.	∴ A is a rogue.

In asserting that the man is a rogue the opponent intends to avail himself of certain presumptions which might follow from the fact that the man was a rogue. The proof that A was not a thief might imply, to untrained minds, that he was a good man, or the disproof might be such as it was not easy to counteract the effect of. Hence if the man can be proved to be a rogue, it is assumed that a presumption against the validity of the disproof is established, or that to prove him a rogue is to prove him a thief. The fallacy, nevertheless, is apparent, in that we may say that all thieves are rogues, but not that all rogues are thieves. The fallacy, then, is in assuming the convertibility of the two conceptions, " rogue " and " thief," and that his assertion contradicts the proposition involved in the original statement.

"The fallacy is the great resource of those who have to support a weak case. It is not unknown in the legal profession, and an attorney for the defendant in a lawsuit is said to have handed to the barrister his brief marked, 'No case; abuse the plaintiff's attorney.'" In all the attacks on a person, or his character, when the question regards a doctrine, the fallacy is the same as in the case of the above attorney. Thus if I praise a man's poetry or his philosophy, it is no refutation of him to show that his life has been bad, or that he has lost his mind.

De Morgan mentions a good instance: "If a man were to sue another for debt, for goods sold and delivered, and if defendant were to reply that he had paid for the goods furnished, and plaintiff were to rejoin that he could find no record of that payment in his books, the fallacy would be probably committed. The rejoinder, supposed true, shows that either defendant has not paid, or plaintiff keeps negligent accounts; and is a dilemma, one horn of which only * contradicts the defence. It is the plaintiff's business to prove the sale from what *is* in his books, not the absence of payment from what *is not*; and it is then the defendant's business to prove the payment from his vouchers."

The observations of Whately and Mill are well worth quoting in this connection at some length, since they furnish so clear an exposition of this fallacy. Says the former: "Various kinds of propositions are, according to the occasion, substituted for the one of which proof is required. Sometimes the particular for the universal; sometimes a proposition with different terms; and various are the contrivances employed to effect and to conceal this substitution, and to make the conclusion which the sophist has drawn answer practically the same purpose as the one he ought to have established. I say 'practically the same purpose,' because it will often happen that

---

* De Morgan in placing "only" in the position which it occupies in the sentence makes his statement liable to a fallacy of accent.

some *emotion* will be excited—some sentiment impressed on the mind—such as shall bring men into the *disposition* requisite for your purpose, though they may not have assented to, or even stated distinctly in their own minds, the *proposition* which it was your business to establish. Thus if a sophist has to defend one who has been guilty of some *serious* offence, which he wishes to extenuate, though he is unable distinctly to prove that it is not such, yet if he can succeed in *making the audience laugh* at some casual matter, he has gained practically the same point.

"So also if any one has pointed out the extenuating circumstances in some particular case of offence so as to show that it differs widely from the generality of the same class, the sophist, if he find himself unable to disprove these circumstances, may do away with the force of them by simply *referring the action to that very class*, which no one can deny that it belongs to, and the very name of which will excite a feeling of disgust sufficient to counteract the extenuation; *e.g.*, let it be a case of peculation, and that many *mitigating* circumstances have been brought forward which cannot be denied; the sophistical opponent will reply, 'Well, but after all, the man is a *rogue*, and there is an end of it;' now in reality this was by hypothesis never the question; and the mere assertion of what was never denied *ought* not, in fairness, to be regarded as decisive; but practically, the odiousness of the word, arising in great measure from the *association of those very circumstances* which belong to *most of the class*, but which we have supposed to be *absent* in this particular instance, excites precisely the *feeling of disgust* which in effect destroys the force of the defence. In like manner we may refer to this head all cases of improper appeals to the passions, and everything else which is mentioned by Aristotle as extraneous to the matter in hand (ἔξω τοῦ πράγματος).

"In all these cases, as has been before observed, if the fallacy we are now treating of be employed for the apparent establishment, not of the *ultimate* conclusion, but, as it very commonly happens, of a premise, then there will be a combination of this fallacy with the last mentioned (undue assumption).

"For instance, instead of proving that 'this prisoner has committed an atrocious fraud,' you prove that 'the fraud he is accused of is atrocious;' instead of proving, as in the well-known tale of Cyrus and the two coats, that 'the taller boy had a right to force the other boy to exchange coats with him,' you prove that 'the exchange would have been advantageous to both;' instead of proving that 'a man has not a right to educate his children or dispose of his property in the way *he thinks best*,' you show that 'the way in which he educates his children or disposes of his property is not *really the best ;*' instead of proving that 'the poor ought to be relieved in this way,' you prove that 'they *ought to be relieved ;*' instead of proving that 'an irrational agent—whether a brute or a madman—can never be deterred from any act by the apprehension of punishment,' as, for instance, a dog from sheep-biting, by fear of being beaten, you prove that 'the beating of one dog does not operate as an *example* to *other* dogs,' etc., and then you proceed to assume as premises, conclusions different from what have really been established," you commit the fallacy of ignoratio elenchi. But it is in a modified form, because it appears less as a refutation than as an attempted confirmation of some position. They can, however, be conceived in the usual form by supposing that the thesis to which it is assumed the conclusions are opposed is suppressed. Besides, one of them, the instance about relieving the poor, might be considered a case of converse accident. But as something is proved with the assumption that it is identical with another position, while it is in reality opposed to it, we have the *ignoratio elenchi* in the converse form.

"A good instance of the employment and exposure of this fallacy occurs in Thucydides, in the speeches of Cleon and Diodotus concerning the Mitylenaeans ; the former, over and above his appeal to the angry passions of his audience, urges the *justice* of putting the revolters to death, which, as the latter remarked, was nothing to the purpose, since the Athenians were not sitting in *judgment*, but in *deliberation ;* of which the proper end is expediency. And to prove that they had a right to put them to death, did not prove this to be an *advisable* step."

Mill observes that "the works of controversial writers are seldom free from this fallacy. The attempts, for instance, to disprove the population doctrines of Malthus have been mostly cases of *ignoratio elenchi*. Malthus has been supposed to be refuted if it could be shown that in some countries or ages population has been nearly stationary; as if he had asserted that population always increases in a given ratio, or had not expressly declared that it increases only in so far as it is not restrained by prudence or kept down by poverty and disease."

Dr. Johnson's refutation of Berkeley's idealism by kicking against a stone is a similar fallacy. And so are all cases of appeal to consequences supposed to contradict a given assertion before proving that such a contradiction exists. In such instances the opponent may accept the consequences, unless the contradiction between them and his assertion is first proved.

In addition to the general form of the Ignoratio Elenchi, there are several special forms which it is important in a treatise of Logic to consider. The valid process, of which the *ignoratio elenchi* is the invalid, is called the *argumentum ad rem*. The special invalid forms or cases of evasion are the *argumentum ad judicium, argumentum ad populum, argumentum ad hominem, argumentum ad ingorantiam*, and the *argumentum ad verecundiam*.

The *argumentum ad judicium* is an appeal to general or universal belief, and so is based upon the common judgments of mankind. The dictum of such an appeal is the admitted or assumed truth of what all men everywhere believe. The controversialist appeals to this maxim because he supposes it is admitted and that it contradicts some conclusion which an opponent is trying to maintain. Thus if I deny the existence of an external world, of spirit, or of an unseen world, it would be an *argumentum ad judicium* to show that all men have everywhere believed in their existence. This universal belief

MATERIAL FALLACIES 251

may create a presumption or make it necessary to consider the matter seriously, but it does not prove it.

The *argumentum ad populum* is an appeal to public opinion, or to the passions and prejudices rather than to the intelligence of people.

The *argumentum ad hominem* is an appeal to the practice, profession, or principles of the person to whom or against whom an argument is directed. It is an effective method of silencing an opponent, but it does not prove the case.

The *argumentum ad ignorantiam* is an appeal to a man's ignorance in order to produce conviction upon his inability to dispute the case.

The *argumentum ad verecundiam* is an appeal to authority, or an accepted body of doctrines.

These several forms of *argumenta* are essentially the same in their principles and their import. Four of them appeal to certain admitted or assumed principles which are supposed to prove the case because they are assumed to contradict the opposite, which is the position to be disproved by the proof of its alternative. But in no case, unless we except the *ad hominem* instance, are we assured either that the *dicta* upon which we depend are admitted by an opponent, or that they are necessarily contradictory to the point in question. They are thus evasions of the issue.

But it is important to remark that they are not always irrelevant or illegitimate merely because they are evasions. There are circumstances in which it is perfectly legitimate to use them; only we must not suppose that this legitimacy implies that they are methods of real proof. Although they have a proper application they are not *argumenta ad res*. It is important to take this fact into account in order not to infer, from their fallacious nature as arguments to the point, that their illegitimacy either impeaches the proposition in question or excludes them from a certain relevancy for another purpose. They are invalid only as proofs or disproofs of a matter in discussion, but they are not invalid as means of establishing a contradiction between two propositions. Common discourse assumes that a man is refuted if we show that he has contradicted himself and that he must accept a given conclusion if the opposite contradicts his profession or his practice. But this is not the fact. It does place him in a position that compels him to choose between the two contradictories, but it does not decide which of the alternatives he must select. Hence the charge or proof of a contradiction in a man's discourse is no disproof of his assertion, unless he still holds to its contradictory. If he denies the contradiction he may hold both alternatives. Hence the several *argumenta non ad res*, in merely proving a contradiction somewhere, are fallacies of *ignoratio elenchi*, in the relation of assuming that they prove anything. But we must distinguish between this and their valid use for establishing a contradiction. An illustration in the case of the *argumentum ad hominem* will make this position clear. We quoted the instance of incomplete disjunction by the ancient Greek philosophers, who sought to prove the impossibility of motion by trying to limit our conception of it either to a change where a thing is, or a change where it is not. Tradition has it, says De Morgan, that the originator of this disjunction called in a physician to set a dislocated shoulder, and the physician turned his argument upon the philosopher to prove that his shoulder was not hurt. He argued that the shoulder must be put out of place either where it was, or where it was not. But as it could neither be put out of place where it was, nor where it was not, it could not be dislocated at all. This is an excellent case of the *argumentum ad hominem*, both in its legitimate and its illegitimate relation. It is an admirable exposure of the absurdity of the Greek philosopher's argument, but it neither disproves the impossibility of motion nor proves its existence. Nor is it a refutation of the assertion which is imputed by inference to the philosopher, namely, that his shoulder was out of place. It only establishes a contradiction between his philosophic doctrine about motion and his present belief about the dislocation of his shoulder. The philosopher would have only to say either that it was not a case of motion, or that his shoulder was not displaced, in order to indicate that his argument or position was not overthrown, while admitting that the reasoning of the physician was correct. Nevertheless he would not escape the charge of a contradiction somewhere, and although his assertion is not disproved by the *ad hominem* argument, he is under obligation to explain the contradiction or to give up one of the alternatives. This is the value of the *argumentum ad hominem*, and of the other similar forms of appeal to admitted principles.

2. FALLACY OF NON SEQUITUR.—As already indicated this is generally called the *fallacia consequentis*, or False Consequent. It arises in connection with the conclusion, and not in connection with the premises. It is, therefore, the introduction of new matter into the conclusion, which is not contained in the premises. There is no special necessity for subdividing it into distinct forms, except that one class has received a separate name for the sake of particular convenience, and perhaps because of its peculiar frequency. If we must distinguish them at all, it must be into the *common non sequitur*, and the *non causa pro causa*, or *false cause*, often called the *post hoc, ergo propter hoc*, fallacy. The form in which the fallacy usually occurs can be represented in the following manner:

> All men are rational.
> Socrates is a man.
> Therefore Socrates is noble.

It is evident that this conclusion cannot follow from the premises unless we regard "noble" as identical with "rational," which it is not intended to be. The fourth term is here in the conclusion. De Morgan's illustration of the fallacy is less simple. It is:

> Episcopacy is of Scripture origin.
> The Church of England is the only Episcopal church in England.
> Therefore the church established is the church that should be supported.

It is evident that nothing has been said about supporting the church in the premises, and hence it does not follow from them. The fallacy is determined wholly by the presence of new matter in the conclusion. It closely resembles the formal fallacies of illicit major and minor. The difference is that in the latter the addition is *quantitative*, while in the *non sequitur* it is *qualitative*.

The literal meaning of *non sequitur* would apply to any fallacy whatever, because the fallacy means that the conclusion does not follow from the premises. But technically logicians mean or should mean by this particular term that the conclusion does not follow from the premises, although they are true. The *petitio principii* vitiates the conclusion because of false premises; in the *non sequitur* the premises are not disputed, but are admitted, at least for the sake of the argument. It is important to observe in this connection that both fallacies are possible at the same time and in the same syllogism. We may question the premises and so charge a *petitio principii* upon the conclusion; or we may say that even if the premises are true the conclusion does not follow, in which case we impute a *non sequitur* to the reasoning. Therefore whenever we can make the error turn upon false assumptions in the premises, we charge the former fallacy against the reasoning, and whenever it turns upon false assumption in the conclusion, independently of the premises, we charge the latter fallacy.

Very frequently the fallacy of *non sequitur* is due to apparent cases of immediate reasoning, which are in reality enthymemes. Thus if we were to say "History is authentic because mankind has accepted its statements," or "Philosophy is useless because it bakes no bread," we might be charged with a *non sequitur* on the ground that the conclusion was not included in the premise. But since the argument is an enthymeme we can complete it in the usual way, so that the conclusion after all might be included in the terms of the suppressed premise. Thus the major premise of the first enthymeme is "Whatever mankind has accepted is authentic," and of the second, "Whatever bakes no bread is useless." When these are supplied we find that the conclusion is valid unless we can impeach the premises, but to question them turns the fallacy into a *petitio principii*. We thus discover that what may be regarded as a *non sequitur* in one relation may be a *petitio principii* in another; what is not involved in one premise may be begged in the other.

We have therefore to be careful in deciding when a fallacy is a *non sequitur* alone. The pure and simple form of it occurs when both premises are admitted, either in reality or for the sake of argument. In such cases as we have just indicated it coincides with the *petitio principii*, and may be reduced to it. But when it occurs in its pure form this cannot be done.

The fallacy of False Cause, or *non causa pro causa*, is the mistake of imagining a necessary connection where there is none, or of confusing a causal connection with a mere coexistence or sequence. It occurs when we argue that a certain thing is the cause of another when we find them occurring together. Thus if we were to argue that a change of the weather was due to the occurrence of a new or full moon, because they coincided, or because the former immediately followed the latter; or if we attributed a pestilence to the occurrence of a comet; or a death in the family to an eclipse of the sun, we should be committing this fallacy. The Latin phrase, *post hoc, ergo propter hoc*, indicates the manner in which the conclusion is drawn, and upon what it depends. "When things are seen together," says De Morgan, "there is frequently an assumption of necessary connection. There is, of course, a presumption of connection: if A and B have never been seen apart, there is probability (the amount of which depends upon the number of instances observed) that the removal of one would be the removal of the other. It is when there is only one instance to proceed upon that the assumption falls under this fallacy; were there but two, inductive probability might be said to begin. The fallacy could then consist only in estimating the probability too high." But a probability is no proof. The inference may be a deductive fallacy, however great the probability, and in spite of the inference being inductively legitimate. No number of mere coexistences or sequences contains the statement of the cause of phenomena, and we are not entitled to infer it from them. *Necessary* connection is not involved in the mere *fact* of connection. If it were, I might argue that night was the cause of day, or *vice versa*, because we find that one invariably precedes the other; or I might argue that the flight of birds was the cause of springtime, because it accompanies the latter.

If we analyze the cases of *non causa pro causa*, however, we shall find that they too may coincide with a *petitio principii*, and perhaps they should be classified with that form of fallacy. Thus we might say that the inference that night was the cause of day was a non *sequitur* when drawn from their invariable connection, but when we complete the syllogism by supplying the suppressed premise the major premise would be, "All that precedes day is the cause of it." The minor premise would be, "Night precedes day," and the conclusion would follow as involved in the major, although not in the minor, premise. But we may charge the major premise with begging the question, and hence, as before, this case, which appears a *non sequitur* in relation to the minor premise, is a *petitio principii* in relation to the major premise. All *post hoc, ergo propter hoc*, fallacies can be reduced in this way, and hence it might seem best to include them as a species of begging the question. But as they usually occur with an enthymeme where the conclusion in such cases is not included in the premise, the conveniences of controversy make it best to regard the fallacy as a *non sequitur*, although it is one which coincides with a *petitio principii*, or may so coincide with it. But the most perfect form of *non sequitur* will occur when both premises are unquestionable.

3d. General Observations.—The first observation to be made regarding the fallacies which we have just considered is that they are not always distinct from each other. This is apparent in the fact that the last two often coincide, and that the *non causa pro causa* may be resolved into a *petitio principii* when the suppressed premise is supplied. A similar reduction

MATERIAL FALLACIES 257

might be possible with some of the others. For example, take the fallacies of Accident, and in particular the illustration of the use of pine wood:

Pine wood is good for lumber.
Matches are pine wood.
Therefore matches are good for lumber.

We gave this as a fallacy of Accident. But in fact it may be resolved in two other ways at the same time, which may show why the fallacy of Accident occurs. In the first place, the major premise is an indefinite or general proposition, and we have already learned that such propositions are very frequently particular in their real import. This is, in fact, the real meaning of the statement. It is not true, nor would it be intended to assert, that *all* pine wood, that is, *all* forms of it, are good for lumber, but only that *some* pine wood is good for lumber. But thus to convert the real import of the major premise into a particular proposition, making the syllogism IAA of the first Figure, prevents the distribution of the middle term, so that the fallacy would virtually be a formal one. Many of the fallacies of Accident can be so reduced. But it is only because we are viewing the premises in their *quantitative* signification instead of their *qualitative*. For it is true, qualitatively, that "*All* pine wood is good for lumber," that is, in substance, but not in every form, and hence the case of Accident can be brought against this conception of it. But interpreting the case mathematically, what would be regarded logically and qualitatively a fallacy of Accident becomes formally and quantitatively an illicit middle.

In the second place, since we suppose the material meaning of the major premise to be that "*Some* pine wood is good for lumber," we impeach the truth of the proposition universally, and it is upon its universal truth that the conclusion depends. Hence, in considering the premise or premises doubtful we can regard the fallacy as a *petitio principii*. There are thus two fallacies, one formal and the other material, which can be

17

imputed to this syllogism, besides that of Accident. In fact we can make it a fallacy of Accident only upon the supposition that the major premise is *universally* true of the *essentia* of pine wood, but not of its *accidentia*, while the predicate of the conclusion would connect with one of its accidents what had been connected in the major premise only with the essence. But aside from this interpretation either an illicit middle or a *petitio principii* can be imputed to the syllogism.

Perhaps a similar resolution of Composition and Division could be made, because the premises in syllogisms committing those fallacies are capable of a double interpretation. It is sufficient to suggest the possibility, and the actual achievement of it can be left to the student. And it will not be necessary to say more on the close relation between the *non sequitur* and the *petitio principii* in many cases, after having shown that the two may be applied to the same conclusion, but in different relations, one of them indicating assent to a premise, but not to the inference, and the other indicating that one of the premises vitiates it. The two will not coincide when they are imputable, one of them only to false premises and the other only to a false inference.

One more remark, which has been alluded to, it is important to make. The imputation of a fallacy in the reasoning does not necessarily imply that the proposition in the conclusion is a false one. In many cases the falsehood of the proposition and the existence of the fallacy go together; but it is not always the fact, and we must learn to recognize this fact because although a proposition may be true, it may lead to error to have it connected falsely with another proposition as proof when that proposition may not be true, or when the conclusion is not an inference from it. We commit a fallacy when we suppose that an error in reasoning is a sufficient disproof of a proposition. The fallacy thus committed is an *ignoratio elenchi*. All that the existence of a fallacy can establish is a mistake in the mode of proving a proposition, unless it serve as the means of discovering the actual error in our propositions. We usually discover the error, in fact, before we find why or how it has been committed. But the fallacy in reasoning is an error growing mainly out of an attempt to deduce one connection of terms from another, and so will not always be an index of material errors of fact. They are, of course, accompanied by error somewhere of a material kind frequently enough, but not necessarily implying it where the ordinary mind assumes it; we require to be on our guard against committing a fallacy when imputing one to others. We must always distinguish between the error in fact which we may first perceive, and the error in reasoning to which such a discovery may have led us.

Another important remark is that it is not necessary to put the argument into the form of a syllogism in order to discover what the fallacy is. We have only to observe the manner of *substituting one term for another.* Most frequently in actual discourse arguments are either stated in the form of enthymemes, or the premises are so expressed as to effectually conceal the Mood and Figure of the syllogism, and we are left entirely to depend upon the manner in which we use certain terms. Then, since in enthymemes we can construct them into syllogisms, at least, of the first or of the second Figure, as we please, in one of which the same matter may be valid which is invalid in the other, and since the three Figures can be reduced to the first at pleasure, it will not be necessary to consider the form in detail, but only how we substitute one term for another. If the fallacy be a formal one, it will be most easily detected in some cases by observing the form of the argument, but in some cases this is not necessary. Besides formal fallacies are not so often committed as the material. When any doubt, however, exists about the nature of an illegitimate inference, it is best to throw the argument into the form of a syllogism, and then ascertain its relation to the general rules.

But in many, if not in most instances of material fallacy, we can determine the error by observing the two or more senses in which a term is used without stopping to consider whether the form of the syllogism is expressed correctly or not, because it may either be *thought* in a different manner from the expressed relation, or it may be materially what it is not formally. Thus in the following syllogism:

> White men are Caucasians.
> The Germans are Caucasians.
> Therefore the Germans are white.

we should be guilty of a formal fallacy of illicit middle. But since we may have stated the major premise in a form in which it was not thought, namely, in the inverse form, the proposition being a definition, the reasoning may be in the form of the first instead of the second Figure, as it is stated, and hence perfectly valid. We cannot, of course, always rely upon this method of dealing with an argument, but in cases of material reasoning we either use the first Figure most commonly, unless we are proving a negative, or our data can be so easily reduced to it, that we can generally depend upon the mere form of substitution of one term for another in order to determine the nature of the fallacy. Thus if we try to prove that a man should give alms to a particular person on the ground of his duty to be charitable, we commit a fallacy of Accident, because we argue from the genus "charity" to a particular case of it where an accidental or differential circumstance may modify the obligation. Similarly with the fallacies of Quantity, if I argue from the effect of a forest in producing a thick shade to a similar effect from a single tree I commit the fallacy of Composition. We have seen how the *non sequitur* may be imputed without considering, at least in some cases, more than the statement which is assigned as its ground, although a completion of the argument by supplying a suppressed premise may convert it into a *petitio principii*. This is only an illustration of observing whether the conclusion is deduced from the given data or not. In the *ignoratio elenchi* we never require to construct the syllogism, but only to observe whether the conclusion evades the question or not, or whether the assumed contradiction is a true one or not. Since all material fallacies, with the possible exception of the *petitio principii* in some cases, are a modified form of Quaternio Terminorum, we have only to see whether terms are used throughout an argument in an identical sense or not, in order to determine the nature of the fallacy committed in any particular case.*

* On Fallacies consult De Morgan: Formal Logic, Chapter XIII.; Mill: Logic, Book V., especially Chapters V., VI., and VII.; Whately: Elements of Logic, Book III.; Hamilton: Lectures on Logic, Lecture XXIII.

CHAPTER XIX.

QUANTIFICATION OF THE PREDICATE

A TREATISE on Logic is hardly complete that omits an account of recent doctrines regarding what is called the "quantification of the predicate." The usual expositions of the subject are confined to the forms left by Aristotle, and which, sufficing for practical purposes, are best adapted to the actual usages of language. But language does not always express explicitly what thought involves implicitly, and hence many logicians have felt it necessary to correct this defect by an *ideal* scheme of logical doctrine which might enable us better to understand logical processes in their pure forms, and then to modify this scheme to suit the exigencies of defective usage. Sir William Hamilton, Professor De Morgan, and George Bentham, all about the same time conceived the propriety, or at least the possibility, of modifying logical doctrine by the "quantification of the predicate." This view we shall proceed to explain, with its importance in practical reasoning.

We have already shown what the *quantification of the subject* is, although it has not been stated under that name. But to quantify it is only to say whether the whole or the part of it is taken into account. Its quantification refers to its distribution or non-distribution. Hence to quantify the predicate is to state whether the whole or only the part of it agrees with the subject, or differs from it. We have seen that the proposition "Men are wise," is an indefinite one so far as its explicit statement is concerned, and that we cannot deal with it logically or with any degree of assurance as to the results unless we first know whether it means "*all* men" or "*some* men are wise." This is quantifying the subject in order to bring the proposition into a definite form for logical use. Thus far the ordinary Logic proceeds, but no further. But why not also quantify the predicate in a similar manner, in order to evade the equivocations incident to its indefinite forms? In the above proposition it is said to be undistributed, because nothing is stated or implied to indicate whether the whole or only a part of its extension is taken into accôunt. It is, therefore, perfectly indefinite. But in some cases, in spite of this mode of statement, we happen to know that the meaning of the predicate is definite; that it is identical with the subject in its quantity. As already explained this is the case with all definitions in which subject and predicate are convertible terms. Supposing that we define man as a rational animal, we can say with equal truth "All men are rational animals," and "All rational animals are men," and so with any other definition. The same happens to be true of the proposition "White men are Caucasians." We can convert it simply into "All Caucasians are white," but only because we happen to know that whiteness and Caucasian are convertible terms. According to the formal laws enunciated regarding the form of such propositions, it would have to be converted *per accidens*: so also even with definitions. But as the form of a definition and that of an ordinary assertion cannot be distinguished in general usage, why would it not be an improvement in the science of Logic to treat the predicate as we have treated the subject, and to state explicitly what is implicitly involved in it? The answer to this question will appear in the sequel, after we have shown how the quantification can be effected and what are its consequences to the ordinary forms of reasoning. Jevons's exposition suffices for the purpose.

"In the proposition 'All metals are elements,' the subject is quantified, but the predicate is not; we know that all metals are elements, but the proposition does not distinctly assert whether metals make the whole of the elements or not. In the quantified proposition 'All metals are *some* elements,' the little word *some* expresses clearly that in reality the metals form only a part of the elements.* Aristotle avoided the use of any mark of quantity by assuming, as we have seen, that all affirmative propositions have a particular predicate, like the example just given; and that only negative propositions have a distributed or universal predicate. The fact is, however, that he was entirely in error, and thus excluded from his system an infinite number of affirmative propositions which are universal in both terms. It is true that 'All equilateral triangles are *all* equiangular triangles,' but this proposition could not have appeared in his system except in the mutilated form, 'All equilateral triangles are equiangular.' Such a proposition as 'London is the capital of England,' or 'Iron is the cheapest metal,' had no proper place whatever in his syllogism, since both terms are singular and identical with each other, and both are accordingly universal.

"As soon as we allow the quantity of the predicate to be stated the forms of reasoning become much simplified. We may first consider the process of conversion. In our discussion of the subject it was necessary to distinguish between simple conversion and conversion by limitation. But now one simple process of *simple conversion* is sufficient for all kinds of propositions. Thus the quantified proposition of the form A, 'All metals are some elements,' would be simply converted into 'some elements are all metals.'" The quantified form of A, "All metals are all elements," would be simply converted into "All elements are all metals;" and so on with all propositions. We could simply proceed upon the rule that *whatever we do with one term we could do with the other.* Their meaning is made definite by their explicit quantification.

"The doubly universal proposition is of most frequent oc-

---

* If the ordinary canon about the signification of *some*, as previously defined (p. 116), is to be enforced here, Jevons is wrong in saying that the word denotes *only* a part. It merely asserts distinctly a part, and does not assert or even imply anything about the whole. But it does indicate that we are not to assume anything about the whole of the predicate. If we adopt its use with the implication that it denotes *only* a part, we should have to define it so.

currence, as in the case of all definitions and singular propositions. I may give as instances, 'Honesty is the best policy,' 'The greatest truths are the simplest truths,' 'Virtue alone is happiness below,' 'Self-exaltation is the fool's paradise.'

"When affirmative propositions are expressed in the quantified form all immediate inferences can be readily drawn from them by this one rule, that *whatever we do with one term we should do with the other term*. Thus, from the doubly universal proposition, 'Honesty is the best policy,' we infer that 'what is not the best policy is not honesty,' and also, 'what is not honesty is not the best policy.' From this proposition, in fact, we can draw two contrapositives ;* but the reader will carefully remember that from the ordinary unquantified proposition A we can only draw one contrapositive. Thus if 'metals are elements,' we must not say that 'what are not metals are not elements.' But if we quantify the predicate thus, 'All the metals are *some* elements,' we may infer that 'what are not metals are not *some* elements.' Immediate inference by added determinant and complex conception can also be applied in either direction to quantified propositions without fear of the errors noticed under those heads."

The quantification of the predicate adds *four* more propositions to those of the quantified subject, A, E, I, O, and Thompson employed new symbols for them, U and Y for the affirmative, and the Greek letters η and ω for the negative. U and Y represent the predicate as distributed in the affirmative propositions, and η and ω as undistributed in the nega-

---

* Jevons here uses the term *contrapositive* in a sense which is different from the usual definition of it, and to denote two distinct processes. We wish logicians could adopt it to denote the inference to complementary propositions, and limit the use of the term *contraversion* to what is generally called contraposition. Instead of speaking of "two contrapositives," therefore, Jevons might say a contrapositive and a contraverse. The conditions are different for inferring that "what is not honesty is not the best policy," from those in which we can infer that "what is not the best policy is not honesty," except with a quantified predicate. See also Antithesis (p. 169).

tive. The following table represents the forms of the eight propositions:

U All S is all P
A All S is some P
I Some S is some P
Y Some S is all P.
} Affirmative propositions.

E No S is any P
η No S is some P
ω Some S is not some P
O Some S is no P.
} Negative propositions.

The existence of these eight propositions makes it necessary to extend the number of valid moods from 24, or rather 19, omitting the weakened conclusion, to 108, without the fourth Figure, and 144 with that Figure. Sir William Hamilton did not include this Figure in his table and notation, nor did Thompson, to whose table we add the moods of the fourth Figure:

TABLE OF MOODS OF THE SYLLOGISM.

	Figure I.		Figure II.		Figure III.		Figure IV.	
	Affirmative.	Negative.	Affirmative.	Negative.	Affirmative.	Negative.	Affirmative.	Negative.
I......	U U U	E U E / U E E	U U U	E U E / U E E	U U U	E U E / U E E	U U U	E U E / U E E
II.....	A Y I	η Y ω / A O ω	Y Y I	O Y ω / Y O ω	A A I	η A ω / A η ω	Y A I	ω A ω / Y η ω
III....	A A A	η A η / A η η	Y A A	O A η / Y η η	A Y A	η Y η / A O η	A A Y	O Y η / Y O η
IV.....	Y Y Y	O Y O / Y O O	A Y Y	η Y O / A O O	Y A Y	O A O / Y η O	Y Y A	η A O / A η O
V......	A I I	η I ω / A ω ω	Y I I	O I ω / Y ω ω	A I I	η I ω / A ω ω	Y I I	O I ω / Y ω ω
VI.....	I Y I	ω Y ω / I O ω	I Y I	ω Y ω / I O ω	I A I	O A ω / I η ω	I A I	O A ω / I η ω
VII....	U Y Y	E Y O / U O O	U Y Y	E Y O / U O O	U A Y	E A O / U η O	U A Y	E A O / U η O
VIII...	A U A	η U η / A E η	Y U A	O U η / Y E η	A U A	η U η / A E η	Y U A	O U η / Y E η
IX.....	U A A	E A E / U η η	U A A	E A E / U η η	U Y A	E Y E / U O η	U Y A	E Y E / U O η
X......	Y U Y	O U O / Y E E	A U Y	η U O / A E E	Y U Y	O U O / Y E E	A U Y	η U O / A E E
XI.....	U I I	E I O / U ω ω	U I I	E I O / U ω ω	U I I	E I O / U ω ω	U I I	E I O / U ω ω
XII....	I U I	ω U ω / I E η	I U I	ω U ω / I E η	I U I	ω U ω / I E η	I U I	ω U ω / I E η

QUANTIFICATION OF THE PREDICATE 267

In this table the columns marked affirmative and negative represent respectively the affirmative and negative conclusions which the moods of the quantified predicate will give. It omits the cases of weakened conclusion. It is interesting to remark that the negative conclusions are twice as many as the affirmative. The large number of both of them adds very much to the difficulties of remembering those that are valid and those that are invalid. It would seem that instead of simplifying the process of reasoning the quantification of the predicate very much complicates it. In some cases it certainly does simplify it, but in so many other cases it complicates it that little is to be gained by the system. But its theoretical principle should be considered.

Hamilton's statement of its value is brief and to the point. He says the fact "that we can only rationally deal with what we already understand, determines the simple logical postulate—*To state explicitly what is thought implicitly*. From the consistent application of this postulate, on which Logic ever insists, but which logicians have never fairly obeyed, it follows, that, logically, we ought to take into account the *quantity*, always understood in thought, but usually, and for manifest reasons, elided in its expression, not only of the *subject*, but also of the *predicate* of a judgment. This being done, and the necessity of doing it will be proved against Aristotle and his repeaters, we obtain, *inter alia*, the ensuing results:

1. "That the *preindesignate* * terms of a proposition, whether subject or predicate, are never, on that account, thought as *indefinite* (or indeterminate) in quantity. The only indefinite is *particular*, as opposed to *definite* quantity; and this last, as it is either of an extensive *maximum* undivided, or of an ex-

---

* Hamilton employs the terms *predesignate* and *preindesignate* to denote the two subdivisions of Particular propositions or terms. "Predesignate" denotes what we have called *General* terms and propositions, as "Man is wise," where we may mean *all* or *some*, the quantity being indifferently expressed. "Preindesignate" denotes the ordinary particular proposition, as, "Some men are wise," where the subject is definitely indefinite.

tensive *minimum* indivisible, constitutes quantity universal (general) and quantity *singular* (individual). In fact, *definite* and *indefinite* are the only quantities of which we ought to hear in Logic; for it is only as indefinite that particular, it is only as definite that individual and general, quantities have any (and the same) logical avail.

2. "The revocation of the *two terms of a proposition* to their *true relation;* a proposition being always an *equation* of its subject and predicate.

3. "The consequent reduction of the *Conversion of Propositions* from three species to *one;* that of Simple Conversion.

4. "The reduction of all the *General Laws of Categorical Syllogisms* to a *Single Canon*.

5. "The evolution from *one canon* of all the *species and varieties of Syllogism*.

6. "The abrogation of all the *Special Laws of Syllogism*.

7. "A demonstration of the *exclusive possibility of Three syllogistic Figures*, and (on new grounds) the scientific and final *abolition of the Fourth*.

8. "A manifestation that *Figure* is an *unessential variation* in syllogistic form; and the consequent *absurdity of reducing* the syllogisms of the other figures to the first.

9. "An enouncement of *one Organic Principle for each Figure*.

10. "A determination of the true *number* of legitimate *Moods;* with

11. "Their *amplification* in number (*thirty-six*).

12. "Their numerical *equality* under all the figures; and

13. "Their *relative equivalence*, or virtual identity, throughout every schematic difference."

The remaining points of modification and advantage are mainly of interest to advanced scientific Logic, and need not be repeated here. Even some of those we have quoted are not of any apparent importance to practical Logic. But all of them present a formidable number of consequences imputed to the quantification of the predicate. The first four and the eighth are important simplifications of reasoning, and the remaining are at least interesting modifications. But there are two or three difficulties in the theory which Hamilton seems not to have noticed, and which go far to offset all the advantages of quantifying the predicate.

The first of these is an incident in the use of the word *some*. This term in the quantification of the subject, according to the Square of Opposition, must mean *some, and it may or it may not be all*, and must not mean *some and only some*. If it meant the latter, the assertion of I would always prevent the assertion of A. Hence, for the purposes of Logic it must take the former meaning, so that A may remain indeterminate and possible upon the assertion of I. But in the quantification of the predicate, if *some* mean an *indefinite part, and it may or not be all* of the predicate, the first thing to be remarked is that the predicate in reality remains as undistributed as before, and the second is, that the antitheses or complementaries of such propositions are no more possible than before, a possibility which was said to be consequent upon the quantification of the predicate. On the other hand, if we use *some* to mean *a part, and only a part*, we obtain the quantification of the predicate in affirmative propositions at the expense of the regular rule about its use in the quantification of the subject; and in negative propositions a serious ambiguity arises which it is hard to overcome. Thus propositions η and ω may each have two meanings. "No S is some P," may mean either that S is not *any* portion of P, or that it is not a *certain indefinite* portion of P. So in the proposition "Some S is not some P." This ambiguity appears most distinctly when we come to draw the antitheses or complementaries of such propositions. Thus in proposition U, "All men are all rational," we can immediately infer, as in exclusive propositions, "All not-men are not rational." When the use of *some* in the predicate means a part, and only a part, the complementary can be inferred that all which is not the subject is not that portion of the predicate. But the terms indefinitely considered would imply that it was not the other portion also. Thus, "All metals are some elements," ought to give "All not-metals are not some elements," and so it will, if we mean the particular *some* identified with the subject. But the complementary may mean not *any* "some" of the elements, which cannot possibly be true. These ambiguities are a serious difficulty in the way of the Hamiltonian theory.

There is a second important difficulty which shows how little the accepted rules of syllogistic reasoning can be relied upon to determine the validity of the new cases introduced by the quantification of the predicate. It is remarked by Keynes. In ordinary reasoning the distribution of the middle term in at least one of the premises, and the retention in the conclusion of the same quantity of the major and minor terms as in the premises, are a guarantee of valid inference. But in the mood AUA of a quantified predicate and of the second Figure, the observance of these conditions is no security against a fallacy, as will be apparent in the following syllogism and its symbolic representation in Fig. 26.

All P is some M
All S is all M
∴ All S is all P or some P.

The circles will bring out the fallacy more clearly.

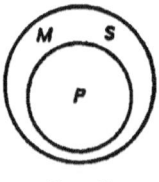

FIG. 26.

It is evident from this representation that we cannot infer from the premises that *all* S is either all or some P. All that we could infer is that "Some S is all P," which is Y or "Some S is some P, which is ω;" and yet neither illicit middle, nor illicit major, nor illicit minor, is committed in the propositions which, it seems, ought to follow, namely, "All S is all P," etc.; as a consequence of this, Keynes finds it necessary to add the

QUANTIFICATION OF THE PREDICATE   271

following new rule to the list already given for determining valid syllogisms : *If one premise is U, while in the other premise the middle term is undistributed, then the term combined with the middle term in the U premise must be undistributed in the conclusion.*

But this addition of a rule only complicates, instead of simplifies, the process of reasoning as we desire to use it in practical life ; and the immense number of valid moods has the same effect, while the ordinary rules of reduction to the first Figure, and the common practice of reasoning in this Figure makes the matter as simple as it can be made, taking into account the nature of language and the defects of expression in our thought. These, then, with other incidents, are defects in the practical importance of the quantification of the predicate.

Nevertheless there is one consideration of great interest and importance in it, which, if we cannot make as much practical use of it as might be desirable, does explain the simplicity of mathematical reasoning—the use of exclusive propositions and definitions, and the habitual tendency of ordinary minds to use predicates as if they were definitely quantified. Hamilton's dictum is a correct one, namely, that ready reasoning is greatly facilitated, and rendered less liable to fallacy by stating explicitly what is implicitly thought. This means that we should state definitely what is definitely thought, and not state it indefinitely. If I mean that " *All* men are wise," " *All* governments are good," I should say so, and not conceal a subterfuge under the indefinite "Man" or "Government," etc. But in such propositions the predicate remains as indefinite in extent or quantity as the subjects just mentioned. If only I knew in such statements whether the subject was the whole or the part of the predicate, I could know better how to use them. Thus if I mean by the first proposition that "All men are all the wise," I know first that "All who are not-men are not wise," and second, that the two terms are perfectly convertible. This is precisely what occurs in exclusive propositions and definitions. In the former class the predicate is definitely, and, we might say, formally, quantified by the use of *only* qualifying the subject, and which indicates that the quantity of the predicate is not greater than that of the subject. In definitions the predicate, although not *formally*, is *materially* quantified, in being made identical with the subject, and so convertible with it. The effect of such propositions upon the reasoning is to make it valid in some instances where there appears to be a distinct formal fallacy. Take an illustration with an exclusive proposition :

> Only virtue is praiseworthy.
> Courage is praiseworthy.
> Therefore courage is a virtue.

This appears from the form of the propositions to be a syllogism in AAA of the second Figure, and therefore invalid because of the undistributed middle, and yet in spite of this we cannot resist the impression that the inference is correct. The reason for this is that the virtual quantification of the predicate in the major premise totally modifies the reasoning as compared with the ordinary rules. We can therefore explain the case in two ways. First, since the subject and predicate are made coextensive by the use of *only*, the major premise is equivalent in meaning to its converse, namely, "What is praiseworthy is virtue," and this change turns the syllogism, in spite of its present form, into AAA of the first Figure, which is valid. We may conceive that this is the order of thought, in spite of the order of expression, and the right to so conceive it depends wholly upon the definitely quantified predicate. Second, since the quantification of the predicate in the major premise, making it coextensive with the quantity of the subject, has the effect of distributing it, we have a syllogism of the second Figure which distributes the middle term at least once in the premises, in spite of the affirmative character of the propositions. This distribution enables us to draw the inference which appears in the conclusion. In the case of definitions the illustration would be the same.

QUANTIFICATION OF THE PREDICATE

Common reasoning often treats propositions as if they were definitions or mathematical propositions, and it is because the thought of the person reasoning is conceived as representing a certain identity between subject and predicate which may not be expressed, but is in mind. Hence the disposition to substitute them for each other. Now this substitution can be done with impunity whenever the predicate is made quantitatively equal to the subject, and so the need of distinguishing between propositions which intend to distribute and those which do not intend to distribute the predicate. If common usage could adopt a symbol which could serve this purpose many logical difficulties would be overcome. The *theory* of pure Logic would thus find its conditions supplied in practice. But as it is, all propositions, except definitive, exclusive, and mathematical propositions, are alike in their form, while some of them in their matter, or the way in which the mind *thinks* them, may be quantified, so that the syllogisms containing them cannot be judged by the formal laws of reasoning until we know what content they are supposed to have. Thus we may say, "Houses are residences," and formally the predicate is not quantified, or is not distributed, and cannot be used with an affirmative minor premise in the second Figure. But if we have in thought the notion that the predicate is identical with the subject it is quantified, and the reasoning is altered. This, then, in brief, is the advantage of quantifying the predicate. It indicates explicitly whether *all* or *only a part* of it is taken into account.

It must be observed, however, that in propositions I and Y no special advantage is gained, because they leave the predicate undistributed, except that we know definitely from their form how to understand them. The chief importance attaches to the distinction between A and U, which, if the quantification were practicable in most cases, would enable us to know the extent of the predicate as well as that of the subject. But even if not of much service practically, the theory shows very clearly what can be done with definite propositions, and how the common mind often acts in its reasoning, when the trained logician would impute a violation of the formal laws of Logic to it. Besides, it explains a common tendency of the mind to carry on the substitution of one term for another without stopping to consider their distribution. The substitution in an unqualified manner is, no doubt, an error, but it is rather an error of assumption in many cases than of inference.*

* References on the doctrine of the Quantification of the Predicate are as follows: Hamilton: Lectures on Logic, Appendix V.; De Morgan: Formal Logic; Thompson: Laws of Thought, Part II., Sections 77-79; Mill: Examination of Sir William Hamilton's Philosophy, Chap. XXII.

NOTE.—The remarks in this chapter which seem to imply the reciprocal and convertible coincidence in quantity of subject and predicate in exclusive propositions must not be misunderstood. In as much as negative exclusive propositions are very seldom used at all, I have limited my statements to the affirmative form. Affirmative exclusive propositions resemble mathematical propositions in the fact that the predicate, although of an affirmative instance, is definitely quantified or distributed, and can be simply converted. But this convertibility into a universal is not reciprocal with the subject and predicate; that is, we cannot convert and reconvert them as often as we please unless we retain the exclusive form with every reconversion of the universal. The illusion which they give rise to comes from the actual distribution of the predicate and the fact that the extension of the subject is at least equal to that of the predicate. That is to say in the exclusive propositions mentioned the extension of the subject must be equal to, and may be equal to, that of the subject, but it is not distributed, while the extension of the predicate cannot be greater and may be equal, but is distributed. Consequently in the first degree of conversion they are, like mathematical propositions and definitions, capable of simple conversion. As already indicated they are really inverted universals and their complementary opposites are the contraverse of the real universals. In negative exclusive propositions the predicate is distributed without reference to the exclusive particle, and hence it has no function practically except to imply the complementary opposite. The quantification is the same as in the affirmative form, but conversion is not applicable because the distribution is the same as in particular negatives, and as we have learned propositions in O are not convertible. But in the affirmative forms not only are the propositions capable of simple conversion, but also are equivalent to the complementary opposite in which both subject and predicate are distributed and which represent a more distinctly mathematical proposition.

CHAPTER XX.

MATHEMATICAL AND OTHER REASONING

THE principles involved in the quantification of the predicate, and for that matter, also of the subject, will help us to understand the certitude and exemption from error displayed in mathematical reasoning, and also the difference between this and other reasoning. We shall define it more carefully after illustrating it and pointing out its relation to the principles discussed in the preceding chapter.

The first important observation to be made is that *in mathematical reasoning all propositions are U or E propositions, or their equivalents;* that is, are *universally and definitely quantified in their subject and predicate.* The effect of this fact is that which was remarked by Hamilton in the eighth observation we quoted from him (p. 268). It makes subject and predicate quantitatively equal, so that we can dispense with the variations of Figure in Syllogisms, in so far as the validity of the reasoning is concerned. All the Figures are valid alike, and from what has been said about the nature of the propositions in mathematical reasoning the Moods will be only three, namely, UUU, UEE, and EUE, all of which are valid in all the Figures. In the following illustrations the sign of equality serves both as the copula and the sign of *equivalence*, and the negative symbol previously explained (p. 140) denotes *inequality* as well as negation. The letters denote the various terms as before, only they stand for numbers or quantities. Mathematical reasoning, then, will appear in the following forms indifferently:

1st Fig.	2d Fig.	3d Fig.	4th Fig.
UUU	UUU	UUU	UUU
M = P	P = M	M = P	P = M
S = M	S = M	M = S	M = S
∴ S = P	∴ S = P	∴ S = P	∴ S = P.

276   ELEMENTS OF LOGIC

1st Fig.	2d Fig.	3d Fig.	4th Fig.
UEE	UEE	UEE	UEE
M = P	P = M	M = P	P = M
S × M	S × M	M × S	M × S
∴ S × P	∴ S × P	∴ S × P	∴ S × P

If we turn to the table of Moods with the quantified predicate we shall find that these two are valid in all four Figures, a fact which is due to understanding definitely that the predicate is limited to the subject, or made equal to it in extension. The process for all four Figures can be represented symbolically by the following circles :

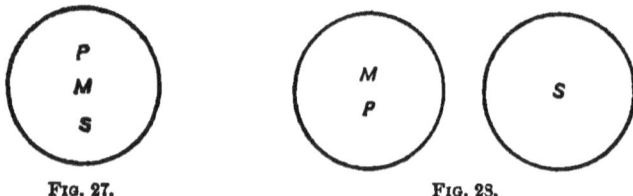

FIG. 27.   FIG. 28.

Fig. 27 represents the Mood UUU, and Fig. 28 the Mood UEE. Mood EUE would be the same as Fig. 28, only the letters S and P would change places. It is very evident from this how simple mathematical reasoning is, and how little regard we need to pay to Figure in the form of presenting it. If all reasoning were the same, it would be less exposed to formal fallacies, and undoubtedly it is the tendency of common minds to construe their reasoning mathematically, that explains both their liability to formal fallacies and the necessity of being on our guard for the existence of material fallacies in it when it is formally correct. The untrained mind, accustomed to deal with its terms and their import in a mathematical sense, very easily confuses with them other terms and propositions having the same form as the mathematical, and is at a loss to discover its liability to error.

The illustrations and their symbolic representation enable us to define mathematical reasoning and to distinguish it from the species which might be called *logical*, but for the fact that all forms of reasoning are logical. It could, however, be called "logical" in the sense in which we have distinguished between the "mathematical" and the "logical" genus. But without pressing this meaning, mathematical reasoning is that *which deals purely with quantity*. It may be called *pure quantitative* reasoning, in distinction from the second form, which may be called mixed *qualitative* and *quantitative*. Mathematical reasoning is based upon relations of quantity which expresses no variations or differences except in amount; quanto-qualitative reasoning is not only based upon relations of quality but of quantity also, and in addition to differences of degree or amount, represents differences of *kind*. This distinction between the two processes merely means that the conceptions employed in mathematical reasoning either deal exclusively with the abstract ideas of quantity, or deal with the concrete objects, only in numerical terms of time, space, force, etc., or commensurable quantities. They are not viewed as a group of qualities, but as individual wholes. When we say "all men," "all animals," "all citizens," we speak of a number of individual persons as individuals, and do not take into account their various differences. They may be of different sizes, different color, or of different powers, but the "all" does nothing but describe them numerically. When the predicate is thus quantified, the identity between subject and predicate becomes purely a quantitative identity, and so the terms are *equated*. But when we say "man," "animal," "citizen," etc., we think of *attribute wholes*, so that any identity between them and a predicate is qualitative, and in my judgment is not of the nature of an equation at all. Its real import may be expressed in one of two ways, according as the predicate is an *attribute* or a *class* term. We have already spoken of the two kinds of judgment, the *intensive* and the *extensive* (p. 123). The former, as for example, "Man is wise," expresses by its predicate that a certain group of qualities called man is *accompanied* by the quality wise, or that amid that group of qualities will be found one denoted by the term "wise." The so-called "identity," or "agreement" between the subject and predicate in such cases, is not one of quantity. The relation should rather be called that of *connection*, a relation that cannot be dealt with quantitatively, except as it may be common or invariable. The extensive judgment, as "Man is an animal," expresses by its predicate a distinct relation of identity between the subject and predicate, namely, that the two have certain common qualities by which they may be classified together. But no relation of quantity appears in this case any more than in the intensive judgment. In both, however, there is nothing said to prevent any number of other and *different* beings from being connected with, or included in, the predicate. The relation, therefore, is one which is based upon the real or the possible coexistence of resemblances and differences between certain objects connected as subject and predicate. This is only to say that in what we have called qualitative judgments and reasoning the conceptions with which we have to deal are *genera and species, essentia and accidentia*, or *conferentia and differentia*. These conceptions do not appear as such in mathematical reasoning. They prevent that definite quantification, especially of the predicate, which is necessary to make the reasoning mathematical.

But there is a class of judgments, intensive or extensive, just as we choose to regard them, in which the identity between subject and predicate is such as to preclude the admission of any other than the existing subject in the same connection. They are such judgments as "Virtue is goodness," "Quadrupeds are four-footed animals," "Honesty is the best policy," "Government is social organization," etc. Here there is a qualitative connection, or identity of qualities, expressed by the propositions. But it is an absolute identity. The terms are either synonymous or represent definitions. The relation expressed is not that between genus and species, or the connection between essentia and accidentia, or conferentia and differentia, but between genus and genus, essentia and essentia, conferentia and conferentia, accidentia and accidentia, or differentia and differentia. But such a relation coincides exactly with the *quantitative* relation, and, so far as reasoning is concerned, can be made convertible with it. Hence, in all such cases the reasoning with such propositions is as assured as that in mathematics, because their quantitative meaning coincides with their qualitative, or their qualitative with their quantitative, and so can be substituted in its stead. It will be self-evident from this fact that the whole process might be greatly simplified if all terms could be made to represent a similar connection between the subject and predicate.

The peculiarity of the relation between the two terms in these two classes of propositions and in all mathematical judgments might justify the use of the name *Traduction* for this species of reasoning, in contrast with ordinary Deduction on the one hand, and Induction on the other. It does not differ essentially, however, from Deduction, as the principle is the same. Derived from *trans* and *duco*, to lead over, it might denote the substitution which is characteristic of all reasoning that involves a predicate in its propositions quantitatively identical with the subject. Jevons alludes to the use of this term in connection with a species of syllogistic form and reasoning which we have yet to consider. But he does not employ it to denominate the essential characteristic of mathematical reasoning, which can carry on its substitution without regard to the question of Figure, as we have already seen. All its propositions are definite, and so are its terms. Even those conceptions which denote a part of a whole are definite. Mathematical conceptions are not qualified by the indefinite *some*, which may denote any portion whatever less than the whole, but they are expressed by some definite fraction of the whole, which is equivalent to a universal notion in its import, whenever a portion of some larger total is to be reckoned with. Even the unknown quantities of Algebra are no exception to this principle. They always represent definite quantities, which are called "unknown" because they may be used for any fixed number we please. In all instances of such terms, however, having a mathematical import, we have for our propositions an equation or equations in which the process of transition from one term to another is substitution or traduction. Mathematical reasoning thus embodies in a perfect form the principles most clearly enunciated in the doctrine of the quantification of the predicate, although implied in the quantification of the subject. It is simply the substitution of one term for another which involves it, because they are quantitatively identical at the same time that they may be necessarily connected in another way.

The principle thus established can be used to explain a form of reasoning which Jevons regards as irregular, and yet valid in spite of its real or apparent violation of the formal laws of the syllogism. The first illustration chosen by him is not formally incorrect, but is regarded by him as irregular. It is as follows:

    The sun is a thing insensible.
    The Persians worship the sun.
    Therefore the Persians worship a thing insensible.

The only apparent irregularity in this example is the use of a part of the predicate of the minor premise in the conclusion, and the consideration of the other part of it as the middle term. But logically I can see no objection to this process, nor a reason in it for regarding the syllogism as in any way irregular. It is true that, grammatically considered, the subject of the major premise is "the sun," and the predicate of the minor premise is "worship the sun," so that being AAA of the first Figure, they occupy the place of the middle term. The difference between them might seem to make a fourth term, and therefore a fallacy of Quaternio Terminorum. This would undoubtedly be the case if we attempted to draw the conclusion, "The Persians *are* a thing insensible." But the whole matter is altered when we transfer the term "worship" to the conclusion, because it signifies that the real middle term of thought is "the sun," and the reasoning becomes perfectly valid, and is no exception to the ordinary form of syllogistic inference. But it is not a case of mathematical substitution, except as a part of the apparent middle term is transferred to the conclusion.

The next instance is somewhat similar, and yet is as different in other respects. Jevons is correct when he asserts that a great deal of our reasoning is done after this manner. The following is his illustration :

The Divine Law commands us to honor kings.
Louis XIV. is a king.
Therefore the Divine Law commands us to honor Louis XIV.

The peculiarity of this instance is that it seems to be a case of AAA in the second Figure, namely, an example of illicit middle, and yet valid in spite of the fact. As before, we substitute in the conclusion a part of the predicate in the minor premise, and this leaves the term "king" as the middle term. Why, then, does the reasoning impress us irresistibly with its validity, although the form is against this feeling? If we stated the case as follows there would be no difficulty in rejecting the conclusion :

Louis XIV. commands us to honor kings.
The Divine Law commands us to honor kings.
Therefore the Divine Law is Louis XIV.

The fallacy is palpable in this instance. Why is it not so in the former?

Two explanations can be given to the case which will answer the question. The first is a reduction of the thought expressed to the proper Mood and Figure. Thus the minor premise of the original example can be interpreted as meaning "Kings are those whom the Divine Law commands us to honor." This is precisely the same thought as before, and it gives AAA of the first Figure, with the conclusion, "Louis XIV. is he whom the Divine Law commands us to honor." In the premise and conclusion, therefore, we have simply the passive for the active form of expression, while the logical import is precisely the same as before. In this way we answer our question by showing that logically the apparently invalid syllogism may really be valid by being of an actually different form in thought from what it is in expression.

But there is a second way of dealing with the case. In the proposition "The Divine Law commands us to honor kings," it is evident that the predicate is not distributed, namely, that other things might command the same. But in such statements we usually quantify or universalize certain terms helping to constitute the predicate. Thus in making this assertion we are most likely to mean *all kings*. Now, since a part of it is substituted in the conclusion, this quantification of the term "king" has the effect of distributing it, that is, of indicating that Louis XIV. is included in the general class as soon as he is affirmed to be a king. The reasoning, therefore, becomes a simple case of substitution of Louis XIV. for "king," which is already included by implication or assertion in the whole class. If we had distinctly said or thought that "the Divine Law commands us to honor *some* kings," and then asserted that "Louis XIV. is a king," we should have had no inclination to draw the conclusion that Louis XIV. is to be honored, and would have very quickly perceived the fallacy of attempting to do so, simply for the reason that it would appear to be a case of undistributed middle.

But it remains to show whether the case can be resolved by the regular laws of the syllogism. If we regard the universalizing or distribution of the term "king" as bringing the case under the quantification of the predicate, we should have, or should seem to have, an instance of UAA in the second Figure, which we have found to be formally valid, because of the distribution of the middle term. But it can be shown, in spite of its appearance, that the reasoning is, after all, not of the second Figure. Thus if the minor premise were negative, "Louis XIV. is not a king," we should have, according to our supposition, UEE, which is valid according to the table. But it requires only a glance at the conclusion, that "The Divine Law does not command us to honor Louis XIV.," to see that it is fallacious, because, so far as the assertion is concerned, the

MATHEMATICAL AND OTHER REASONING   283

Divine Law may command us to honor all other persons as well as kings. If we examine the fallacy carefully we shall find that it is the same as AEE of the first Figure. Let it be represented by the following diagram: We observe in this representation that "kings" are included in those whom the Divine Law commands us to honor, and also that Louis XIV. is so contained, but excluded from the class "kings." The representation is thus precisely like that which symbolizes the fallacy of AEE in the first Figure. The placing of "Divine Law" in a circle is only to indicate that the entire predicate is not distributed. For considering the fallacy we require only to take the other circles into account.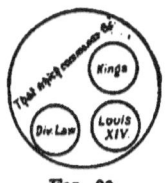

Fig. 29.

But the fallacy, being shown to be equivalent to that of AEE in the first Figure, creates the suspicion that the original syllogism, with the affirmative premise, is really AAA of the first Figure, although it appears to be more nearly related to the second Figure. The next diagram will illustrate this supposition. In this representation the relation between the terms is that of the first Figure, and it also indicates what is implied in the proposition that "the Divine Law commands us to honor kings," namely, that it may command us to honor others also. But it brings out most distinctly what is involved in the quantification of a term and the relation implied by it between this term and some other conception in the argument, namely, that a substitution can take place in any particular instance included under a class. It is done in this case without regard to the *apparent* Figure of the Syllogism, and in so far resembles the process of mathematical or quantitative Logic. But the reasoning is neither a case of traduction nor one of the ordinary syllogistic form as it is expressed. It is what older logicians called the *complex* syllogism, and it has generally been maintained that it could not be reduced to the regular form. But this we shall show can be done, and the process may bring out more clearly the dif-

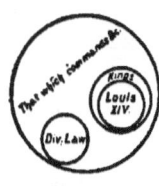

Fig. 30.

ference between mathematical and ordinary qualitative reasoning.

We have shown that the reasoning cannot be of the second Figure, although it appears to be. It remains to show that it is of the first, and so to confirm the suspicion created by the nature of the fallacy involved in supposing it to be of the second figure. The two diagrams show the complexity of the relations to be dealt with. In the first place, under the ordinary interpretation of the proposition "Divine Law" is the subject, and is distributed, and "commands us to honor kings" is the predicate, and undistributed. This relation is represented by the largest circle and the small one containing "Divine Law," and merely denotes that other sources of the same command might exist, in so far as the proposition says nothing to the contrary. But the predicate is a complex one, and it is not upon *the whole of it that the reasoning depends*. This is evident from *the substitution of a part of it in the conclusion*. The part which is not found in the conclusion, then, is the real middle term. This is the word "kings." "Commands us to honor," is not a part of the middle term, but of the minor or major term, as the case may be. "Kings" is the term upon which the reasoning turns, and as the circles represent it, is to be included in the larger, because other persons may be included under the same command. This relation is not excluded by the universal quantification of the word, so that it cannot be conceived as a predicate, because its universal quantification in that case would imply its *coextension* with the subject; which would imply that the divine law could not command us to honor any one else. But since the proposition does not exclude honor to others besides "kings," and "kings" is the middle term and distributed, the only possible way of representing the fact is *to conceive it as subject*. But to conceive it as subject makes the proposition the major premise, and the other proposition, "Louis XIV. is a king," the minor premise, and the form becomes AAA of the first Figure, as we have already asserted it must be, in order to make the reasoning valid.

But in order to effect this transformation a change in the order of the statement is required. Since the object of thought is mainly the middle term, as in such cases it always is, there is no objection to the process which explains how the mind really treats it, although the form of statement appears to make it different. The reason, however, for the change is in the peculiar nature of the proposition, together with the quantification of the term "kings." If we observe the two propositions we shall discover that only one of them connects the subject and the predicate by means of the ordinary *copula*. But this is of minor importance compared with the fact that the two terms so connected are, one of them an individual or singular and the other a class term. This makes the proposition quantitative, or an *extensive* judgment, its intensive meaning being implied. We have already explained how all propositions may be either intensive or extensive, according as we choose to view them. This double conception of them generally enables us to proceed in an argument without regard to the other fact, which we remarked of them, namely, that *the quantity of intension is represented in the reverse order of that in extension*. In other words, according to Hamilton and others, judgments of extension represent the subject as contained in the predicate, and judgments of intension represent the predicate as contained in the subject. Practically, however, this inversion of the order of comprehension has no importance so long as the judgments are convertible from the intensive to the extensive, and *vice versa*, by the mere thinking of those qualities and without a change in the order of the terms. This is because the propositions usually chosen for illustration are simple ones. But the proposition, "The Divine Law commands us to honor kings," is not a simple one in the ordinary logical sense. In the first place, it is an *intensive* judgment. The predicate, "commands us to honor kings," is not a class term, but expresses a quality inherent in the "Divine Law." But nevertheless, as explained before, it can be immediately transformed into an extensive judgment by saying, "The Divine Law is that which commands," etc.

This is the form in which the diagrams represent it. But at the same time it is apparent, both from the reasoning and from the diagrams, that the inference does not depend upon this feature of its intension and extension. *The quantitative element which determines the reasoning is in connection with the term "kings."* It is this that renders the proposition logically complex. There is a double quantification of conceptions. But being an intensive proposition, with the assertion or implication that "kings" is to be taken in its extension, and because the whole reasoning depends upon the latter, we may ignore the relation of intension and extension expressed between "Divine Law" and "commands," etc., because this is reproduced in the conclusion, and reconstruct the proposition formally so as to express the way in which the mind deals with it. In other words, we transform it from its intensive form into an extensive judgment like the other premise in the syllogism, and having the quantitative conceptions implicitly used in the process explicitly stated in the order to indicate the conformity of the reasoning to the regular laws of the syllogism. The proposition "The Divine Law commands us to honor kings," becomes, considering that "kings" is universalized in its meaning, "All kings are those whom the Divine Law commands us to honor." The Mood and Figure is apparent from this construction, and it is brought out by observing that the judgment is an extensive one, with its quantity, so far as it affects the reasoning, in one of the terms, and not in the whole predicate.

Now, as we have remarked, a large amount of our ordinary reasoning is of this kind, namely, is reasoning with propositions involving complexities of intension and extension that require analysis to bring them under the regular rules. As indicated, we do this by making both premises *extensive propositions*. Even if one of them is intensive *it must be capable of an extensive conception or expression which indicates the proper relation of inclusion or exclusion between it and the other premise, and this is making it in effect extensive.*

That this is a common state of matters in reasoning can be

seen by more examples than the one we have been discussing. We shall give some, both in the valid and the invalid form. Take the first:

Men show a disposition to respect the brave.
Leonidas was brave.
Therefore men show a disposition to respect Leonidas.

This instance is precisely like the previous one, and appears to be of the second Figure, but is resolved into the first by the same process, and for the same reasons as before. The major premise becomes "the brave are those whom men show a disposition to respect." That the reduction is dependent upon the peculiarly intensive nature of the proposition is apparent from the following instance which has the same form, but does not resolve so easily:

Men show a disposition to *be* noble.
Lincoln was noble.
Therefore men show a disposition to be Lincoln.

This is manifestly absurd. But when reconstructed it gives the conclusion, "Lincoln is one of those whom men show a disposition to be," which is rational enough. Why does it not appear so in the first case, where we seem to proceed as before?

The answer to this is found in the difference of import between the use of *be* and a *transitive* verb. "To honor kings," "to respect the brave," are intensive conceptions whose extension can be brought out only as in the diagrams, *by reversing the order of expression*, as, "Kings are to be honored," "The brave are to be respected." But "to be brave," "to be a king," may be intensive or extensive, as we choose to regard them, and without any alteration of order. In the former the two qualities can be made to coincide only by the inversion we have indicated, and hence such intensive conceptions or propositions must be transformed in order to show the formal process of reasoning actually involved in their use.

An instance of an invalid form may be produced in order to confirm in a negative way the reduction we have made :

All oxen are animals.
The Levitic law permits us to use oxen for food.
Therefore the Levitic law permits us to use animals for food.

If in the conclusion we mean "*some* animals," the reasoning is valid, but if we mean "*all* animals," as we have explained to be usual in such propositions, the inference is not valid. We have distributed the term "animals" in the conclusion when it is not distributed in the premises. But it is noticeable that the form of the syllogism is that of AAA of the first Figure, and yet is not valid except on the condition that we say or mean "*some* animals" in the conclusion. When reduced it becomes in reality AAA of the third Figure, which is a case of illicit minor term. Thus,

All oxen are beings which the Levitic law permits us to use for food.
All oxen are animals.
Therefore all animals are beings which the Levitic law permits us to use for food.

Here the illicit minor is quite apparent. But if we had said "Some animals are beings," etc., the conclusion would be valid, and it would explain why to have said or meant "some animals" in the first instance would have been valid reasoning. As is usual in the third Figure, it does not matter which of the propositions is taken for the major premise. The conclusion will be substantially the same in all cases. We placed them in the above order solely for the convenience of getting the conclusion in an order that would dispense with conversion for comparing it with the original.

The first of the cases can be solved also according to the same principles. The proposition, "The Persians worship the sun," can be reduced to "The sun is the object which the Persians worship." This will make the syllogism AAA of the third Figure instead of the first, and the conclusion will be, "The object which the Persians worship is a thing insensible."

But there is no necessity for this reduction in this instance, because the middle term is *singular* and the predicate of the major premise is made indefinite or left unquantified by the particle "*a*," which prevents it from being universalized in the conclusion, and which has the effect of equating it with the subject in the minor premise. Were it universalized in the conclusion we should have an instance of the same fallacy as that which we have just exposed. But the case is an interesting one as showing the use of substitution whenever *singular* terms are employed, and the peculiar influence which may be exercised by the particles "a" or "an," and "the," the former denoting an individual which is at the same time a part of a class and the substitutive equivalent of the term which it defines. The latter often denotes that the term which it agrees with is quantified universally, or that a certain *definite number* of objects is considered, which enables us to resort to substitution. They do not, however, alter the form of a proposition, and do not always produce the effect described. The meaning, as affected by them, is a material factor of the proposition. But it is important to consider it for the reason that it may make the reasoning actually valid in cases where, tested by purely formal laws it would appear invalid.

In all these instances of complex syllogisms we have clear illustrations of an apparent resemblance to mathematical reasoning in that there is at least an apparent disregard of the Figure of the syllogism. But with the reality of this disregard denied the illusion is dispelled, and we have illustrations of reasoning in which traduction is not possible unless the terms used are singular. The contrast, therefore, between purely quantitative reasoning, in which the subject and predicate are identical in extension, and qualitative reasoning combined with the quantitative, where there is a disparity of extension between subject and predicate could not be better brought out than in these cases. And as much of our reasoning is, perhaps, of the type we have just been considering, we see the liability to fallacy incident to it because of its variation from the form where certitude and assurance are guaranteed.

CHAPTER XXI.

THE LAWS OF THOUGHT

THE Laws of Thought do not require any elaborate treatment in an elementary treatise upon Logic, but the manner in which they have been assumed, or in which they underlie all our reasonings makes it necessary to state them and their meaning very briefly. We have already explained what a "law of thought" means, in our statement that it denotes the uniform way in which we think and must think. In all our reasoning we take these laws for granted. They are conditions of our reasoning and of the relation expressed between subject and predicate, antecedent and consequent, in propositions. We do not require to announce them as premises in our processes of transition from proposition to proposition, because they are either universally assumed without question, or they are the conditions of the formal and material truth of the data themselves, which it is not the business of formal Logic to investigate. Besides, they are of that axiomatic nature which renders it necessary to admit them before we could construct an objection or an argument against them. We do not require, therefore, to investigate them to determine their validity, but only to state what they are, their meaning and their functions.

The Laws of Thought may be divided into two classes, the *Primary* or *Fundamental*, and the *Secondary* or *Derived*. The primary laws are those which regulate all thought, whatever, whether of Conception, Judgment, or Reasoning. The secondary are simply those modified forms of the primary laws which are formulated in a particular way to suit the contingencies of syllogistic reasoning. We shall consider them in their order.

1st. The Primary Laws.—As defined, they are the fundamental laws of all thinking; that is, of conceiving percepts, concepts, and judgments in relation to each other. There are four of these general laws, *The Law of Identity*, *The Law of Contradiction*, *The Law of Excluded Middle*, and *The Law of Sufficient Reason*.

1. THE LAW OF IDENTITY.—This law expresses the right of the mind to affirm that a thing is identical with itself. Thus, A is A, or "Whatever is, is," that is, any existence is equal to itself. This is the usual form of statement for the law. But the law is the general principle at the foundation of all affirmative judgments, whether the subject and predicate are quantitatively or qualitatively equal or not. Hence there are two kinds of identity, *absolute* or *total*, and *relative* or *partial* identity. Absolute identity is represented by the truistic or tautological proposition, where both the form and the matter of the terms in subject and predicate are the same. Relative or partial identity is that of the ordinary proposition where there is a difference of extension between subject and predicate, or where the form of the terms makes it possible to identify them with others also. Thus to illustrate both instances: "Man is man," "Animals are animals," are cases of absolute identity in which a conception can be affirmed to be equal to itself. But of partial identity we have "Man is an animal," "Horses are quadrupeds." In these instances there are certain elements of identity, but the terms are not convertible with each other; that is, the propositions cannot be converted simply.

From what has just been said it might be inferred that exclusive propositions and definitions, as well as synonymous terms, are illustrations of absolute identity. This is true; but it is identity of *matter* and not of *form*. The terms are such as can be used in other than truistic propositions.

2. THE LAW OF CONTRADICTION.—This law Hamilton observes should be called the law of *non-contradiction*, because it denotes that an object cannot be affirmed to be what it is not. It is sometimes defined as denoting that a thing cannot exist and not exist at the same time; or that it cannot be affirmed to be one thing and its opposite or contradictory at the same

time. This law is the complement of the law of identity, and may be said to be the same law for thought as the law of impenetrability is for matter. It is the principle which determines the relations and inferences in the Square of Opposition and the drawing of negative conclusions in the syllogism.

3. THE LAW OF EXCLUDED MIDDLE.—This law denotes that of two contradictions only one can be true. Thus of the propositions, "Man is an animal," and "Man is not an animal," only one can be true. They are supposed to represent a completely dichotomous division of all objects into two contradictory classes, and to achieve this the subject and predicates must be so related that if the law of identity applies to the relation in one case, that of contradiction applies in the other. Hence only one can be true, and the other must be false. The law is a combination of the first two laws, those of identity and contradiction, and is at the basis of disjunctive propositions and syllogisms.

4. THE LAW OF SUFFICIENT REASON.—This law is briefly defined as denoting that every phenomenon, event, or relation must have a sufficient reason or cause for being what it is. There is some dispute about the right to regard this law as a law of reasoning, and it is certain that it seems more appropriate an assumption or postulate for the physical sciences than for the logical. But it nevertheless dominates certain modes of thought which are occupied, not with the identity or non-identity of objects or concepts, but with their *connection*. We shall not discuss the merits of this question, since it does not belong to the elementary plan of the present work to do so. Hence, we shall only explain the meaning attached to the law by those who regard it as a law of thought.

The law has been formulated to mean that we should affirm or infer nothing without a reason or ground. This reason or ground is the condition upon which the truth of a proposition or the reality of an event depends. The condition may be called the antecedent and the resultant the consequent. Hence the law will appear to determine the relation expressed in hypothetical propositions and syllogisms. But as these are reducible to the categorical form the law must either be reduced to that of identity, or be found to apply to categorical judgments. Perhaps it is the dependence of all individual truths upon general truths or principles that represents the application of this law as well as that of identity in all ordinary propositions and reasoning. If so, it can have an independent place in Logic. But not intending to decide this question, *pro* or *con*, we must be content with recognizing what is frequently regarded as a law of thought, and what certainly expresses a relation of dependence quite fundamental to our mental processes.

2d. The Secondary Laws.—These are simply derivatives of the primary laws of Identity and Contradiction. They are often called the axioms of Logic, and are simply formulas for justifying the inferences in formal reasoning. They are four, two of them for immediate inference and two of them for mediate reasoning. After what has been said in previous chapters they may be stated without further explanation.

1. If two concepts agree in one relation, they may be stated to agree in the converse relation.

2. If two concepts differ or contradict in one relation, they will disagree or contradict in the converse.

These rules regulate the process of immèdiate inference or Conversion, and their principle is simply assumed in every case of convertible and non-convertible terms.

3. If two terms agree with one and the same third term, they agree with each other.

4. If of two terms one agrees and the other disagrees with one and the same third term, they do not agree with each other.

It must be observed, however, that the latter two axioms are applicable in their purity and in an unqualified sense, only to mathematical syllogisms, or reasoning with universally or definitely quantified subjects and predicates. Hence for ordinary syllogisms they have to be modified by the following rule or law :

5. The terms which agree or disagree in the conclusion

must have no greater or the same distribution as in the premises, and they can agree or disagree only when the middle term is properly distributed in the premises.

This last law is to provide against the commission of the fallacies of Illicit Middle, Illicit Minor, and Illicit Major terms, when the third and fourth laws have been conformed to in their qualitative relations. The fifth law or rule specifies the quantitative conditions affecting the conclusion.

CHAPTER XXII.

INDUCTIVE REASONING

1st. The Nature of Induction.—It has been usual to define Induction in a manner contrasted with Deduction. But there are some peculiarities in connection with the meaning of the term which must be considered before we define the process with which we are at present concerned. Usage has not given the term a uniform conception, such as belongs quite generally to the word "Deduction." All classes of thinkers are tolerably agreed in regard to the process of deduction, whether they are logicians or scientific investigators. It is assumed to be the process of finding the proof of a particular truth in a general principle or proposition already containing it, explicitly or implicitly. Induction is often described as the reverse of this process, namely, as inferring general truths from the particular. But this is not the only conception of the term, and it is because this conception is not the only one in use that there is so little agreement about the nature, functions, and importance of induction. It will, therefore, be necessary to examine the several imports of the term in order to make possible a true conception or theory of the process denoted by it.

There are three different applications of the term "Induction," which are generally assumed to mean the same thing. To explain what they are we have to produce the usual divisions of the subject, which are the so-called kinds of induction. They are "*Perfect Induction*" and "*Imperfect Induction.*" The first meaning of the term applies to the first kind, and the other two are modifications of what is implied in imperfect induction. We would not suspect a difference of meaning from this general fact alone, but if we examine carefully the illustrations chosen to describe the nature of the process as thus distinguished into different kinds, we shall discern very clearly the great differences of real meaning attaching to the term. Thus "Perfect Induction" is simply *an enumeration of the particulars which form a class.* It is the process which characterized the method of Socrates in reaching his definitions, and which Aristotle remarked was a new method compared with the argumentation of his predecessors. An example of perfect induction is the following: "Mercury revolves on its axis ; so do Venus, the Earth, Mars, Jupiter, Saturn, and Neptune. But these are all the planets, and therefore all the planets revolve on their axes." Although this is stated in the form of reasoning, it is not reasoning at all. This fact is apparent in the nature of the conclusion, which is that "All the planets revolve on *their* axes," not on the axis of Mercury, although the same thing would be true if we had said "on the axis of Mercury." But the special proof of its not being a case of reasoning is in the fact that the so-called conclusion is merely a universal statement of what had been enumerated in detail in the premises. We are supposed to have observed the individual fact that "Mercury revolves on its axis," and then again that "Venus revolves on its axis," and so on throughout the entire number of planets. Hence when we say, "All the planets revolve on their axes," we but universalize our particular observations—we use the terms "all planets" as an economical device to avoid repeating the proper name of each planet. But we do not infer anything, or reason from one proposition to another. We do not establish any new connections of thought by the process, as we do in syllogistic reasoning, as already explained, but we only *generalize* what we had observed in detail. It is precisely the same with all enumerations of individuals or particulars into a whole or class with a general name denoting those enumerations only. They may be called "Inductions" if we choose so to name them ; but they are not reasoning. They are only *generalizations* as opposed to or distinct from reasoning, while the term "Induction," as now used by logicians, denotes a process of inference or reasoning of some kind. It is quite generally agreed since the time of Bacon that the so-called "Perfect Induction" is not properly called "Induction" because it is not a mode of reasoning. It has been the name for the Socratic process of obtaining universal conceptions and definitions. But the contingencies of the growth of knowledge and the demand for a method which would take the place of the Aristotelian Logic suggested the term "Inductive" as opposed to "Deductive," and the rejection of "Perfect Induction" on the ground that it was not ratiocinative in its nature, implied that Induction must be a process of reasoning in order to compare it with Deduction.

This second general meaning is the more important of the two, and was called "Imperfect Induction" because *the conclusion contained more than the premises.* Thus if I had inferred that "All the planets revolve about their axes," from the mere fact that one of them did so, I should have drawn an inductive inference. I should not in this case have merely generalized the particulars of my observation or experience, but have conjectured or inferred that what was true of one case would turn out to be true of all the objects known upon other grounds to belong to the same class. But this conclusion has no definite certainty such as the mind desires, and hence to give this conjecture greater probability I must vary my observation of facts in connection with the several planets, and find whether they agree or disagree with my supposition. If, for instance, I observed that certain of them presented an absolutely invariable appearance, such as a particular spot always in sight and in the same place, the fact would be at least a presumption against the supposition of the planet's axial revolution. On the other hand, if the spot presented certain regular changes of position and periodical disappearance and reappearance, the fact would be in favor of the hypothesis. This mode of repeating and varying observations or experiments in the case of the experimental sciences, according to certain methods, which are called the "Method of Agreement," the "Method of Difference," the "Method of Concomitant Variations," etc., has been called the *Inductive Method* in general, as a mode of ascertaining certain truths in a manner quite distinct from the ordinary syllogistic and deductive reasoning. This is the third meaning of the term, with which a theory of Induction has to reckon. It remains for us to select which of them we are to deal with under the heading of this chapter. It is the confusion of all three under the same term that leads to the uncertainties about the process itself.

It will conduce to clearness in the discussion of Induction if we sketch briefly the history of the general meaning of the term, and indicate specially the implication carried along with it which is no necessary part of its import as describing a logical process. This latter fact is not sufficiently taken into account by logicians when treating of the subject, and yet it involves grave consequences to their theory of Induction.

The most general meaning of the term "Induction," as used, especially by popular writers, since the time of Bacon, is *that of a process which adds to our knowledge*. The accusation which Bacon and his admirers have brought against the "deductive method," or Aristotelian Logic, was that it could not give us our premises, and therefore neither assured us of our data for reasoning, nor, when we were assured of them, could it add anything to our knowledge. In other words, as we have seen, our conclusion depends wholly upon knowing our premises, and if we already know the premises, the conclusion adds nothing to what we know. The deductive method is, therefore, useless for giving us knowledge, say its opponents. It only manipulates in various ways that which we are supposed to have in an implicit, if not an explicit, form. How, then, do we get the knowledge we have? How do we ever make any additions to our general knowledge, or to the premises with which deduction deals and which it assumes?

This was the question which Bacon attempted to answer, and he employed the term "Induction" to define the method of acquiring new data and principles of truth. The Aristotelian method was rejected as useless, because it could never advance beyond what was already given, and hence, in adopting the new method, which was to contrast with the old, the term "Induction" took on the meaning which that contrast implied, namely, *that process which produced an increment to the knowledge which we already have at any given time.* If we take the first meaning of the term, that describing the Socratic induction, and called "Perfect Induction," we shall see that it involves to some extent this implication; not that the generalization at the end of our observations expresses any addition to our previous knowledge, but that the enumeration of individual experiences or observations, or perhaps better, the accumulation of them, is such a process. It is quite as apparent that the inference from one or more known facts to a greater number is an addition, or the necessary step to such an addition. But if it be called "induction" it must be regarded as quite of a different nature from the process which we have just mentioned. It nevertheless resembles it in the one quality of providing an increment to existing knowledge. And again, the "methods of Induction" which seek by repeated observations and experiments to verify a supposition or conjecture already made, represent means of adding to previous knowledge. But they represent also something more than an inductive inference. They are complicated with direct experience and observation, and with deductive principles, assumptions, and inferences, so that they are not *simply* cases of inductive reasoning, although they are the proper means of adding to our knowledge and widening our generalizations.

But Logic does not immediately treat of increasing our knowledge, or of the material means for applying scientific methods. It has to do with *thought or reasoning* and *with the manner in which one idea is inferred from another*, not with the complicated methods of verifying this inference after it is made. Hence in so far as "Induction" is a logical process in contrast with Deduction, we must confine the term to a certain kind of reasoning, and not extend it to all modes of adding to our previous knowledge. We, therefore, choose the second of the three meanings explained, as the proper one to represent "Induction" in so far as *formal* Logic has to deal with it, and which is the only meaning that will enable us to contrast the process with Deduction or consider it exclusively as a process of reasoning. This is very clear in the case of the so-called "Perfect Induction," which we have found to be simply observation and generalization, and not properly reasoning at all, and which is as much a condition of Deduction as of Induction.

The third form is a process of verification of conjectures, or probable inferences already made, and although such additional inferences may be connected with these various methods of verification, the methods themselves are not pure inductive inferences, as is admitted by more than one writer on Logic. But to these points we shall return again, when they may be treated more fully. It has been important here only to fix upon the meaning which the term "Induction" is to have when comparing it with deductive reasoning, and to establish the fact that the notion or implication so common with popular writers, and tacitly admitted by logicians themselves, namely, that "Induction" is *any* process of adding to what we already know at any given time, is not the main conception with which we have to do when considering it as a form of reasoning. It does imply addition; but it is by way of inference, not by observation or verification.

Having fixed upon the second of the three meanings as the proper one for the term, so far as formal Logic is concerned with it, we may indicate, before defining the process more carefully, the three methods and their characteristics by a terminology to some extent new. The first we shall call *Observation and Generalization*, the second, *Induction*, and the third, *Verification* or *Scientific Method*. We use the term "Induction," therefore, to describe a process of inferring one truth from another in a manner somewhat different from Deduction, and are now prepared to define it more accurately.

We have already defined Induction as reasoning from the particular to the universal in contrast with Deduction as reasoning from the universal, to the particular. It is also frequently defined as reasoning from effects to causes, from the known to the unknown, from the actual to the possible and probable. From two of these conceptions it is often supposed that the process is the inverse of Deduction, and this is true in some respects, but not in all. For instance, to argue from the known to the unknown, or from the actual to the possible and probable, is not the inverse of Deduction, for if it were, we should be obliged to regard the latter as reasoning from the unknown to the known, or from the possible to the actual, a process which is the very opposite of the real case. Induction is only the inverse of Deduction in certain respects, and these are with reference to the *extension* of the conceptions involved. In one we narrow our conceptions as we proceed to the conclusion, and in the other we widen them. In other respects we cannot say that the two processes are the inverse of each other, but only that they are different from each other. Thus in Deduction we argue from the known to either the known or the unknown ; to the known, if the conclusion is an actually known fact, but not seen in all its relations to general principles until these relations are enunciated ; and to the unknown, if the conclusion happens to be a truth implicitly contained in the premises, but not explicitly realized in the consciousness of the reasoner or the hearer until stated. But it is never a process of reasoning from the unknown to the known. Both Induction and Deduction, therefore, appear to be reasoning from the known to something else. But there are two differences between them. In the first place, the known data are different as syllogistic matter. In Deduction the known is either a universal principle in the abstract, or certain universally known concrete facts which will enable us to affirm the same thing of all individual instances included under the universal. In the second place, it is not strictly true in either of them that we argue *to the unknown*. In Deduction we argue to what is necessarily included in the nature of the premise, whether known or unknown, explicit or implicit. In Induction we argue to the possible and probable from known *facts*, not principles, and so to more universal truths, which may in their turn become deductive data *when verified*. These facts explain why it is best to define Induction as proceeding from the particular to the general or universal, or from effect to cause. The process gets the last conception of its nature from the frequency with which it is occupied in determining the causes of known phenomena, which can never be ascertained deductively because more than one cause may produce the same known effect, and when the cause is known we can directly know what the effect will be. But as our knowledge begins with matters of fact we usually have to argue to their probable causes by inductive inference or conjecture, no doubt after some accumulated experience and observation, and then verify our inference by additional methods.

It is sometimes said that Induction is reasoning from particular to particular, and some writers on Logic admit or assert that nearly all the reasoning of common life is of this order. It is certainly true that we do reason from one particular instance, or set of instances, but when we examine into the case more closely, the result or conclusion in its real meaning is a universal broader than our original premise and containing both what we had reasoned from as a particular and what we had reasoned to as a particular. This, I think, will be clear, when illustrated, to all who study the process. But in order to include such cases in the definition, Inductive reasoning may be defined as *reasoning from what is known in a certain fact, or facts, to the possible or probable truth of the same thing in other facts where it has not been observed or proved, or is reasoning from actual to necessary connection.*

As an illustration of this inductive reasoning we may take the case of discovering gravitation by Sir Isaac Newton. The story about his having been moved to the discovery by a falling apple is probably legendary, but it may be used as if it were a fact because it is to the purpose and might have been employed by him as an illustration. But supposing his attention to have been arrested by the fall of an apple while reflecting on the position of the heavenly bodies, he had before him several facts. There is, first, the fixed and suspended position of the apple before it breaks loose from the tree. Then there is its falling to the earth under the influence of its weight. The supposition that it is attracted by the earth is another fact accounting for the weight of the apple. A fourth fact is that the moon, the sun, the planets, and other heavenly bodies are suspended in space, somewhat in the same relation to the earth as the apple which we are considering. But it is not known that any force from the earth is exerted to hold them in their position, or create in them a tendency to fall toward it. But assuming their relation to the earth and the attraction which caused the apple to fall, we may infer the possibility that the same force of attraction extends to the moon, sun, etc., inasmuch as we can see no reason in the nature of space to prevent this action. As in the case of the moon, for example, the resemblance between it and the apple, in regard to relative positions, was such that it was natural to infer from the attraction exerted upon the apple that the moon was similarly affected, and so the other planets, by reason of their likeness to the moon in qualities concerning the matter at issue. But there was one circumstance in the case that prevented the inference from being verified by observation. It was the fact that the moon and the planets, with their satellites, did not fall toward the earth, or that each did not fall to the body toward which it gravitated. In the case of the apple there was no difficulty, because as soon as it was released from its support on the tree, the phenomenon of its fall was an observable fact, and attraction was presumed to be the cause. But the moon did not perceptibly move toward the earth, although it was suspended in space, and without visible support. Observation, therefore, could do nothing directly in producing or verifying the belief that attraction was exerted upon the moon, but the belief was an inference from resemblances of relation and material qualities to a resemblance in one other quality already known of the apple and the earth. Thus far the inference, however, expresses no other degree of certitude than a possibility, and it remains to ascertain how a greater degree might be given it in connection with the same or the same kind of reasoning.

There was another known circumstance in connection with the case which may have helped to suggest the inference, and which certainly gives it greater probability. The motion of the planets about the sun, and of the satellites about their respective planets in elliptical orbits, was an admitted fact at the time of Newton. Now he knew that such motion involved the existence of a tendency on the part of all moving bodies to keep in a straight line unless drawn from it by some other force than that of their impulse, and also that bodies moving in circular and elliptical orbits tend to fly off from the centre unless held in their place by some external force. As the planets were moving in an elliptical orbit, and the satellites about their central bodies, there must be this tendency to fly off from their centres, and unless they deviated from this line there must be some force to sustain them in their place. Gravitation, therefore, came in as the complementary centripetal force to counterbalance the action of centrifugal tendencies. The existence of some such influence was presumed in the nature of the case, but that it should be the same gravity that pulled the apple to the earth was not proved by that fact, and hence it remained a probable inference of greater or less degree, according as the nature of the case would make it. The inductive nature of the inference lies precisely in this fact, that it is not necessarily conclusive, but only that the cause inferred is adequate to the effect, and that the circumstances render it probable that the supposition is true, or that it has more in its favor than any other hypothesis. The mere possibility of some other conditions to the same effect prevents the case from being proved by the circumstances which occasioned the inference. But it will increase in probability with the number of incidents consistent with it, or which it aids in making intelligible. In the case before us the probability of gravitation was greatly increased by the mere circumstance that a centripetal force was needed, and the conjecture extending to the moon the agency which caused the apple to fall precisely supplied this want. In getting a clear notion of what the inductive inference was, however, the student has only to remember that it consisted in the extension to the various planets of the same force known or supposed to control the movements of the apple, and under circumstances which prevented the action of such a force from being an observed fact.

Any number of similar illustrations might be chosen. For instance, if we find that two or three gases are compressed into liquids under certain degrees of temperature and pressure, we might infer the same of other gases not yet so compressed. The conjunction of a solar eclipse with the dark of the moon would suggest the inference that the moon was the cause of the phenomenon. The known fact that intervening bodies cast a shadow is the basis of supposing that the same effect will take place when the moon is between the earth and the sun. Our ignorance of the exact position of the moon would make the inference merely conjectural, but a knowledge of the fact that its latitude and longitude corresponded exactly with the position of the sun at the place would be equivalent to a demonstration of the inference, or would make it so highly probable that a very peculiar combination of circumstances would be required to weaken it. But it must be noticed that the circumstance rather verifies than instigates the inference, and becomes the basis of predictions of the eclipse. Again, the peculiar change in the shape of sun spots leads to the inference both of the rotundity and the axial revolution of the sun. They move across the visible plane of the sun's surface, and appear elliptical or elongated as they approach the edge of the sun. The known fact of such elongation in connection with spherical bodies would suggest sphericity in the sun in connection with the elongated character of the spots, and the conjunction of this form in the sun spots with their motion across the sun's surface suggests axial revolution according to well-known facts. And again, the frequent conjunction of a certain kind of cloud and rainfalls will lead me to suppose that this kind of cloud is a cause, and that it will be invariably accompanied with rain. To Benjamin Franklin certain resemblances between lightning and electricity led to the inference that they would resemble each other in conduction over a wire and in charging a Leyden jar. The identification of the two, and hence the verification of his supposition, was the result of an experiment suggested by this inference, and is too well known to be repeated.

These illustrations suffice to indicate the nature of inductive inferences, and it remains to show the syllogistic form which they assume. The premises consist of certain facts and assumptions, and the conclusion, of something wider than the known facts in the premises. In representing them, however, we cannot follow the usual order of the Figures in the syllogism without too much preliminary explanation. We take first, therefore, the form most frequently adopted.

2d. The Form of the Inductive Syllogism.—The usual form of inductive reasoning is that of the third Figure. My own opinion, however, is that it can be stated in all the Figures. But we take first the Third Figure, because it is undoubtedly the one in which the inductive inference occurs most frequently. Thus:

> Mars, Venus, etc., revolve around the sun.
> Mars, Venus, etc., are planets.
> Therefore the planets revolve around the sun.

In deductive reasoning we have learned that this would be a case of illicit process of the minor term, because it is undistributed in the minor premise, but distributed in the conclusion. But we must remember that, although the *inference* or *reasoning*, deductively considered, is false, the conclusion may be true as a matter of fact. Only we have no right from the nature of our premises to assert it. This indeterminate nature of the fact leaves it open to conjecture, hypothesis, or inductive reasoning to infer that *all* the planets revolve about the sun, because two or three of them do; that is, that the other planets being like the known planets in many respects, will prove to be like them in this other characteristic. It is, of course, not a *necessary* inference from the data given, but only a possible or probable one, as the case may be.

The matter is sometimes put in the following manner in order to avoid the appearance of a deductive fallacy:

> A, B, C (magnets), attract iron.
> A, B, C represent all magnets.
> Therefore all magnets attract iron.

Here the inference is undoubtedly wider than the known facts. But having used the phrase "represent all magnets," we distribute the term magnets, so as either to bring the case under the rules of the quantification of the predicate, and, therefore, deductive reasoning, or we create one of those intensive propositions with the quantification of one of its terms which modifies the real form of the reasoning, and makes it some other Figure than the apparent one, as has been explained. Besides, it is to assume the conclusion, if we thus express the case in the premises, and so far from having our knowledge increased by the inference, as it should be by Induction, it remains the same as in the premises in such cases. Hence I do not think this form of expression conducive to the proper representation of the inductive form of inference. We must not assume as known in the premises what we are to infer in the conclusion. Hence the better form for showing the nature of the inductive inference is such as the following, where nothing is assumed in the premises to anticipate or necessitate the conclusion.

> All magnets attract iron.
> All magnets are attracting bodies.
> Therefore all attracting bodies attract iron.

We may know the premises to express facts and yet they may not include the conclusion drawn from them. But in this instance the agreement or connection between the two predicates and the same subject awakens the supposition that they, the predicates, are as essentially connected with each other. This is, of course, the thing to be proved. But it is possible in the nature of the circumstances, and will be probable according to the extent of our knowledge regarding the nature and action of electricity. It may even reach the stage of demonstration. But the first degree is that of possibility or probability suggested by certain known resemblances.

In the second Figure the reasoning might be represented as follows:

   Magnets attract iron.
   Loadstones attract iron.
   Therefore loadstones are magnets.

In deductive reasoning this is a case of illicit middle. But for the same reason as before it is possible that the two subjects agree, although this is not necessarily the case. Their agreement in the matter of attracting iron is merely the conception of a single effect which has its cause, and as both objects produce the effect we may reasonably suppose that they are identical in their nature. This is what we express by saying that "Loadstones are magnets."

In this illustration we must not mistake the historical manner in which magnets and loadstones were actually identified for the inductive inference we are trying to illustrate. Historically, "magnets" was only another name for "loadstones," and we came afterward to apply the same name to manufactured articles having the same qualities. But we are here supposing a mind made acquainted with the two things independently of each other, and discovering that both attracted iron in the same way, and hence inferring that they belonged to the same class. In this case, then, we might say either that the loadstone was a magnet, or the magnet a loadstone. We should only intend by it that one of them was either possibly or really more comprehensive than the other.

The difficulty with the second Figure in the inductive syllogism is that it is usually impossible to connect the two subjects by the copula which signifies their identity in a class relation, which is not always capable of being expressed by language which is adjusted to observations already made.

There is a reason for this difficulty which will appear in the sequel. It remains to see whether we can state the inductive syllogism in the form of the first Figure.

Meteors are followed by a train of light.
Rapidly moving bodies are like meteors in their motion.
Therefore, rapidly moving bodies are like meteors in being followed by a train of light (other things being equal).

The conclusion in this example is not meant to indicate an observed fact, but for that very reason it is better calculated to represent the inductive nature of the inference. It is not a matter of experience that "all rapidly moving bodies" are followed by a train of light, and yet we are accustomed to explain the absence of this effect by saying that we mean such bodies as have the velocity of meteors. Our argument is that if they had as swift a motion as meteors they would display the same effects, and the inference that they would do so is inductive, in so far as it supposes true of all kinds of matter what is observed to be true of meteoric matter and bodies. It might not be true of terrestrial bodies, because of their peculiar nature and established connection with the earth. But, on the other hand, the effect would be probable in proportion to the knowledge we have of the effect of friction from the air upon rapidly moving matter of whatever kind, and the temperature necessary to make it luminous, etc.

In regard to the form of this syllogism it will be noticed that the middle term is quite peculiar. We could not state that the subject of the minor premise was contained in the class "meteors," because this would have made the argument deductive, while it is our object either to "prove" their inclusion in that class, or to show that both classes belong to the same general kind of objects on the ground of a common quality observed and a common quality inferred, which shall be as essential to the subject of the conclusion as to that of the major premise. Hence we must, in the first Figure, state our middle term in such a way as to represent an identity between it and the minor term which is only partial, and from which we may infer a more complete identity inductively. Also we must state it so that it represents properly the extent of our actual knowledge about the two terms, so that the conclusion shall contain more than our actual experience of the facts. In the minor premise, therefore, we cannot say that the subject *is* the predicate or middle term, but only that it is *like* it in certain particulars, and then we can infer its identity with the major term so as to complete the conception of the identity between the two classes of objects. Otherwise, as we have remarked, the argument would be purely deductive. It will be interesting to remark that this is the valid inductive inference, the counterpart, and correspondent of the invalid deductive process in Ambiguous Middle.

It will not be necessary for present purposes to represent an inductive syllogism in the fourth Figure. We have intended only to show that so far as Induction is reasoning it must follow the same form as Deduction, and that it differs from Deduction in the character of its conclusion, permitting us to go beyond the premises with our inferences, although affording nothing but a probability of truth instead of a certainty or necessity. But there is a peculiar characteristic which is best illustrated in the example representing an inductive syllogism of the first Figure, and which is our reason for having given it. We have observed the peculiar form of statement necessary to prevent the case from being a deductive syllogism, and in comparing mathematical with ordinary logical or deductive reasoning we remarked that the former was purely *quantitative* and the latter both *quantitative* and *qualitative*. We have now still further to observe that inductive reasoning is purely *qualitative*. Mathematical reasoning we found to be quantitative because it dealt only with quantity, and was not concerned with the qualities which constitute objects, and we chose to call it Traduction because its data could be transposed and substituted without any reference to the Figure of the syllogism. Induction, on the other hand, has nothing to do with quantity, but solely with the *qualities* of things. It reasons from quality to quality, from attribute to attribute. Thus in identifying electricity and magnetism we argue from certain resemblances of effect to an identity of cause, although in a modified form in each instance. We do not have any special class or numerical quantity of instances under which the minor premise can be subsumed, and hence we argue from the known qualities in one or more things to the existence of the same quality in other objects. The number of objects specified in the premises is of no importance. Each premise may represent only a single individual, and the data be sufficient for an inductive inference of some kind. Indeed, when class terms are employed they have no more value than singular terms, except for strengthening our convictions about the fixity of the connection between the subject and the predicate. We do not so much use them for class or general terms as for conceptions *representing the presence of certain qualities*, and it is *from the peculiar conjunction of certain qualities with each other that we infer a similar conjunction where it has not yet been observed*. Thus we are not arguing from a class to a sub-class or individual under it, on the ground of *numerical inclusion* in the larger, but from a quality or qualities observed to the same things unobserved in other objects, or in a modified form concealing their identity. Thus Franklin, in arguing to the identity of lightning and electricity, reasoned from certain known phenomena, the production of sparks by an electric current, to the production of the same by lightning, if the proper conditions were satisfied and his flying of the kite was only a verification of the inference. The number of instances in which he had observed electrical phenomena and lightning had nothing to do with his reasoning, but only in strengthening his conviction regarding the permanence or uniformity of the resemblance between the two sets of phenomena. The logical process was then not concerned with the mathematical aspect of his conceptions, but only with their qualitative import; that is, with the resemblances between phenomena. A verified induction may enlarge the area or extension of general concepts, but it is not done by any logical stress upon the mathematical import of the conceptions employed. It is done only by observing, or inferring, and then verifying a resemblance of qualities. After the general conception has been formed, we may use its quantitative import as a basis for assured deductive reasoning, but only the qualities of objects and their connections can be the basis of induction, and hence we feel justified in calling it purely *qualitative* reasoning.

We must be on our guard, however, about the use of the terms quantitative and qualitative in describing these several kinds of reasoning, because they are used here in a somewhat new and modified sense. Jevons, for instance, and most writers on Logic, speak of "quantitative" reasoning when they mean reasoning with *extensive* propositions, and "qualitative" reasoning when they mean reasoning with *intensive* propositions. The meaning I attach to the terms rather includes this than differs from it essentially, because I have already laid it down as a fact that all extensive propositions have a corresponding intensive import, and intensive propositions an extensive import; so that they may be conceived either quantitatively or qualitatively. But by quantitative reasoning I mean the comparison of terms on the ground of quantitative identity, or quantitative relations of whole and part, and by qualitative reasoning, the comparison of terms on the ground of a qualitative identity, which never permits of additions to form quantitative wholes; and hence quantitative or mathematical reasoning will deal with conceptions in purely numerical relations; qualitative or inductive reasoning only with the qualitative resemblances and connections of phenomena. If, then, we can apply the term Traduction to mathematical reasoning, we may indicate briefly the characteristics of the three forms of inference. Traduction is purely quantitative, the most assured in its certitude and the most free from fallacy. Induction is purely qualitative, the least assured in its conclusions and the least exempt from mistakes of inference. Deduction is both ·quantitative and qualitative, and so combines the characteristics of both Traduction and Induction. It is assured in its conclusions precisely in proportion as the reasoning turns upon the quantitative import of its conceptions and propositions, and it is exposed to fallacies precisely in proportion to the lack of coincidence between the group of qualities constituting the individuals in a class of objects and the mathematical import of the class term denominating them. That is, if the same term has a wider mathematical application than the group of qualities it may denote there is great liability to fallacies of all kinds in deductive reasoning, and hence when the question turns upon the qualitative import of terms we see the probability, or at least the possibility, of error increasing.

3d. Kinds of Inductive Inference and Its Principles.
—The inductive conclusion does not always take the exact form which has been given it in our illustrations. We had in view such a statement of the process as would give a clear idea of what it was in its essential characteristics, and so to separate it from that class of conclusions which were the result of verification or of a surreptitious introduction into the premises of matter which ought first to appear in the conclusion. But there is some confusion about what matters of fact should belong to an inductive inference, and this is generally caused by the mode of representing its syllogistic form. This confusion, however, is not one with which logicians have dealt, or about which there has been any controversy. It is rather the unconscious classification under the head of inductive reasoning of two or more kinds of general conceptions, beliefs, or principles, which are suggested by individual facts much narrower than the truths they give rise to, that has caused the confusion. They are spoken of as inductive inferences without taking into account either the complications and conditions that make some of them appear quite different from others, or the various degrees of certitude that attach to them, making some of them quite assured and others quite conjectural.

We spoke of "the surreptitious introduction into the premises of matter which ought first to appear in the conclusion," if the inference were to be truly inductive, as characteristic of some representations of the process. Hence it is common to illustrate inductive reasoning in the following manner:

The Earth is molten under Vesuvius.
Vesuvius fairly represents the interior of the Earth.
Therefore the interior of the Earth is molten.

The only appearance of induction in this instance lies in the fact that the major premise is a particular circumstance, and the conclusion another supposed to be suggested by it. It also appears to be a syllogism of the fourth Figure, but when stated in another form it appears to be of the third Figure. Thus,

The Earth under Vesuvius is molten.
The Earth under Vesuvius fairly represents the interior of the Earth.
Therefore the interior of the Earth is molten.

But even this is not the completely correct form in which the logical relation of the terms is actually thought, because the only conclusion, inductive or deductive, which can be drawn from these premises is, "That which fairly represents the interior of the Earth is molten." The minor premise, as we have indicated before in propositions of this kind, is an intensive judgment which requires reduction to express the true relation of the terms in it, and hence it would be, "The interior of the Earth is like that under Vesuvius." But this makes the case one of AAA in the first Figure, and also a deductive syllogism, unless we so state the resemblance as not to indicate that it is complete, but only in a matter that supports the probability of a conjunction with the predicate "molten." In stating the likeness in the premises as we do, we practically assume in them all that is found in the conclusion, and so many of the cases which are taken to represent inductive are simply instances of deductive reasoning. The appearance of induction comes only in the *admitted probability* of the conclusion. But this characteristic is due solely to the probability of one or perhaps both of the premises. If the premises, stated in the manner of the above illustration, be proved facts, the conclusion is a necessary one, and not probable at all. The probability of an inductive inference is one which is the result of an inference from established facts, and *so is due to going beyond the premises under the stimulus of certain general principles which may be called the Principles of Induction.* These will be considered in a moment. But when we say in the minor premise of an inductive syllogism that certain things "*represent*" the subject of the major, or "are like" it, without qualification, we assume in it what ought to be an inference in the conclusion from admitted or supposed facts which do not include the inference. The probability of the conclusion must not be borrowed from that of the premise, but from the principles which justify inferences beyond the actual data given, otherwise we should practically be guilty of a *petitio principii*.

Thus if we were asked to give inductive proof for the proposition that "the interior of the Earth is molten," we should state the syllogism as follows:

The Earth immediately under Vesuvius is molten.
The interior of the Earth and that immediately under Vesuvius resemble those conditions in which openings are outlets of what is contained deeper within.
Therefore the interior of the Earth resembles that immediately under Vesuvius in being molten.

Here are two known facts with such resemblances as inevitably suggest a further resemblance, that of a quality which is observed and known in one of the objects, but unobserved in the other. It is also to be noted in this case that the middle term is peculiar. There is no such identity or inclusion as in Deduction. It is only partial, and this is the true characteristic of inductive reasoning. The known facts of the middle term must represent partial identity and resemblances, and the inference must contain the total identity. It is this feature of the inductive syllogism or reasoning ; namely, the purely *qualitative*, and *not quantitative*, nature of the comparison, that lends support to the supposition that a truly *formal* expression of the process is impossible. It is certainly very difficult in many cases. But aside from these questions the important feature of the inductive proof is the statement of certain known and partially agreeing facts. They may wholly agree, or they may only have a partial agreement, but which of the two it really is must be determined by the verification or disproof of the inductive inference. All that we are presumed to *know* in the premises is their partial agreement or actual connection, and we are to infer from them under the circumstances their total agreement or their necessary connection. We are not to assume any data in the premises which should first appear in the conclusion.

It has been necessary to dwell upon this point in order to distinguish clearly between deductive probability and inductive probability, as a condition of fixing upon truly inductive inferences. It has been too common to confuse the probability of a conclusion with its inductive nature, and we need to know exactly the marks which make the inference the one or the other. In deductive reasoning, if one or both of the premises be probable truths, the conclusion will be probable, but it will necessarily follow as an inference from the supposed data. But in inductive reasoning this is not the case. The inference is not a necessary one, although it is always a probable one. Its probability is independent of the question whether the premises are positively known facts or only likely assumptions. It comes from the peculiar nature of inductive principles in connection with its extension beyond the data of the premises. The difference, therefore, in this respect, between deductive and inductive inferences is simply this : In deduction the probability of the conclusion is purely *material ;* that is, determined by the material character of the premises in this respect. In induction the probability is both *formal and material.* But it is not due to that characteristic in the premises. It is due to the manner in which the inference is drawn, and this is the introduction of new matter in the conclusion. Hence inductive inferences are not determined by the mark of probability alone, but by *probability plus an increment of knowledge, or conceptions not necessarily involved in the premises*. This maxim must be constantly kept in mind or the student will confuse with each other two entirely distinct kinds of reasoning.

It will be important to examine this liability to confusion because of its bearing upon the general theory of inductive reasoning. There are, at least, three conceptions of the process, which are not necessarily conflicting, but will be so upon the supposition that each one is exclusive of the other. They are:

(1.) That the inductive inference is from the particular to the particular, or from one or more individual cases to another.

(2.) That it is from the particular to the universal, or an inference which forms a generalization from a particular instance or instances.

(3.) That it is an inference from *actual* to *necessary* connection, or from uniformities of coexistence and sequence to causal relations.

It can be shown that each of these forms of inference ultimately result in generalization, and this is probably the reason that John Stuart Mill identifies inductive reasoning with generalizations from particular cases. But there are some generalizations which only *seem* to be from particular instances, and hence may be confused with inductive inferences, merely because the data we start with happen to be individual facts and the conclusion includes more than these, and perhaps all similar facts, when in the meantime a deductive process has been surreptitiously but unconsciously introduced. This is the case when a probable conclusion in a deductive syllogism is based upon an inductive conclusion as one of the premises, or when our generalization simply states explicitly what was implicitly involved in the conceptions arrived at by induction. Let us examine the application of these principles in the example which Mr. Fowler presents as an illustration of inductive reasoning, as follows:

(*a*) I *observe* that these two bodies (though of unequal weight) reach the bottom of the receiver at the same moment.

(*b*) This fact must be due to some cause or combination of causes (Law of Universal Causation).

(*c*) The only cause operating in this instance is the action of gravity.

(*d*) Therefore the fact that these two bodies reach the bottom of the receiver at the same moment is due to the action of gravity operating alone.

(*e*) But whenever the same cause or combination of causes is in operation, and that only, the same effect will invariably follow (Law of Uniformity of Nature).

(*f*) Therefore whenever these two bodies, or any other two or more bodies (even though of unequal weight) are subject to the action of gravity alone, they will reach the bottom of the receiver at the same moment, or, in other words, will fall in equal times.

The first remark here is that we have in this instance two syllogisms, one ending with proposition (*d*) and the other with proposition (*f*). The first can be shown to be purely deductive, and the second to contain both a deductive and an inductive conclusion. Whether Mr. Fowler intended that proposition (*d*) should be an inductive inference I cannot say. But it is certain that the only appearance of such a character in the inference is the nature of proposition (*a*), which merely states a fact or phenomenon, and of proposition (*d*), which includes the cause of the phenomenon. In the meantime, however, if we observe closely, all the conditions for making the conclusion deductive are introduced into propositions (*b*) and (*c*). Propositions (*b*), (*c*), and (*d*) form a complete deductive syllogism, with (*c*) as the major and (*b*) as the minor premise; or the mode of statement can so be altered as to make (*b*) the major and (*c*) the minor premise. Let us see how this is the case.

Proposition (*b*) is not an inductive inference, nor an observed fact, although it contains new data not found in proposition (*a*). It is a deductive inference from the assumption that all phenomena must have a cause, and the observed fact of the two falling bodies, which is taken to be a phenomenon. Now proposition (*b*) becomes a premise for another syllogism, and it is assumed again that the action of gravity is the only operating cause in the case, from which it immediately and necessarily follows that this instance of falling bodies has gravity for its cause. The causal influence of gravity, instead of being directly inferred from the observation of a fact, is assumed in the premise, and so makes the conclusion deductive.

The only proper way to state the inductive inference in such cases is as follows :

I observe that two bodies of unequal weight fall through equal spaces in equal times.
I observe that the action of gravity is isolated at the same time.
Therefore the isolated action of gravity is the cause of the phenomenon ; that is, of the falling through equal spaces in equal times.

In this form of statement we merely assert the *observed coincidence* of two facts, the falling of the two bodies and the isolation of gravity, and from the coincidence we infer the cause, or from their actual connection we infer their necessary connection. This is not involved in the data known, although we may have some foreign reason for making the inference. But to assume that gravity is a cause in the case, and more especially that it is the only cause, is to introduce in the premises what we find in proposition (*d*), and so make the inference deductive.

In the second syllogism the inferences seem to be somewhat different. They are that the same two bodies will *always* fall in the same manner under the same conditions, and that *all other* bodies will do so. Here we have two extensions beyond the observed fact. But we must deal with only one of them at a time. We take the first, that these two bodies will always fall through equal spaces in equal times, and consider it in relation to the premises from which it is drawn.

As before, we have a deductive instance from the very fact that the universality of the case is assumed in proposition (e). The universality of the connection must be suspected or drawn from individual cases in order to be inductive. Proposition (e) really states what should be the inference, and since it states it as a premise the conclusion only deduces it in another form, but does not induce it. That the same effect will invariably follow the operation of the same cause, and that alone, is the truth to be inferred and proved. But if it be stated in one of the premises its presence in the conclusion cannot be the result of an inductive inference. To make its universality in time an inductive act of reasoning, we should compare the same effects with each other in several different times, and then from the observed fact that a difference of time had exercised no influence upon the result, infer that this incident alone never would do so, and that with the changes of time, the effect would always be the same. But a grave doubt may exist about the inductive nature of this inference, because to infer that a certain event will always happen under the same conditions, if we have reason to believe that those conditions are its cause, or necessarily connected with it, may well be claimed to be an *a priori* or deductive inference on the ground that the universality of the causal connection is implied in the fact of its existence at all in a particular case and that difference of time is always implied in universality. If this be so, then to infer that an event will *always* occur under the same circumstances that were once its cause, is only to state explicitly what is implicitly involved in the idea of their being the cause in the first place.

A complete theory of Induction would require us to discuss this question at length. But in an elementary treatise we cannot be expected to do so, and we must therefore be content to announce the conclusion we have adopted for such cases. It is that all such generalizations as we have mentioned, namely, the inference that a given event will always happen under the same circumstances, are deductive when they are from a causal connection to its universality, but inductive when we reason from coincidences and sequences to causes. To infer a causal connection in a single instance of observed coincidences is of the nature of a universal generalization, not because it is inferred from a single incident, but because the universality of the causal relation is involved in the idea of its necessity, and hence no new conception is involved in the generalization. But to infer the causal connection from the mere fact of actual connection is different, and so is inductive. Consequently Mr. Fowler's first inference is deductive because it only states explicitly what is implied in the idea of necessary connection in the observed case.

These observations show how we are to dispose of the second inference in proposition (*f*) of Mr. Fowler's illustration. We find there that he has argued from the instance of two bodies of unequal weight to *all* bodies of equal or unequal weight. Inequality of weight is shown by the observed case not to be a material circumstance, but for all that we know about them the different qualities of other bodies may be precisely the circumstances which will prevent the formation of a true or perfect middle term for the syllogism, and prevent the expected effect in the case of experiment. These differences are not included in the premises, and hence can only be included with a certain degree of probability in the conclusion, according as the principles of induction determine it. For this reason, therefore, we agree that the inference from the specified case to the idea that *all* bodies will fall through equal space in equal times, is inductive, because it is a generalization involving an increment in the conclusion not contained in the premises. To suppose it true of *all* bodies is to infer it of other conditions than those in which it was first inferred.

We are now prepared to examine the kinds of inductive inference. They may be divided, first, into two species (1) the *statical* and (2) the *dynamical* inference. By a statical inductive inference we mean an inference to the actual existence of certain qualities, coincidences, or sequences, from their observation in given cases. This class we subdivide into two subordinate species determined by the different characteristics mentioned in the definition. There are, first, those which infer resemblances or differences, or the existence of certain qualities from the partial resemblances and differences of observed cases. The second class consists of those which infer the repeated or continued existence of observed coincidences and sequences. The two classes often coincide, so that it may be difficult to decide whether the inference is to resemblances and differences, or coincidences and sequences. But the one or the other aspect predominates frequently enough to use the difference for distinguishing different forms of the same kind of reasoning. By a dynamical inductive inference we mean an inference from *actual* to *necessary* connections. The observed incidents may be of coincidences, or sequences, and the inference in this case must not be to the mere probability of their recurrence, but to the probability that one of the circumstances is the cause of the other. Whenever we infer, therefore, a particular cause from the mere occurrence of a phenomenon, we are reasoning inductively. Reasoning to the existence of resembling qualities may be called *attributive* inductive inference; to coincidences and sequences, may be called *connective* inductive inference; and to causes may be called *causal* inductive inference. The causal inference, however, may be implied along with the others, and hence its distinction from them may not be absolute. But it is often so prominent a feature of the inference as to serve for a criterion of its nature. In general, the terms statical and dynamical comprehend the two characteristics determining inductive inferences, namely, reasoning to unobserved facts, whether qualitative or connective, and the causal nexus. The following outline summarizes results :

Inductive Inferences { Statical { Attributive = Unobserved identity and differences. Connective = Recurrent coincidences and sequences. Dynamical = Causal = Necessary connections.

Another classification of inductive inferences can be made upon the basis of the kinds of generalization involved in them. Directly or indirectly, primarily or ultimately, all inductive inferences result in a generalization, explicit or implicit, and hence we may be able to distinguish them by this characteristic. But it is not all generalizations that are inductive inferences. We have seen that those of "perfect induction" are not cases of reasoning at all; so also with the universalizing of the causal connection. Hence we must assign the characteristics which mark the generalizations of inductive inferences. They are (1) inferences from one particular case to another, or from species to species. This results in forming or implying a larger class, comprehending both particulars, and so identifying them in a higher genus. Thus if I infer the qualitative identity of magnets and loadstones, of electricity and magnetism, of potassium and metals, from resemblances of qualities or of effects produced by them, I am forming a higher genus, or as in the case of potassium and the metals, widening the last class. In chemistry, biology, and zoölogy this inference is very frequent. It is very common in the classification of animals, and especially in investigations involving the doctrine of evolution. There are, then, (2) inferences from part to whole, from particular to universal, or from the individual to the class, from the species to the genus. This form, however, needs explanation. It might be better to define it as an inference from *accident to essence*, or from *differentia to conferentia*, only we must not assume that the basis of the inference is *known* to be an accident or a difference. It is merely an observed fact that a certain quality is present, and when we infer that it will be found to characterize the class we infer that it will be a common quality, and we thus pass from the *relation* of accident or difference to that of essence or conferentia. Thus if we observe that very frequently ruminants are horned, and infer that all ruminants are horned, we do not extend the genus as a class term, but we merely increase the number of qualities to be taken as the essence or conferentia of the class; we infer that what might be a mere accident is probably an essential property. Or if we infer from the conjunction of a single or several east winds with rainfall that all east winds must be so characterized, we infer that rainfall is an essential, not an accidental, accompaniment of east winds. We do not widen the genus, but we increase the conferentiæ.

It is important to remark the difference between these two classes of generalization. The first class widens the genus, and the second does not. The first does not change the conceived character of the quality or relation, but only its extent, and the second does so change it. A further important difference is, that in the first class the resemblances between the particulars are actually observed, and it is the identity of the cause or quality producing them, when that identity of kind is otherwise not apparent, that constitutes the object of the inference. In the second class, the whole genus is known, not inferred, from the individual or species; but it is merely a sufficient number of common qualities, or conferentiæ to determine the group of individuals named. The resemblances, however, are not *observed*, but *inferred*. The coincidence of the essentia with a quality which might be an accident is observed, and then its universal coincidence with the essentia inferred, supposing that it is a general instead of an accidental characteristic. This is why we may be said in some cases to pass from the individual to the universal, or from the species to the genus. It is only when the genus is already known by a definite essentia or conferentia, that this inference can be made. Otherwise we should have to depend upon "perfect induction," or the addition of individual instances in order to form a generalization. It may be that observation forms the first generalization; but afterward the process may be materially aided by the inference we have just discussed.

The two generalizations we have just considered are statical. When we come to the third and dynamical we have to be cautious in our judgment of their inductive nature, as we have already observed in our examination of Mr. Fowler's example. But there is, nevertheless, a third class of generalizations that are inductive. They are (3) inferences from actual to necessary and universal, or causal connections. The basis of these inferences must always be the coincidences or sequences of two or more phenomena, and the increment supplied by the inference must be either the causal nexus or the universality of the coincidence and sequence, or both. But they are very much complicated. We sometimes directly infer that a given phenomenon is the cause of another, from the fact of their connection, and then we infer that it will *always* be so. But in the last instance we may imply that the conditions are absolutely the same for all cases, or that if they are, the events will always happen in this way. If this be our assumption the inference is due to a subsumption under the universal law of causation, and becomes deductive, or merely an explicit statement of what is involved in the idea of causation, as already explained. On the other hand, if we infer that a particular event will *always* be the cause of another, we may do so either on the ground that it is the *known* cause in this instance, or that its causal relation is the result of a real or an implied inductive inference. In the former case the generalization is like the previous one, a mere enunciation of what is implied by the idea of cause. In the latter case the generalization is a probable one, but might be considered a deductive inference with a probable conclusion, on the ground that the supposition of the particular case being inductively an instance of causality, was the minor premise of a syllogism, in which the universal law of causation was the major premise. The generalization could be inductive only upon the assumption that it was only another way of making the transition from observed facts of coincidence or sequence to necessary connection, or that, instead of supposing tacitly or openly the existence of absolutely the same conditions, we inferred that a particular one of them was sufficient in the future to produce the effect. This last, however, is really based on the assumption that the event is not known as a cause in the first place, but that we are still dealing with it merely as a coincidence or sequence.

Let us take an illustration. I observe in one or more instances that the dew falls on clear, cool nights. If I infer that clear, cool nights will always cause a fall of dew, I evidently have a generalization with an increment of conception not involved in the data known, and one which, in meaning, is, that *all* clear, cool nights will produce this effect. But whether the generalization is inductive or not will depend upon the question whether the inference turns upon the universality of the causal nexus in such cases, or the probable similarity of all other clear, cool nights with the one observed. If it turn upon the universality of the causal nexus, this may be merely a statement of what is implied in supposing that it was the clearness and coolness of the night that caused the dewfall in the observed cases, so that the inductive inference will lie in the mind's implicit or explicit transition from the fact of coincidence in the first place to the idea of causal connection, and not in universalizing it. On the other hand, if it turn upon the probable similarity in other respects of all other clear and cool nights to those observed, the inference may be inductive, on the ground that it is not from the assumption of identity in all the conditions to the effect, but is an inference to the single efficiency of the clearness and coolness. Of course, if we suppose at the outset that the clearness and coolness of the night were the *sole* causes of the dew falling, the generalization involving all such nights is only an explicit enunciation of that idea, and not a new conception. But usually the generalization in fact means the selection of those qualities out of all that are present, as the *sole* cause of the phenomenon. In this light the inference might be regarded as inductive because it adds to the observed possibility of a number of conditions the conception that a particular condition is the sole one and will always be the only one.

These illustrations show the limitations under which causal generalizations are to be regarded as inductive, and they all practically resolve themselves into the one rule that dynamical inferences are inductive only when the transition is from coincidences or sequences to causes. The mere transition from individual instances to the universal in cases involving the causal condition is not necessarily inductive, *because it is complicated with the assumptions about reasoning from cause to effect which is usually spoken of by logicians as deductive.* It is, therefore, necessary in such cases to find some other criterion than the mere generalization as a mark of what the nature of the inference is, and this is to observe whether the case involves any transition, explicit or implicit, from merely actual to necessary connection.

The possibility of a confusion of deductive with inductive inferences where the causal nexus is involved, is brought out admirably by Mr. Venn in a passage of his " Empirical Logic," which we quote in illustration.

" A man is bitten by a cobra. We have known or heard of many other such cases, and they all proved fatal. We conclude with some confidence that XY, the present sufferer, will die ; as A, B, C, the former ones, are all supposed to have died. Here in these few words we have had all the requisite facts put before us, and we also have the inference from them.

" Now, since we are looking, in the spirit of logicians, at the existence of this belief, which we know will inevitably arise in every normal mind, we proceed to exercise what Hume calls ' our sifting humor,' by beginning to press a series of questions. We start by asking the observer *why* he believes in the approaching death of XY ? To this question two distinct answers might readily be given. Some would say off hand, ' Because every one who is so bitten always dies ' ; others would say, 'Because A, B, C, whom we know to have been previously bitten, have all died.' When these answers are expanded into proper shape they would stand respectively as follows :—

" *Deductive.*—All men who are bitten die : the man XY is bitten : therefore XY will die.

" *Inductive.*—The men A, B, C, were bitten and died. The man XY has also been bitten. Therefore XY will die."

This illustration is only to say that the inference " XY will die" may be either inductive or deductive, according as the premises do or do not include it. In the second form it is assumed to be inductive because there is a transition from one particular case or set of cases to another, so that the known and inferred instance may ultimately be included in a wider generalization than those already observed. *But if this individual inference be a deduction from the tacit assumption that "all persons bitten by the cobra will die," the induction originally was not from the particular known cases to this one, but was made from the known instances to this universal assumption by supposing the connection to be a necessary one in any case.* If anyone wishes to object to the cogency of the reasoning in such cases, it must be on the ground that the assumption made in the original instance was not proved, or that the inference is only an inductive one. But whether it is inductive or deductive must be determined by first deciding what are the real premises in the case.

This is sufficient to recommend caution in regard to the nature of inferences in particular cases respecting the causal nexus and generalizations involving this idea. The present treatise does not require a complete exposition of the conditions and limitations of induction in this respect. The student must therefore be satisfied with the ordinary criterion, that causal generalizations and inferences, to be inductive, must be transitions from what is conceived as actual coincidences to necessary ones.

The Principles of Induction are the next subject of consideration. They are suggested by the demand to know why or upon what ground we make the generalizations just discussed, or inferences extending beyond the known data of the premises.

The principle supposed to lie at the foundation of inductive reasoning is sometimes announced as the *Law of Universal Causation*, and sometimes the *Law of the Uniformity of Nature*. There is an important difference between these two principles which we shall have to notice, but it is not such as to affect seriously the theory of induction. The latter form of statement is adopted by those who feel doubtful about the existence of any such necessary connection as is implied by the notion of "cause." The difference between the two conceptions can be brought out by their definition.

The Law of Universal Causation means that *every event must have a cause, and every cause must have an effect.* This implies that the mind is not satisfied with the mere existence or occurrence of a phenomenon, but must know upon what it depends, the ground for its being what it is. The principle is usually spoken of as an *à priori* one, which means that we must assume it in all our thinking and can give no proof of it. As Logic is not concerned with proving that it is a priori, or with the speculations centering about the use of the term, we do not enter into these discussions, but must content ourselves with recognizing that Logic and logical processes take the law for granted.

The Law of the Uniformity of Nature means that *the phenomena of the world present a certain uniformity of qualities or occurrence that enables us to conceive it as a system of them having a definite unity and order.* The law differs from that of causation in not representing a necessary assumption or principle of the mind, but in expressing *the observed facts of experience.* We must think of events as having a cause, whether they occur regularly or not. But the discovery of their actual uniformity, whether of resemblance or occurrence, is a matter of actual observation and experience. The law is therefore called *empirical,* in contrast with the term *à priori* describing the former law.

In regard to their relation to Induction, the two laws may be said to be complementary of each other, as we shall explain. But it is important to remark that they are not exclusively principles regulating the process of induction. They are both equally related to deductive reasoning. Thus the first law, the Law of Universal Causation, is only another statement of the Law of Sufficient Reason, which we have discussed in the chapter on the Laws of Thought. It is true that some writers would limit the Law of Sufficient Reason to inductive processes, but the present writer sees no ground for excluding the same law from a determining influence on the deductive inference. The main distinction is that the law is applied in a somewhat different way in the two processes. But it is not important in this treatise to discuss the question, nor even to take any positive attitude upon one side or the other of it. It is sufficient to know the general opinion of logicians, that the law of causation is an accepted principle affecting inductive inferences.

The manner in which the law affects induction is this: There is no reason in the fact itself why we should go beyond the data of the premises. When we infer something which has not been stated in the premises we must know why we are so disposed to act, or upon what ground we thus go from the known to the unknown, from the particular to the universal, etc. The only ground or reason assignable is that there is, in the circumstances involved, a phenomenon requiring explanation, and that the most probable one in the case is to be found in the inference drawn. Thus I observe a very marked resemblance between certain phenomena of sound and light, and I know that in the case of sound this phenomenon is connected with its nature as vibrations. This resemblance requires explanation because every event must be supposed to have its cause. In this special instance the resemblance of the phenomenon, and its known connections in the other, are taken as facts justifying the inference that undulations in light will account for a phenomenon which is essentially the same as that which is accounted for in the same way when occurring in the case of sound.

But the mere rationality of the demand for an explanation of a resemblance, a difference, a coincidence, or a sequence, is not a proof that the inference or conjecture is the right one. It only explains why the mind seeks to go beyond the known facts, and guarantees only that the phenomena shall have *some* cause, but does not indicate even with a degree of probability what particular cause it shall be. Another law must come in at this point to determine what inference shall be the probable one. This is the Law of the Uniformity of Nature, or the actually observed frequency with which certain phenomena have been connected in our experience. This law decides the greater probability of any given inference both by itself and in relation to other possible inferences at the same time. The occurrence of a resemblance between two bodies, say of electrical effects in the case of electricity and magnetism, on a single occasion, might justify under the law of causation the supposition of their possible identity. But owing to an equal possibility that the resemblance was only accidental, the inference would be a weak one, unless we were able to give it some probability from the peculiar nature of the resemblance; and this characteristic is often a factor in such cases, but has no rules for determining when it does and when it does not give probability. But if we frequently, and under all sorts of varying circumstances, have observed this resemblance between the two phenomena, the probability of their identity is greatly increased. The law of causation determines the possibility of the identity, and even the likeliness that the same effect has the same cause; but the law of the actual uniformity of nature determines the probability that the cause is to be found in these particular bodies. The frequency with which peculiar resemblances and coincidences, differences and sequences occur affords a probability not only of their reoccurrence in the same connection, but also of the fact that this cause will be found in some of the known causal agencies in connection with the objects manifesting the phenomena.

But in addition to the two general principles thus enunciated, there are two subordinate principles or canons which are determinative of the process, and which are, in a measure at least, corollaries of the general law of the uniformity of nature, and are mainly the means of determining the legitimacy of the inductive inference. I shall call them the *Principle or Canon of Agreement*, and the *Principle or Canon of Difference*. These are practically the same as the Methods of Agreement and Difference so denominated by logicians generally, only I wish here to distinguish them as organs of discovery, from their frequent and actual application in connection with deductive assumptions, as verifications of inductive inferences already made. This latter use of them we shall comment upon again.

The Principle of Agreement can be briefly defined as the principle which determines the probability of a given identity or connection on the ground of the actual frequency of certain resemblances or coincidences under varying conditions. Or more simply still, the agreement of two phenomena in respect of the qualities producing them, or in respect of the connection in which they occur is a criterion of their cause. Thus, to illustrate, if A, B, C, and D, E, F, resemble each other in a certain marked phenomenon or quality known to be characteristic of each object, the probability that they are identical in their nature, structure, functions, etc., is proportioned to the degree of resemblance, the frequency of the phenomenon's occurrence, and the assurance we feel about the relation between particular causes and effects. If A, B, C, and D, E, F, represent events associated together, the probability that they are necessarily connected is proportioned to the frequency with which they occur together under varying conditions. "For example, bright prismatic colors are seen on bubbles, on films of tar floating upon water, on thin plates of mica, as also on cracks in glass, or between two pieces of glass pressed together. On examining all such cases they seem to agree in nothing but the presence of a very thin layer or plate, and it appears to make no appreciable difference of what kind of matter, solid, liquid, or gaseous, the plate is made. Hence we conclude that such colors are caused merely by the thinness of the plates, and this conclusion is proved true by the theory of the interference of light. Sir David Brewster beautifully proved in a similar way that the colors seen upon mother-of-pearl are not caused by the nature of the substance, but by the form of the surface. He took impressions of the mother-of-pearl in wax, and found that although the substance was entirely different the colors were exactly the same. And it was afterward found that if a plate of metal had a surface marked by very fine grooves, it would have iridescent colors like those of mother-of-pearl."

It should be remarked, respecting these examples, that they also contain illustrations of the Canon of Difference, as all cases of agreement actually do which have any conclusiveness at all. But the important thing to be observed in them is that the inference is first suggested by a number of resemblances where they might not be suspected; that is, in conjunction with differences that make the resemblances significant.

The Canon of Difference lays the emphasis upon the differences between groups of phenomena, and so infers a causal connection from the separation of certain phenomena from others. It means that, if two phenomena are constantly isolated together from other groups which remain invariable, the separated phenomena may be taken as necessarily connected in the relation of cause and effect; that which is known to be the antecedent being the cause, and the consequent the effect. An example of this method is given in the case quoted from Mr. Fowler, in which the isolated action of gravity is inferred to explain the equal velocities of bodies falling through equal spaces in equal times. Ordinarily, bodies of different weights, where this difference is considerable—say lead and feathers—show marked differences of velocity in falling, and it was inferred that gravity did not affect all bodies equally. Here was a case of differences, however, which did not take into account the uniformly accompanying fact of resistance from the air. But the separation of this influence and the concomitant isolation of gravity and of the falling bodies gives rise to another inference, and the very opposite of the previous one, upon the basis of the difference between this phenomenon and others. The inference is conclusive in proportion to the certitude that the conditions are as we suppose them. If we had not felt in isolating gravity that there could be no other disturbing factor at the time, our inference would have been liable to the same error as that which had been based upon the actual differences of velocity in falling bodies.

The Principle of Difference is complementary to that of

Agreement, and is often, if not always, found in connection with it, except that the inference does not turn upon the agreements, but upon the differences. But the differences are noticeable directly in proportion to the invariability of other phenomena without those differences, and hence such circumstances are a great indirect help to the inference. In the example quoted to illustrate the Principle of Agreement, it was certain differences between the instances that made the inference drawn a possible one. The differences between the natures of the various substances while the iridescence was a common quality showed very distinctly that the peculiar kind of substance had nothing to do with the effect, and this negative inference had as important a part in determining the inference as the agreement of the various substances in reflecting light. So, in the example quoted from Mr. Fowler, the agreement between the retarding influence of the air and the decreased velocity of the lighter body, and the uniformity of opposite effects without the isolation of gravity, are as necessary to the inductive inference as the variation from the usual order when gravity is isolated. But the explicit ground of the inference depends in the one instance upon the agreement, and in the other upon the difference, between the phenomena.

It must be remarked regarding the Principle of Difference, that it is most frequently applicable in cases of experiment, and experiment belongs more properly to *verification* rather than *discovery*, or to confirming the inference rather than first suggesting it. This is not exclusively the case, however. The same remark is not so true of the Principle of Agreement, which is more frequently the incident of discovery, although it can be made, and often is, the instrument of verification and experiment. It is when the two principles are combined with observation and experiment artificially applied, and with deductive principles assumed or proved, that they become means of verification rather than the means of originating the inductive inference ; and it is this fact which makes it so difficult, if not impossible, to distinguish as accurately as is desirable between pure inductive reasoning and scientific method or verification. They are regulative of inductive inferences pure and simple, when the reasoning is suggested by them alone, and not complicated with assumed or known facts or principles involving an application of the Laws of Identity and Contradiction. As principles, however, they are only modifications of the Uniformity of Nature.*

* For general references on the nature of induction, the following works may be consulted : Mills: Logic, Bk. III., Chaps. I. to V. inclusive. Hamilton : Lectures on Logic, Lects. XVII. and XXXIII. Venn : Empirical Logic, Chaps. XIV. and XV. Jevons: Principles of Science, Bk. I., Chap. VII. ; Bk. II , Chap. XI. (the whole book should be carefully read). Fowler ; Inductive Logic, Chap. I. Ueberweg : System of Logic and History of Logical Doctrine, Sections 127, 128, and 129, pp. 470–490.

CHAPTER XXIII.

SCIENTIFIC METHOD

I. THE NATURE OF SCIENTIFIC METHOD.—We have hinted at the distinction between inductive reasoning and scientific method, and perhaps we have implied that the latter is limited to the verification of the former. So far as scientific method is identified only with what is usually called "Inductive Method," this is perhaps the case. But there is a broader use of the conception, which involves also the application of deductive principles, which, besides being distinguished from purely inductive reasoning, may appear in combination with the inductive, or be a deductive process solely. This broader use of the conception gives rise to two subordinate forms of logical procedure. They are usually called:

(1) The Method of Discovery.
(2) The Method of Instruction.

The method of discovery is occupied with the acquisition of knowledge, and is usually identified with Induction or the Inductive Method, as it is generally called. The method of instruction is occupied with the communication of knowledge. But the name purports to include more than the mere imparting of truth once discovered, and hence it is meant to be identified with what is called the Deductive Method, which aims to prove as well as to impart truth. But the two methods are not wholly independent of each other, as the sequel of the present discussion is intended to show. They may be combined in certain stages, both of discovery and instruction. This requires us to recognize the divisions only for provisional purposes, and to indicate processes which do not exactly coincide. In certain features the two methods are entirely distinct from each other. It is when they are both implicated in the discovery and establishment of the same truth, that they together form the perfect application of scientific method. We shall call them respectively the *Inductive* and the *Deductive* Methods, and examine the deductive first in order.

II. THE DEDUCTIVE METHOD.—The deductive method has to do with the communication and proof of existing knowledge. The knowledge may exist as a positive acquisition, or be positively known, only by the person imparting it, or proving it, and so involve something of discovery to the person receiving it. But this fact does not prevent the mental processes of the receiver from being deductive as well as those of the imparter. The object of the method is to assure the truth of the matter concerned, to give it more than a probable value. The method comprehends three distinct processes: *Definition, Division,* and *Probation.*

1st. Definition.—Definition is the process of making clear all the conceptions entering into the thesis or proposition to be proved. It unfolds their *intension*. The first thing, of course, is to know what is to be established, and the method of proving it begins with the definition of the elements of the judgment or thesis enunciated for proof. The nature and principles of definition have already been discussed, and need not be repeated here. The process itself does for the conceptions of a proposition what proof does for the proposition itself.

2d. Division.—Division is the process of rendering our conceptions more definite in respect to their relations to each other. It is complementary to Definition, and so unfolds the *extension* of a notion. It is to some extent implied in Definition, but can go beyond that by showing the exhaustive nature of our ideas, and the amount of truth involved in the thesis to be proved. This subject has also been duly treated in its proper place, and requires no further mention.

To illustrate what is meant by Definition and Division, as aids to instruction and proof, suppose it is required to establish the truth of the proposition, "Governments are useful." I must first define what "Government" is, and what "utility" is. In so doing I indicate the characteristics constituting them, conferential and differential. But this does not tell me how much is involved in the proof of the thesis. I must divide the two terms into their species; "government," for instance, into monarchic, oligarchic, and democratic, or other such species as the convenience of the argument may require. This will show more distinctly what is involved in proving the main thesis. So also if "utility"·be divided into species, such as economic, artistic, scientific, etc. The two processes, definition and division, bring out more clearly than one of them alone, the conferentia and differentia, genus and species, or qualitative and quantitative aspects of our conceptions.

3d. Probation.—Probation is the process of proof; or the statement of certain truths which render the thesis a necessary conclusion from them. The thesis is the proposition to be proved. The truths which prove it are the known facts and principles which may constitute the premises, and the thesis will be the conclusion. These determining truths may be axioms, postulates, proved propositions, or any truth which the person to whom the probation is made may accept. Their acceptance is the condition of their proving anything. We must observe, therefore, that probation is a material as well as a formal process. It is always syllogistic, but, as Hamilton observes, the converse is not necessarily true; namely, that all syllogistic reasoning is probation. The object in probation is to prove the material truth of the thesis, and not to conduct merely a formal process of reasoning. We must therefore enunciate some facts or principles accepted by the person to whom the probation is made, and then bring the thesis under it in such a way that it will be a necessary consequence of what is already admitted. In this way we give absolute certainty to our proposition, provided we avoid the usual fallacies in syllogistic reasoning.

To illustrate, suppose I am required to prove that the sum of the angles in a triangle is equal to two right angles. I must first define the terms involved, namely, "angle," "right angle," "triangle," etc., and then I must announce the fundamental principles that are either assumed or admitted in the process of proof. There may be such axioms and postulates as that things equal to the same thing are equal to each other, all right angles are equal to each other, and all triangles are *essentially* identical in their properties, etc. If then I can show, either by observation or proof, that the sum of the angles of the triangle is equal to some quantity which is known or admitted immediately to be equal to two right angles, the major premise involved in the axiom mentioned insures the conclusion I am required to draw. The proposition may not carry with it its own evidence, but the truth of some prior proposition or propositions may prove it. The following diagram and argument illustrates the whole method. The the-

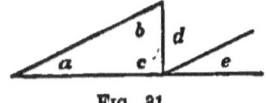

Fig. 31.

sis to be proved is that $a + b + c = 2$ right angles. It is assumed that c and $d + e$ are right angles and equal to each other. Hence $c + d + e = 2$ right angles. By construction or previous proof, b is taken to be equal to d and a to e, so that $a + b = d + e$, a right angle; $a + b + c$, therefore, equals $c + d + e$, and by assumption $c + d + e = 2$ right angles. Hence whoever admits the previous steps and conditions must admit that $a + b + c = 2$ right angles.

The process would be quite similar to this in any other proposition, such as "civil law is necessary to the preservation of order." If this statement did not evince its own truth, we should be obliged to announce some acceptable truth which carried the given conclusion with it. Thus we might show that restrictions of individual liberty were necessary to this end, and this assertion might be admitted without proof, but on its own transparency. Then, if the definition of civil law showed it to be the only restriction adapted to such an end, the truth of our proposition must follow as an inference, whether the person to whom the proof was directed could fully see the contents and meaning of the conclusion or not.

It is not always essential that the deductive method should take an explicit syllogistic form, because one of the premises may be so apparent that the conclusion would follow upon the enunciation of a single conditioning truth, or it may follow as an immediate inference from the conditioning proposition. Usually, however, a syllogism is implied. In disproof, which involves the same general method, we state or assume the contradictory of the thesis, as in probation we state the conditioning proposition. These serve to measure the truth of a given assertion by its agreement or its contradiction with truths already known or accepted. No guessing, or probabilities, based upon the actual uniformities of nature, enter into the case, except as they might have originally determined the proposition which is assumed as proof. The proof depends upon the acceptance of the principles, and not upon the method by which they were originally suggested. We take up the inductive method next.

III. *THE INDUCTIVE METHOD.*—After what has been said about the meaning of the term " induction," it might seem objectionable to use the term for describing a method which is to be sharply distinguished in some of its features from the process of reasoning going by that name. But as long as we are careful to distinguish between induction or inductive reasoning as a mode of inference, and the process usually called " the inductive method" and comprehending the principles of verification, there will be no serious reason for rejecting the traditional use of the term ; especially as the method is mainly, if not wholly, occupied with the acquisition of knowledge. It is often identified with scientific method, which, as we have explained, is really a combination of inductive and deductive processes in many of the cases assumed to be purely inductive. Because of this fact and general usage, we shall treat, in this section, of "the inductive method" as usually conceived, and without trying to push the distinction between inductive reasoning and the complications of induction and deduction in the process of verification into any new departures of phraseology. Were we discussing the subject for any other than elementary purposes, we might be tempted by innovations. But they are not necessary here, and hence we shall employ the expression to include all the steps involved in the passage from facts to general ideas and their verification. The process so named includes an exposition of the preliminary methods or inductive inferences, the four inductive methods and the verification therein involved, along with the frequent application of deductive principles which give greater assurance to truths first suggested by induction. In other words, we use the expression here to indicate all the processes involved in the acquisition and verification of knowledge that is not deduced solely from previous conceptions. These processes may be reduced to two general classes, somewhat merging into each other in actual experience. They may be called *Acquisition* and *Verification*, primary and secondary conditions, or subsidiary and confirmatory conditions of new inferential knowledge. Their import and the extent of their application will appear on examination. It must be observed that we are here dealing only with *new inferential* truth, and not with what may be acquired by non-logical processes. In scientific method, however, we shall be permitted to mention the conditions or data upon which such inferences are based. Some of them are included under the subsidiary processes to induction, as they are called by writers generally.

1st. Acquisition, or Primary Stages of Ascertaining New Facts and Truths.—These are so called because they are the first conditions of the truths which turn out to be acquired by a method which is not deductive, or which does not represent in the first stages of its conception any derivation from established knowledge. In the acquisition of knowledge in general, there are of course more processes involved than mere inference or reasoning. Indeed, the first and most important means of attaining such are entirely prior to any form of reasoning whatever. They are the common processes of Perception and Introspection, by which we simply observe facts of experience, and perhaps group together into classes by comparison and generalization. But these are processes whose analysis and investigation belong more properly to Psychology. Yet it is not possible to understand their relation to the properly logical processes of the mind unless we mention them, and indicate the manner in which they either condition the inductive inference, or confirm it when made. We therefore begin with a brief consideration of these processes as furnishing the data of scientific method.

1. OBSERVATION.—Observation is simply the perception of facts. It denotes *to watch for*, and merely involves what is contained in perception and introspection. The data from which to make inferences must first be known, and observation is the means of ascertaining them. The amount of our knowledge is determined by its accuracy. Those who are good observers will discover more data for inductive inferences than those who are not. Observation may be of two kinds, spontaneous or casual, and voluntary or rational. The former requires no distinct effort on the part of the person observing; the latter requires an act of attention and is frequently, if not always, employed to verify some expected or anticipated phenomenon. The progress of science depends much more upon rational than upon casual observation.

Experiment is generally mentioned as a subsidiary process to induction. This is often the case in regard to particular kinds of induction, but it is not a condition of all inductions, nor of the process in its earliest forms. It is, however, a very important means to increasing the data for inductive inferences. It is not an instrument which can be employed independently of observation, but is only a help to that process. It is the means both of varying and multiplying observations. It is the artificial production or repetition of natural phenomena for the purpose of insuring the accuracy of first observations or of correcting them. It is therefore one of the incidents of observation in its highest forms. "Mr. Mill distinguishes between the two processes by saying that in observation we *find* our instance in nature; in experiments we *make* it by an artificial arrangement of circumstances. When, as in astronomy, we endeavor to ascertain causes by simply watching their effects, we *observe* ; when, as in our laboratories, we interfere arbitrarily with the causes or circumstances of a phenomenon, we are said to *experiment."* Experiment may also be used as a means of verification as well as acquisition, and so also can observation. But of these functions we shall speak in the proper place. It is sufficient to remark of the two that they occupy, as subsidiary processes, the position in inductive method that is held by definition in deductive method.

There are two rules for regulating the use of observation as an instrument of scientific discovery, in so far as it is merely a subsidiary process to inductive inferences. They are :

(*a*) The observations should be precise.

(*b*) They should be relevant or material.

The first rule enjoins accuracy, and the second the disposition to seek and see what is essential and important.

2. CLASSIFICATION.—This process does the same for inductive methods that Division does for the deductive, and it is also the reverse of Division. It is the process of reducing phenomena to systematic groups on the ground of their resemblances and invariable connections. It is the means of keeping together all relevant and material circumstances in connection with objects and phenomena, and hence is of great help in determining the extent of the inferences we draw on particular occasions, as well as assigning the limits within which they can be drawn. It is the reduction of phenomena to their classes or *genera*, as Division reduces genera to species. But the process need not be dwelt upon at length in this treatise. We turn to the third and more important process.

3. HYPOTHESIS, OR THE FORMING OF HYPOTHESES.—An hypothesis may be briefly defined as a *supposition*. It is an inference from certain facts to others, or to the cause of those which are matters of observation. This account of it identifies it with the inductive inference, and this is our intention.

Some would distinguish it from "induction." But the distinction, when examined, only turns out to be one between a supposition expressing a mere possibility and one expressing a probability of such a degree as to set aside the legitimacy of other suppositions, at least for the time. An hypothesis simply supposes, in its primary stage, a possible fact or cause to explain phenomena, and may turn out to be tenable or not, according as further observation may determine. But an "induction" is supposed to represent a firmer mental allegiance than hypothesis, and so something more approaching to validity. But we insist that in their nature they are the same, and that at most they can differ only in degree. We, of course, use the term "induction" in this comparison as denoting a particular kind of inference suggested by observed facts, and including more than they contain. Hypothesis we use as a convenient term for representing this function, of whatever degree, in the application of scientific method, and so indicating the mental act which must be preliminary to all verification, and intermediate between the observation of facts and the proof of whatever guesses, surmises, or probable inferences we may make.

The term is sometimes compared with the term *theory*, and a distinction drawn between them. An hypothesis is called a supposition, meaning that it represents a mere possibility, and a theory is called a *verified* hypothesis. But this distinction is only an arbitrary one, except as it implies the difference between the lowest and the higher degrees of probability in various inferences. The fact is, also, that the terms are often used interchangeably. We speak indifferently of the "Darwinian hypothesis," and the "Darwinian theory;" the "nebular hypothesis," and the "nebular theory." Also we say "undulatory theory" of light and sound, but never "hypothesis," although the conception is precisely that of a conjecture or supposition awaiting satisfactory verification. On the other hand, the term theory is sometimes used to comprehend a body of ascertained truth, or a number of related ideas brought together to represent a system of truths. Thus we say "the theory of the universe," "the theory of the solar eclipse," "the theory of heat," "the theory of equations." It is this conception which has given rise to the notion that a theory is a verified hypothesis. But their actual interchangeability in many crucial and important instances, and the fact that they both represent an inductive process having indistinguishable degrees of probability, are sufficiently cogent arguments for making them identical in all essential features, and so using one of them for describing the whole process incident to the attainment of new conceptions which require verification. An hypothesis, then, we regard as a supposition or inference from given data to their cause, their principle of unity, or their identity. Although the term is usually employed to denote a very low degree of probability in the supposition, this does not make the act of forming it different in kind from more assured inductive inferences, and so we used it merely to denote the stage of thought immediately antecedent to verification.

To illustrate: When Franklin observed certain resemblances between lightning and electricity, his supposition that they were identical was an hypothesis. His experiment with the kite was of the nature of a verification. The supposition that the sun's heat is supplied from the falling of meteors into the sun, is an hypothesis. So also is the vulcanic theory of the molten character of the earth's centre. Some such inference or supposition from given facts must first be made before any steps can be taken to give it greater probability or positive proof. In some cases, of course, the circumstances give so great a probability to the first inference that it scarcely requires verification. But the inference is nevertheless an hypothesis, if we are to take the term in its broadest sense, to denote what is less than a positively ascertained certainty.

The formation of hypotheses must be subject to certain conditions affecting the right to make them. It is important to observe that there can be no absolute rules regulating their legitimacy, because nothing can determine their positive validity or invalidity except the process of verification. Hence there must be a wide range of liberty in forming them. Nevertheless, there are certain proximate rules regulating their formation and determining the degree of presumption in their favor, which ought to be observed by everyone entering into scientific study. Some of the more important of them may be briefly discussed.

(a) *An hypothesis should not be inconsistent with known facts and causes, but should be required by new phenomena that are not explained by existing suppositions.*

We have no right to introduce any new hypothesis when the facts, forces, or causes already known are admittedly capable of explaining the phenomena under consideration. If there be anything new and unaccounted for, there can be no objection to new hypotheses; or if the known causes account for only a part of the effect, the new supposition is admissible. But we must in all such cases be assured that the old suppositions are not adequate to the effect, and that there are new phenomena to be explained. Thus, for example. if the law of supply and demand be a fully accepted one as regulative of prices, we should be careful about introducing any other influence as determinative of them. Or, if the material resources of a country are known to be especially rich and fertile, it will not be legitimate to assume that its prosperity is due to government interference with industrial action, until we have shown phenomena which cannot be accounted for by any other means. For example, we often see protection appealed to as the cause of industrial wealth where the impression conveyed is that this influence is the sole cause of it. But although it may be a factor, as long as other causes are known to be active it is illegitimate to introduce protection as the *only* supposition required to account for the effect. We must first be assured of an unexplained increment. Or, to give another illustration, if it be known that petrifactions actually occur, and that animal and vegetable life may leave traces of their existence in deposits of sand and rock during the present period of the earth's development, it is unnecessary to suppose the existence of some "materia pinguis," fatty matter, or "lapidifying juice" to account for the traces of fossils which do not represent present forms of organic life. It is easier to assume that nature is uniform where precisely the same kinds of effects are discernible, and in this case the known fact of death or the extinction of life, in many instances, has only to be added to the suppositive fossil deposits in past ages in order to account for the difference of genera and species which is observed. The instance only proves that we must first exhaust the explanatory power of known causes before resorting to new ones.

There are several rules which may be regarded as mere corollaries of the one just discussed, or at least as closely related to it. Such as that an hypothesis should be as simple and as free from complexity as possible ; or, that it should not be arbitrary, but should be relevant to the phenomenon and according to actual experience. An illustration of unnecessary complexity is the persistence of some astronomers in the theory of epicycles and eccentrics after the simpler hypothesis of Copernicus had been proposed. An example of arbitrary assumptions is that of the Italian philosopher, mentioned by Fowler, who sought to reconcile the Aristotelian and Platonic theory of the rotundity of heavenly bodies, with the supposition of Galileo, proved by observations of the telescope, that the moon was of a rough surface, by imagining that the hollow parts were filled with transparent crystal, which would permit the same appearances of light and shadow as those we observe, and would observe without this substance. We may reconcile *facts* by a new hypothesis, but we should not undertake to reconcile *theories* or suppositions when the simpler of the two may explain the facts or cover all that is known of them. The difficulty, however, in all such cases is to distinguish between the facts and the hypotheses, because suppositions which are believed for a long time, and which rendered a large body of phenomena intelligible, come to be considered as facts, and so give rise to attempts to reconcile them with new theories apparently in conflict with them. What is needed is a criterion to distinguish between facts and hypotheses, or verified and unverified theories. We turn next to the second rule.

(b) *An hypothesis should permit the application of deductive reasoning, or of the inference to consequences which are capable of comparison with the results of observation.*

This is the first of Jevons's rules, and seems to have been suggested to him by the observations of Hobbes and Boyle. It differs from the first rule we have formulated only in the requirement that the hypothesis be capable of deductive inferences consistent with the facts of observation, and so is regarded by Jevons as merely another form of the rule which he regards as the sole test of the legitimacy of the hypothesis, namely, *conformity with observed facts*. The simplicity of this last mode of statement, however, is so great that illegitimate hypotheses might appear admissible for lack of the definiteness which is necessary to make the rule useful.

The importance of this rule is, that an hypothesis will not be capable of verification unless deductive inferences can be drawn from it, and the results compared with experience and observation. The theory of the transparent crystal filling the hollow spaces of the moon's surface, which was mentioned above, is of this nature. Perhaps the old theory of Phlogiston is of the same kind, since no deductions could be made from it. Jevons states the case in the following manner: "We can only infer what would happen under supposed conditions by applying the knowledge of nature we possess to those conditions. . Hence, as Boscovitch truly said, we are to understand by hypotheses 'not fictions altogether arbitrary, but suppositions comformable to experience or analogy.' It follows that every hypothesis worthy of consideration must suggest some likeness, analogy, or common law, acting in two or more things. If in order to explain certain facts a, a', a'', etc., we invent a cause A, then we must in some degree appeal to experience as to the mode in which A will act. As the laws of nature are not known to the mind intuitively, we must point out some other cause, B, which supplies the requisite notions, and all we do is to invent a fourth term to an analogy. As B is to its effects b, b', b'', etc., so is A to its effects a, a', a'', etc. When we attempt to explain the passage of light and heat radiations through space unoccupied by matter, we imagine the existence of the so-called *ether*. But if this ether were wholly different from anything else known to us, we should in vain try to reason about it. We must apply to it at least the laws of motion, that is, we must so far liken it to matter. And as, when applying those laws to the elastic medium air, we are able to infer the phenomena of sound, so by arguing in a similar manner concerning ether, we are able to infer the existence of light phenomena corresponding to what do occur. All that we do is to take an elastic substance, increase its elasticity immensely, and denude it of gravity and some other properties of matter, but we must retain sufficient likeness to matter to allow of deductive calculations."

These two main rules are sufficient for all practical purposes. Most other rules are either corollaries of the second, or combinations of this and the corollaries of the first rule. Hence there is no need of multiplying them unnecessarily.

It is important to remark, however, that conformity to these rules does not verify or prove an hypothesis. It only shows their right to be entertained as representing a possibility or probability. They are legitimately made when they explain a given number of facts, but may be set aside by a better one which supersedes the place of the old one on account of superior simplicity and comprehension, or superior deductive capacity in connection with observed facts.

Hypotheses are sometimes advanced for the purpose of deductions which shall be a *reductio ad absurdum*. These correspond to the negative proof of deductive methods, and assume a positive hypothesis already in mind, and in this way conform to the rules we have laid down.

2d. Verification.—Verification is the process of testing hypotheses and inferences to see whether they have been correct or not. After a supposition has been made, or an inference drawn that something is possibly or probably true, the mind naturally looks for other evidence to substantiate them. This evidence may come in the repeated occurrence of the original phenomenon, showing that it is probably not a mere accident of circumstances, or it may come in its being a necessary deduction from some known fact or law, but was not perceived until the event was better known. The verification, then, is intended to give greater assurance than the hypothesis can be supposed to have on the first occasion of its suggestion to the mind. When Newton first thought of the law of universal gravitation, he was not content with its power to explain the single phenomenon which suggested it, but he saw that certain other facts must follow from it, or must inevitably be associated with it. He therefore set about a mathematical calculation to see if the result coincided with what ought to occur in the case, and seeing that it did not, he gave up his theory until new data, some ten years later, were discovered regarding the true distance of the moon from the earth. He then resumed his calculations, and found the result to coincide with his hypothesis, and this was regarded as a verification of it. And so, when the hypothesis that all gases might be compressed into liquids or solids was advanced, it was at least a partial verification of it to have succeeded in compressing hydrogen into a liquid under a low temperature and high pressure.

The process of verification may be of two kinds. The first is that which *increases the probability* of an *hypothesis or inference by the four inductive methods*; the second is that which *establishes the certitude of it by deductive or by observational principles*. The one insensibly passes into the other, as we shall see when considering certain peculiarities of the method of difference; or the two may be combined in such a way that no absolute line of distinction can be maintained, and we can only rely upon extreme instances of their application for making the distinction at all. The manner in which the two methods may be connected, consciously or unconsciously, will be remarked in its place. We must take the several forms of verification according to their simplicity, and not according to the degree of certitude supplied by them. These we reduce to four: *Observation, Experiment, Inductive Methods,* and *Deduction,* and shall treat them in that order.

1. OBSERVATION.—Observation in verification is the same process that it is in acquisition, only it is applied with somewhat different presumptions. When an inference or hypothesis has once been made, observation may come in to ascertain whether it is correct or not, and according as the expectation is realized or not, will the verification be. Thus, when the Copernican theory of planetary motion around the sun was proposed and the explanation of the phases of the moon was accepted, it was argued that Mercury and Venus ought to exhibit similar phases. This was admitted by the advocates of the theory, and when Galileo turned his telescope upon them the inference was verified by the discovery of these phases. Here the observation of a fact was the condition of proving the hypothesis. Whenever the inference concerns the existence of some physical fact beyond the control or production of experiment, observation is the only resource for verification, and the assumption made is, that the realization of expectation cannot be due to chance. The effect of finding the inference turn out true is like that of seeing a prediction realized. It is a proof of the assumptions on which it is made. If I have inferred from observations in a particular region that ague is due to miasma from swampy soil, it will be something of a verification of my hypothesis to observe that the same disease is associated with like conditions in other places, and my certitude regarding it will depend upon the number of my subsequent observations, and the varied circumstances under which the disease occurs.

The certainty which verification by observation gives varies with the nature of the conditions under which it is made. If we are assured of the simple and isolated nature of the conditions, it may amount to positive proof. But if the combination of the conditions is great, the observations may only increase the probability of our hypothesis, and this varies between the lowest and the highest degree, which last in most cases cannot be distinguished from demonstration or certainty, so far as the feelings of the mind are concerned.

2. EXPERIMENT.—As already remarked, experiment is only a modified method of observation. But in the process of verification it is much more valuable than observation. It is the artificial reproduction of phenomena under conditions that isolate them more successfully than we can expect to find them, or at least be assured of, in nature. Besides, it enables us to vary the circumstances under which the effect may occur. Such was the effect of Sir David Brewster's experiment showing that iridescence was due to the form and not the nature of a substance. By taking a wax impression of mother-of-pearl, as before mentioned, he produced the effect where there could be no condition affecting the result but the form of the surface, and hence whatever doubt might have existed in regard to the observation of natural instances was here removed by a decisive experiment. In observing the occurrence of phenomena in nature there may be so many conditions associated with their production that a doubt may often exist regarding the conclusiveness of the verification by observation, because the effect may be due to other than the inferred cause which has not yet been separated from the others. But experiment eliminates many possible conditions of the phenomenon, and so gives greater assurance to our inferences. This certitude may vary from all degrees of probability to positive certitude, the degree depending wholly upon the amount of isolation effected in the conditions of the phenomenon. It is the great means of verifying conjectures in the physical sciences, and has more force than any other method.

It should be remarked, however, that both observation and experiment, or modes of verification, are very likely to be associated either with the inductive methods of Agreement and Difference, or with inductive principles, as verifying instruments. Indeed, it may be said that observation can hardly be separated from them. But the two processes will verify an hypothesis in proportion to the extent of their connection with the above-mentioned methods and principles. If observation enlarges the number of instances illustrating the method of agreement, it increases the probability of the inference, and if it detects a case or several cases of difference, the probability may amount to a proof. Experiment will do the same. It will enlarge the area of application for the method of agreement, and more easily effects an application of the method of difference, so that its verifying power is greater than that of observation. And, of course, if any deductive assumptions, or known facts and principles, are illustrated by the results of the two processes, the verification becomes a demonstrated certainty. Usually some deductive assumption is made, whether of a probable or an assured truth, before observation and experiment are applied. Hence we must always calculate whether our assurance or verification is due more to the methods of observation and experiment than to the realization of assumptions already made under the inspiration of agreement and difference or deductive truths.

3. INDUCTIVE METHODS.—We have already explained the methods of agreement and difference as principles of induction, or grounds of the inductive inference and hypothesis. It remains to show that the same principles may be used for verification, in that they may be the means or conditions of increasing the probability of our inductive reasoning. They renew the inference under circumstances which increase its cogency and probability. They are, however, quite as frequently the source of the inductive inference or hypothesis as the means of verification. They act as verifying agencies, when we can assume that the re-occurrence of a phenomenon in accordance with the conditions involved in these methods fulfils the expectation expressed by our hypothesis, and so the original inference is confirmed and strengthened by the multiplication of the instances involving the effect. The verification consists more in the number or quantity than in the nature or quality of the instances, because the fact of frequency is a better indication of a law of nature. Although we speak a great deal of the uniformity of nature, there is as much

variety at the same time, and it is this variety that acts as a disturbing factor to our expectations and calculations. Besides, this variety and diversity indicate a constant change of causal agencies, so that we may never be assured that an event will have the same connections a second time. But if these connections are repeated under a variety of circumstances and other changes, the supposition suggested by a peculiar incident will be confirmed because of an approximation thus made to a law of nature. This is the Method of Agreement. But if with the isolation of a particular cause we discover a simultaneous isolation of a given phenomenon, we may infer their causal connection. But the verification of this inference by the Method of Difference, as it was suggested by the same method, may come in one of two ways yet to be noticed. We take up the methods separately.

(a) *Method of Agreement.*—This has been sufficiently defined. To understand its application as a source of verification, we have only to recall the illustration of iridescence, where the common fact was this phenomenon in connection with a variety of substances. By increasing this variety of conditions and retaining the common fact of iridescence, I increase the probability that the cause, which was supposed after a few observations to be the form of surface on the bodies, is the true one. The highest degree of probability in this case will be reached where the observations pass over to the method of difference, which we shall notice presently.

Again, I find that after taking a particular kind of food I am ill. But I have taken it along with other kinds at the same time, and my conjecture that the effect may be due to the particular kind supposed will be very weak, until I have repeated the circumstances frequently enough to confirm my belief in the supposed cause. This is greatly strengthened by changes of general condition—climate, temperature, air, and food—which show that this presence or absence is not material to the result. Here again is an instance where the hypothesis was first suggested by the principle of agreement, and then verified or confirmed by the same, on an extension of the instances to which it was applicable ; and it is also noticeable that the verification increases in probability or cogency with the approximation of the case to one involving the method of difference. This approximation to the method of difference consists in the constant elimination of conditions which might possibly be the cause of the phenomenon in question ; that is, the elimination of what appears by this separation *not* to be the cause, until the phenomenon and its causes are left in complete isolation, when the inference becomes distinctly verified by the fact.

(*b*) *Method of Difference.*—Mr. Mill states the Canon of Difference in the following language : " If an instance in which the phenomenon under investigation occurs, and an instance in which it does not occur, have every circumstance in common save one, that one occurring only in the former ; the circumstance in which alone the two instances differ is the effect or the cause, or an indispensable part of the cause, of the phenomenon."

A clear conception of it may be afforded by comparing it with the method of agreement. In the method of agreement, *everything may vary except the phenomena in question ;* in the method of difference, *nothing may vary ; that is, everything may be the same, except the phenomena in question.* In the latter case the probability of the inference is always regarded as greater than in the former. The reason for this is that with the method of agreement it is assumed that several causes may produce the same effect, and hence the probability that any one of them is the true one is diminished. Besides, we are arguing from effect to cause. On the other hand, with the method of difference, when everything is common except the phenomena in question, we have instances in which causes are present and yet the effect to be accounted for does not follow, but something else connected with another isolated circumstance. If these common qualities were causes of the phenomenon, it must accompany them ; but as it does not accompany them they cannot be its causes. Hence, since the common qualities are not the causes, we have no other choice but to accept the antecedent which has been isolated with the phenomenon as its cause. The probability is here proportioned to the assurance we feel that the phenomena have really been isolated. Besides, a part of the argument is from cause to effect, and deductive. That is deductive which proves what is *not* the cause, and the remainder, even if regarded as inductive, has its assurance based upon the probabilities of complete isolation in the phenomena concerned.

A passage from Jevons will furnish illustrations of the method of difference. "Thus we can clearly prove that friction is *one* cause of heat, because when two sticks are rubbed together they become heated; when not rubbed together they do not become heated. Sir Humphry Davy showed that even two pieces of ice when rubbed together in a vacuum produce heat, as shown by their melting, and thus completely demonstrate that the friction is the source and cause of the heat. We prove that air is the cause of sound being communicated to our ears by striking a bell in the receiver of an air-pump, as Hawksbee first did in 1715, and then observing when the receiver is full of air we hear the bell; when it contains little or no air we do not hear the bell. We learn that sodium, or any of its compounds, produces a spectrum having a bright-yellow double line, by noticing that there is no such line in the spectrum of light when sodium is not present, but that if the smallest quantity of sodium be thrown into the flame or other source of light, the bright yellow line instantly appears. Oxygen is the cause of respiration and life, because if an animal be put into a jar full of atmospheric air, from which the oxygen has been withdrawn, it soon becomes suffocated."

These are instances where the method of difference is applied, and it will be apparent on examination that the inference turns upon the elimination of the two phenomena with each other while all else remains the same. But now it requires to be shown how the method is one of verification as well as of suggesting the inductive inference. The inductive inference is from the fact to its cause; the verification is an increase of its probability or its proof. How does the method of difference effect this at the same time that it is the source of the inference?

In answer to this question it is proper to remark first, that some logicians, notably Mr. Venn, regard the method of difference as properly one of verification, and not of suggestion. But this mere expression of opinion is not a proof of it, nor have adequate reasons been given for it. The first incident of its verifying power, however, would be the repetition of the observations and experiments involving its application, so as to show that the phenomena concerned were not accidental, and in this way the inference would be strengthened. But usually the inference seems so cogent that it requires no such repetition to reinforce it. The mind seems convinced positively by a single instance. The inference and its verification seem to occur at once. The reason for this is the existence of certain assumptions that are an invariable part of the method of difference. In the first place, there is the deductive argument involved, which we have observed, and which shows *what is not the cause*. The separation of the common qualities from the phenomenon to be explained settles this fact. In the second place, the assumption made whenever the method of difference is strictly applied and which is, *that there are no other than the isolated circumstances to be taken into account*, leaves no room for doubt, and verifies the case by another implied deductive syllogism with this assumption as its major premise. Of course this may be doubtful, but this does not alter the form of the argument or the method of verification, and hence we see that the whole certainty, or the degree of probability attaching to the inference, is determined by the assurance we feel that there are no other circumstances to be considered in the case. These are the verifying agencies in the method of difference, and they are probably associated with it in all its forms and applications.

There are three other methods which are simply modifications of one or the other of the methods just considered, and which may be briefly considered.

(c) *Method of Concomitant Variations.*—This is the method of difference applied in circumstances where the variations of one phenomenon are constant and simultaneous with another. Thus, all other things being equal or remaining the same, the variations of the altitude of the mercury in a thermometer along with a corresponding variation of temperature, lead to the inference that the temperature is the cause, and the verification by the same method is as before. We may repeat the observation or experiment, or use the reasoning which is incident to the method of difference in order to add more probability to our inference. In so far as the method is one of difference it represents more of verification than suggestion or acquisition.

(d) *Method of Residues.*—This is again another application of the method of difference. If we subtract or eliminate from any group of phenomena what we know is due to a given cause, the residue will be the effect of the remaining conditions. Thus, if we suspect the impurity of a mass of gold, we have only to determine its specific gravity and compare this with what we know of the specific gravity of pure gold. The existence of a difference proves the case. We can thus determine what influence the sun has on the tides by subtracting the influence of the moon from the total effect.

(e) *Joint Method of Agreement and Difference.*—"This method consists in a double employment of the method of agreement and a comparison of the results thus obtained, the comparison assimilating it to the method of difference. We, first of all, compare cases in which the phenomenon occurs, and, so far as we can ascertain, find them to agree in the possession of only one other circumstance. But, though we may not be justified in regarding this inference as certain, we may increase our assurance by proceeding to compare cases in which the phenomenon does not occur." The assurance may not be complete in this case as in the method of difference, but it is greater than in the method of agreement. We quote Fowler's illustration of the method :

"A very thin sheet of light proceeding from incandescent hydrogen is passed through a prism, and it is invariably found (with one exception) that in the spectrum thus obtained there are, in proportion to the intensity of the light, one, two, or more bright lines occupying precisely the same relative position. Moreover, very thin sheets of white light proceeding from various incandescent substances are passed through incandescent hydrogen, and the emergent light is then separated into its constituent elements by a prism. In the spectra thus obtained it is found that there are invariably (with the above-mentioned exception) dark lines occupying exactly the same positions in the spectrum as the lines above mentioned. Hence it is inferred, by the method of agreement, that a sheet of light, whether it proceed directly from incandescent hydrogen itself, or be transmitted through it from some other incandescent substance, will invariably (allowing for the exceptional case) produce these lines. But, if we try the same experiments with any other element than incandescent hydrogen, although we may obtain bright or dark lines, we never find these lines occupying the same positions in the spectrum as the lines in question."

The principles of verification involved in the joint method of agreement and difference are those of the two methods themselves, and require no illustration at length. We turn to the last and most conclusive methods of verifying hypotheses.

4. DEDUCTION.—This method of verification is simply the application of deductive reasoning to show that the conclusion of the inductive process may be a necessary deduction from some other known fact, law, or principle. We have seen what place it has in the method of difference, but it may be applied in many instances to the inferences obtained by the method of agreement, and it may possibly be a contingent factor in all forms of verification. But we shall not insist upon this possibility. It suffices to find some cases where the proof of an inference originating inductively may become deductive and so absolutely assuring it.

This deductive method of verification merely consists in comparing an hypothesis or inductive inference with a known truth, and ascertaining whether they agree or not. If they do not agree, but contradict each other, we have positive proof by the principles of opposition that the inference is false. If they do agree, and the inference bears all the marks of being identical with the known truth, it must be true. Whenever it is possible the method is always resorted to. The certitude it possesses depends wholly upon the certitude of the general principle by which the particular inference is tried. But in all cases where a conclusion by induction can be confirmed by its agreement with some other known truth or principle, its probability is very greatly increased, and if the incidents are of the proper kind the verification may be demonstrative.

The most conclusive illustrations of verification by the deductive method are those where the hypothesis accords with the results of mathematical calculation. Thus the conformity of Newton's theory of gravitation with the mathematical calculations made in regard to planetary motion was universally taken as a proof of the theory. Then again, Newton's Binomial Theorem was at first an inductive inference from a few observed instances, and it was afterward proved deductively by later mathematicians; that is, it was shown by admitted mathematical principles that the co-efficients and exponents inductively inferred by Newton *must* be the ones assumed. Galileo inferred from some observations and experiments that the area circumscribed by a cycloidal curve is three times that of the generating circle or wheel. But he had no means of proving it, although he tried to do so experimentally "by cutting out cycloids in pasteboard, and then comparing the areas of the curve and the generating circle by weighing them. In every trial the curve seemed to be rather less than three times the circle," so that he began to suspect the correctness of his supposition. But afterward Torricelli showed by mathematical calculations that this ratio *must* be true.

The method has applications outside of mathematics. If we know any general principle we may compare it with an assumed or inferred fact and test its validity. Thus if I infer from a few observations of the fact of potatoes driving out wheat as a food, that the cheaper food has a tendency to supplant the more costly, I may deduce this fact from Gresham's law regarding the influence of cheaper upon better currencies, when I know or assume, of course, that money is only a commodity like all other things. Mr. Fowler mentions two instances of the combination of induction and deduction which are to the point.

"In the science of Political Economy, Ricardo's theory of rent, when stated in the slightly modified form that 'the rent of land represents the pecuniary value of the advantages which such land possesses over the least valuable land in cultivation,' is an easy deduction from two principles which are supplied by every one's experience; namely, (1) that land varies in value, and (2) that there is some land either so bad or so disadvantageously situated as to be not worth the cultivating."

Professor Cairnes' work on the *Slave Power* furnishes a remarkable example of the successful application of the deductive method to the determination of economical questions. "The economical effects of slavery are thus traced. We learn from observation and induction that slave labor is subject to certain characteristic defects: it is given reluctantly; it is unskilful; and, lastly, it is wanting in versatility. As a consequence of these characteristics, it can only be employed with profit when it is possible to organize it on a large scale. It requires constant supervision, and this for small numbers or for dispersed workmen would be too costly to be remunerative. The slaves must, consequently, be worked in large gangs. Now, there are only four products which repay this mode of cultivation, namely, cotton, sugar, tobacco, and rice. Hence a country in which slave labor prevails is practically restricted to these four products, for it is another characteristic of slave labor, under its modern form, that free labor cannot exist side by side with it. But besides restricting cultivation to these four products, some or all of which have a peculiar tendency to exhaust the soil, slave labor, from its want of versatility, imposes a still further restriction. 'The difficulty of teaching the slave anything is so great—the result of compulsory ignorance in which he is kept, combined with want of intelligent interest in his work—that the only chance of rendering his labor profitable is, when he has once learned a lesson, to keep him to that lesson for life. Accordingly, where agricultural operations are carried on by slaves, the business of each gang is always restricted to the raising of a single product. Whatever crop be best suited to the character of the soil and the nature of slave industry, whether cotton, tobacco, sugar, or rice, that crop is cultivated, and that crop only. Rotation of crops is thus precluded by the conditions of the case. The soil is tasked again and again to yield the same product, and the inevitable result follows. After a short series of years its fertility is completely exhausted, the planter abandons the ground which he has rendered worthless, and passes on to seek in new soils for that fertility under which alone the agencies at his disposal can be profitably employed.' Thus from the characteristics of slave labor may be deduced the economical effect of exhaustion of the soil on which it prevails, and the consequent necessity of constantly seeking to extend the area of cultivation. From the peculiar character of the crops which can alone be successfully raised by slave labor may be explained the former prevalence of slavery in the southern, and its absence in the northern States of the American Union; and from the necessity of constantly seeking fertile virgin soil for the employment of slave labor may be explained the former policy of the southern States, which was invariably endeavoring to bring newly constituted States under the dominion of slave institutions."

The confirmation of an inductive hypothesis by deductive reasoning from accepted principles is very common, and it varies in degrees of assurance according to the assurance we feel about the principles or premises upon which it depends. As we have already explained, it may still be deductive, although the premise may only be probably true, provided the verification is reasoned out deductively. But we require in such cases to be cautious about mistaking another instance of comparison by agreement for a deductive premise. We may take as proof in some cases merely another instance of the same kind, and so only increase the probability of our inference by enlarging the area of the induction. Hence we must be careful, in attempts at deductive verification, to secure a principle or a group of facts under which the particular case or cases involved in the hypothesis may be placed. The instances, however, where this method is applicable are mostly in Physics, or Mechanics, Astronomy, and the Mathematico—physical sciences in general. They are more rare in what may be called the moral sciences.

IV. *FALLACIES AND ERRORS IN INDUCTIONS.*—Errors in application of scientific method, when it is purely inductive, are not so easily detected as in deduction. There are no fixed rules for determining the degrees of probability attaching to inductive inferences, and hence the errors in them are mostly discoverable only after some contradictory truth has been proved. But it is known that we are liable to mistakes in the process, and hence, in order to avoid them we are required to observe certain conditions as precautions against error, or to keep in mind the sources of error that may appear after the attempts at verification. These sources of error may be divided into two kinds : first, *Errors of Observation*, and second, *Fallacies of Inductive Inference*. The latter is an error connected with reasoning.

1st. Errors of Observation.—These are called errors rather than fallacies because they are not due to any process of reasoning, but to mistakes in the acquisition of facts. They give rise to an error in the conclusion by falsifying the data from which the inference is drawn. This may occur in two ways, and against these contingencies the scientific student should perpetually be on his guard. There are no rules, however, to determine when they may occur. But the two sources of error in observation are as follows :

1. NON-OBSERVATION.—This is simply the failure to observe the facts which determine the nature of the inference or the right to make it. We may fail to observe all the facts, or we

may fail to observe those which are essential to a legitimate inductive inference. The failure may be due to various causes with which it is hardly the business of Logic to deal at length, or farther than to refer in general to the admitted sources of mistake in perceptions; but it is an important fact in scientific method that we are liable to such errors in the acquisition of our data as vitiate the conclusion, by bringing it into conflict with non-observed facts. This non-observation may be of some of the cases or instances involved in, or necessary to, the inductive inference, or of incidents in connection with instances that are observed. We may not be able always to avoid these errors, but we should be on our guard against them.

2. MAL-OBSERVATION.—Mal - Observation is not a failure to see facts, but is mistaken or distorted observation. It may be either sensuous or intellectual. Ill observation by the senses may be due to various causes, such as defect of the organs, indistinctness of the impressions, indirectness of vision in the case of eyesight, etc., so that we may often mistake a thing for what it is not. Ill observation from intellectual causes may be due to mistaking an inference for a fact, to preconceived ideas distorting the appearance of those which are facts, to the concentration of attention which may prevent the distinct perception of all that is in the indirect field of consciousness, etc. Some of the last may actually be a source of error to the senses at the same time. And there may be cases where there are tendencies in the very mental and physical organism to modify the impressions and perceptions which consciousness must receive. The mistaking of an inference for a fact of observation is a very frequent error under this head. Mr. Fowler mentions the instance of the objections at first made to the Copernican system of astronomy. People claimed that the theory was contrary to what they *saw*, when the truth was that what they claimed to see was only an inference from certain facts. Illusions often exhibit the same phenomenon. We misinterpret the data of sense and take the result as a fact.

SCIENTIFIC METHOD 365

2d. Fallacies of Inductive Inference.—It is difficult to treat of fallacies in this field, because there can be no absolute rules laid down to determine the limits of correct inductive inference; that is, to determine when we may and when we may not go beyond the data of the premises. Perhaps the only rules applicable here are those which determine the limits of legitimate hypothesis. But these have no *formal* character, and hence must be capable only of *material* application and limitations. In deductive reasoning the fallacies are mostly fallacies of inference which can be determined by rules. In inductive reasoning what is called an error of inference is never, perhaps, *formal*, but must be *material*, and hence there is no way of determining them until after the process of verifications has been applied and a result established contrary to our supposition. But if errors do not arise from the mode of our inference, they may arise from the assumptions we make in the interpretation of phenomena. Hence there may be what we may call presumptive fallacies in the inductive process, which vitiate the *conclusiveness* of the inference; that is, its probability or certainty, although it may be admissible as representing a conceivable case awaiting distinct verification. Of these fallacies it is sufficient to recognize two general kinds : first, the mistaking an inductive for a deductive inference ; and second, the mistaking of partial for total causes.

1. CONFUSION OF INDUCTIVE WITH DEDUCTIVE INFERENCE.—We may often mistake the degree of proof we have for an assertion. We may have arrived at a truth inductively, and then assume that the process was one of positive proof, or deductive. Or in answering the demand for proof of a proposition, we may quote instances or facts which are only the incidents from which the inductive inference was drawn. They undoubtedly support the inference as an inductive one, but if we assume that they prove it, we assume that we have given deductive reasons, when we have only referred to facts. In deduction this would be called a *petitio principii*. Thus, if I inferred that the moon influenced the weather, or that frost was caused by cool calm nights, and supposed that my belief was confirmed

or proved by the incidents of my observations and experience, I would be committing this fallacy. It is a very frequent error, and in no phenomena is it more frequent than in the confusion of coincidence or sequence with the causal connection. This is the *non causa pro causa*, or *post hoc, ergo propter hoc* fallacy already discussed. We have a right to suppose it possible or probable, but not as proved by the circumstances. Perhaps, however, these should not be called inductive fallacies, so much as fallacies which accompany inductive reasoning. Their association with it justifies the consideration of them in this connection.

2. CONFUSION OF PARTIAL AND TOTAL CAUSES.—This is merely the presumptive error of supposing that what may be one of the causes is the only one. Thus we may infer from a set of observations that the moist air of a given region is unhealthy, when it may be the moist air in conjunction with the temperature, or it may be due to influences of temperature as well as moisture. Or again, the effect may be as much determined by the character of the individual as by the circumstances in which he is placed. The error here, however, is not in inferring a cause, but in assuming that the discovered cause is the only or the whole cause of the phenomenon, and hence in the application of scientific method we need to be as much on our guard against this as against the preceding fallacy.

This fallacy may take several forms, and there are perhaps others of an allied nature. But we do not consider it important to discuss any of them at length in this elementary work. The careful treatment of them belongs to special treatises on scientific method.*

* For a more complete discussion of this subject the student may consult the following references: Jevons: Principles of Science; Fowler: Inductive Logic; Mill: Logic, Books III., IV., and VI.; Hamilton: Lectures on Logic, Lects. XXIV., XXV., and XXVI.; Venn: Empirical Logic Chapters XIV.-XVIII., and Chapter XXIV.

PRACTICAL QUESTIONS AND PROBLEMS.

CHAPTER I.

1. What is the distinction between "science" and "art," and how apply it to Logic?
2. Examine the merits of the several definitions of Logic.
3. Define the logical use of the term "thought."
4. What is Sir William Hamilton's account of "thought?"
5. What is the meaning of the term "law" and such expressions as "laws of nature" and "laws of thought?"
6. Illustrate the use of the term "law" in the sciences.
7. What are the meanings of the term "form" in common usage and in Logic? Also the term "matter?"
8. What is the relation of Logic to the other sciences? Especially to Psychology?
9. What are the divisions of Logic? How define the meaning of each subdivision?

CHAPTER II.

1. What are the elements of logical doctrine? From what points of view are they to be regarded, and why?
2. Define "terms," "propositions," and "syllogisms."
3. What are the "formal" elements of logical doctrine, and how define them?
4. What ambiguity is found in the word "conception?"
5. What is the difference between a "term" and a "concept?"
6. What is the definition of "concept," and to what ambiguities is the term exposed?
7. What is a "percept," and how are concepts formed?
8. What is meant by Perception or Apprehension?
9. What is meant by "attribute" or "individual" and "class wholes?"
10. Explain the difference between mathematical and metaphysical or logical concepts.
11. What is the use of Denomination in Logic?
12. Define Judgment and compare the process with that of Conception.
13. Define Reasoning and compare it with Conception?

Chapter III.

1. What are categorematic and syncategorematic terms?
2. Define and illustrate singular terms. When will terms ordinarily singular become general?
3. Define and illustrate general terms.
4. Distinguish between distributive and collective terms.
5. Examine the following propositions and state the distributive and collective use of the terms:

(*a*) The inhabitants of Germany constitute a nation.
(*b*) "All men find their own in all men's good,
And all men join in noble brotherhood."—*Tennyson.*
(*c*) All standing armies are dangerous to the state.
(*d*) Non omnis moriar (*i.e.*, I shall not all die).
(*e*) All the men cannot lift this weight.
(*f*) All of the regiment was put to flight.

6. Define and distinguish concrete and abstract terms.
7. Give illustrations of pure concrete and pure abstract terms, and show why terms cannot be classified according to the distinction between concrete and abstract.
8. Indicate in the following list the concrete and the abstract terms, and those also, if any, which may be both:

Act.	Ability.	Plato.	Production.
Action.	Presidency.	Solitude.	Warmth.
Agency.	Timeliness.	Dexterity.	Science.
Agent.	Virtue.	Government.	Art.
Beauty.	Excellence.	Library.	Truth.
Man.	Wisdom.	Introduction.	Stone.

9. Do abstract terms admit of being plural?
10. What disposal should be made of general terms in the classification of concrete and abstract concepts?
11. What are the common uses of "concrete" and "abstract," and why?
12. State the views of Wundt on concrete and abstract terms.
13. In the following list of terms show how each term may be taken as concrete and abstract, according to its meaning:

Individuality.	Personality.	Equivocation.	Government.
Society.	Science.	Philosophy.	Institution.

14. Define and distinguish positive, negative, and privative terms, naming the marks of the negative.
15. Define and illustrate nego-positive terms, giving the reasons for recognizing such a class.

PRACTICAL QUESTIONS AND PROBLEMS 369

16. What is meant by *infinitated* conceptions?
17. How can any term be considered as the negative of all others but its synonym?
18. Illustrate absolute and relative terms.

CHAPTER IV.

1. What is meant by the ambiguity of terms?
2. Define and illustrate univocal terms.
3. Distinguish the three kinds of equivocal terms.
4. Distinguish the three causes by which the third and most important class of ambiguous terms has been produced.
5. Explain the ambiguity of the following terms, referring each to its cause:

Chair.	Paper.	Minister.	Volume.	Earth.
Man.	Stock.	Count.	Scale.	Law.
Country.	Air.	Period.	Feeling.	Art.
Sensation.	Glass.	Clerk.	Kind.	Bolt.
Bill.	Peer.	Order.	State.	Star.
Table.	Sense.	Wood.	Service.	End.
Term.	Ball.	Bull.	Subject.	Class.
Letter.	Interest.	Pole.	Age.	Can.
Commons.	Church.	House.	Virtue.	Light.
Mount.	Currency.	Fault.	Lace.	Dip.

6. Explain and illustrate what is meant by the generalization and specialization of terms, and how they give rise to ambiguous ideas.
7. How are the laws of generalization and specialization applicable to concrete and abstract terms?

CHAPTER V.

1. What terms are used to express the same meaning as the terms intension and extension?
2. Define and distinguish what is meant by the intension and extension of concepts.
3. Can abstract terms have both intension and extension?
4. What is the relation between intension and extension, and how may it be formulated?
5. Between what class of terms can a comparison of extension be made?
6. When can a comparison between them not be drawn, and what is peculiar in this respect regarding affirmative and negative propositions?
7. What can be said about the nature and accuracy of the law expressing the ratio or relation between intension and extension?

8. How can the relation be expressed by symbols?

9. What arguments can be produced for the substantial truth of the law?

10. What is meant by denotative, connotative, and non-connotative terms?

11. Examine Mill's doctrine upon this point, and state what modifications are commended by Keynes and Fowler?

12. Select from the following list the terms which belong to the same classes, and arrange them in the order both of the greatest intension and the greatest extension:

Emperor.	Person.	Animal.	Ruler.
Teacher.	Horse.	Dissenter.	Organized substance.
Baptist.	Heavenly body.	Individual.	Lawyer.
Timber.	Christian.	Jupiter.	Alexander.
Planet.	Matter.	Quadruped.	Napoleon III.
Mammalian.	Solicitor.	Being.	Episcopalian.
Frenchman.	Man.	Word.	Greek.

13. Select and arrange the following, with the use of triangular symbols, so as to show both the order of intension and extension:

Animal.	Machine.	President.	Engine.
Tree.	Vertebrate.	Manufactures.	Locomotive.
Gladstone.	Man.	Statesman.	Englishman.
Substance.	Element.	Vegetable.	Iron.
Lily.	Metal.	Steel.	Flower.
Organism.	Being.	Lincoln.	American.

CHAPTER VI.

1. What are the five predicables?

2. Define and illustrate the term "property." What other terms are its equivalents?

3. What is meant by essential property? Illustrate.

4. Define and illustrate what is meant by non-essential properties or "accidents."

5. What is meant by universal and contingent accidents? Also by the term "peculiar" property?

6. What is meant by "differentia" or difference?

7. What is the meaning of the term *conferentia*, and what is the reason for using it?

8. Define "genus" and "species," illustrating them and showing how they are purely relative in their meaning, with two exceptions.

9. What two meanings has the term "genus?"

10. What is meant by *summum genus* and *infima species?*

PRACTICAL QUESTIONS AND PROBLEMS 371

11. What is the analysis of concepts?

12. What is logical division? Illustrate and show what is meant by the *fundamentum divisionis*.

13. Apply division as far as possible to the following list of concepts:

Animal.	Government.	Matter.
Vegetables.	Man.	Vertebrate.
Stones.	Book.	School.
Trees.	Science.	Poetry.
Races.	Furniture.	Metal.

14. What is meant in division by the terms *super-ordinate*, *subordinate*, and *co-ordinate*?

15. What is *dichotomy* and *trichotomy*? Illustrate the "tree of Porphyry."

16. Define and illustrate "partition." What is meant by mathematical and logical partition?

17. Analyze the following concepts by partition:

Metal.	Picture.	Cathedral.	Knowledge.	Ink.
Iron.	Stone.	Religion.	Money.	Book.
Plant.	House.	Literature.	Government.	Ice.
German.	Diamond.	Virtue.	Production.	Wheat.

18. Name and explain the several kinds of definition.

19. Compare the processes of definition and division.

20. What are the rules for correct definition?

21. Give a logical definition of the following terms:

Biped.	Spirit.	Desk.	Action.	Proposition.
Reptile.	Water.	House.	Book.	Literature.
Nation.	Ability.	Agent.	Religion.	Spectacle.
Purity.	Honor.	Imagination.	Club.	Gravitation.
Diet.	Success.	Republican.	Money.	Race.

22. Criticise the following definitions:

(*a*) A member of the solar system is anything over which the sun has continued influence.

(*b*) A chair is an object upon which men sit.

(*c*) An animal is a being which increases in size.

(*d*) Death is the opposite of life.

(*e*) A king is one who exercises regal functions.

(*f*) A gentleman is a man having no visible means of subsistence.

(*g*) Tin is a metal lighter than gold.

(*h*) Government is an association of men.

(*i*) Science is the study of phenomena with a view to a scientific knowledge of them.

(*j*) Man is a bundle of habits.

23. What can be said of the definition of *faith* in the eleventh chapter of the Epistle to the Hebrews?

24. Give examples of indefinable words and explain why they are indefinable.

CHAPTER VII.

1. What is the definition of a judgment or proposition?

2. What is meant by subject and predicate *logically* considered.

3. What are the divisions of judgments, and the difficulties suggested by some of the current divisions? Define and illustrate each.

4. How reduce disjunctive propositions to hypothetical?

5. What is meant by the *quality* of propositions?

6. What is meant by the division of propositions according to *quantity?*

7. Show how the five forms of propositions according to quantity can be reduced to two.

8. Define and illustrate analytic and synthetic judgments.

9. Explain the difficulty of drawing an absolute line of distinction between analytic and synthetic judgments.

10. What are tautologous, pure, and modal propositions?

11. What are the three sources of ambiguity in propositions?

12. Define what is meant by Inverted and Duplex propositions.

13. Illustrate and explain each of the kinds of duplex propositions.

14. State the complementary of the following duplex propositions:

(*a*) None but the wise can be virtuous.

(*b*) All persons except criminals and foreigners are not allowed to vote.

(*c*) Only bipeds have hands.

(*d*) Man alone is not obedient to his instincts.

(*e*) Few persons are as strong to resist temptation as they should be.

(*f*) Only those substances which are not subject to gravity are immaterial.

15. Explain the quantity of exclusive propositions.

CHAPTER VIII.

1. Explain the meaning of "judgments of extension" and "judgments of intension."

2. How show the possibility of expressing a mathematical relation between the subject and the predicate in at least the extensive judgments?

3. How show that intensive judgments have an extensive or quantitative, as well as an intensive or qualitative import?

4. What is the difference between the quantity of extension and quantity of intension expressed by propositions?

5. Symbolize propositions A, E, I, and O, by Euler's diagrams in every

possible form, and reduce them to the simplest expression or representation.

6. How symbolize the quantity of intension in judgments?

7. Define what is meant by the *distribution* of the subject and predicate, and give the rules for it.

CHAPTER IX.

1. Explain the meaning of opposition.

2. What is meant by the terms contrary, contradictory, subalterns, subalternans, subalternate, and sub-contrary?

3. If we assume any one of the four propositions, A, E, I, and O, to be false, what follows in regard to the others?

4. How must we express the opposite of any *singular* proposition?

5. How treat the relation between the two propositions, "Daniel Webster was an American," and "Daniel Webster was not an Englishman?"

6. How are propositions best proved or disproved, and what difficulty do universal propositions present in disputation?

7. Select pairs of the following propositions and arrange them so as to show all the various relations of opposition illustrated by them:

(a) All men are mortal.
(b) Some men are not mortal.
(c) No men are mortal.
(d) Some men are mortal.
(e) Most men are mortal.
(f) All men are not mortal.
(g) Not all men are mortal.
(h) Only men are mortal.
(i) Few men are not mortal.
(j) Only mortals are men.
(k) All men except a few are mortal.

8. Examine the relation between the following propositions and examples of assertion, and state what is implied by a given assertion against an opponent.

(a) One man asserts that all men are wise, and another that they are all ignorant.

(b) Apples are a species of fruit, but they are not an ordinary vegetable.

(c) Free-trade lowers prices and protection raises them.

(d) The leaders of one party assert that Mr. A will be elected president, and the leaders of the other that Mr. B will be elected president.

(e) Mr. X asserts that not a nail was made in this country before 1861.

So far is this statement of X from being true that in 1856 there were 2,645 nail machines in operation in this country with an output of 86,462 tons, and in 1859 as many as 4,686,207 pounds of nails were exported.

(*f*) "Will the educated woman marry? So queried one of our alumnæ in a recent magazine article. The review roll of our alumnæ shows that of 76 ladies who graduated in our classes, 32 have already married."

(*g*) "Great efforts are made to show that a general glut of the market is not a source of evil, and that the appreciation of gold has been the cause of the extraordinary fall in prices. To those who hold this view I would put the question why ivory and whalebone have not fallen in price, but, on the contrary, have steadily risen in price during the last decade."—*Cotter Morrison, Service of Man, Preface,* pp. x.-xi.

(*h*) "An import duty is not a tax, and yet raises the price of articles to the consumer.

"The object of raising prices to the consumer is to enable the domestic manufacturer to pay higher wages to his workmen.

"The duty does not raise prices to the consumer, but, on the contrary, lowers them.

"The manufacturers are enabled to pay higher wages in spite of this fall in prices."

CHAPTER X.

1. What is the meaning of *inference?* Of immediate inference? Of mediate inference?

2. Define and illustrate Conversion.

3. What are the rules for Conversion, and what is meant by the convertend and the converse?

4. How many kinds of Conversion are there? How are they variously applied to propositions A, E, I, and O?

5. What is true of the conversion of definitions, and of singular propositions with a singular predicate, and why?

6. What is Conversion by Negation, to what proposition is it applied, and how? Show whether it is a legitimate form of conversion or not.

7. Explain and illustrate the process of Obversion, and apply it to the four propositions.

8. Explain and illustrate Contraversion or Contraposition, and show why it cannot be applied to proposition I.

9. Explain in what sense Contraversion is a process of immediate inference, if it be so at all.

10. Define and illustrate the process of Inversion, and show to what propositions it is applicable, and to what propositions it is not applicable.

11. What is meant by Inference by Contribution, and what are its divisions? Illustrate each.

12. When will Inference by Added Determinants and by Complex Conception be invalid?

13. State the logical process by which we pass from each of the following propositions to the succeeding one:

(a) All metals are elements.
(b) No metals are non-elements.
(c) No non-elements are metals.
(d) All non-elements are not-metals.
(e) All metals are elements.
(f) Some elements are metals.
(g) Some metals are elements.
(h) No metals are elements.

14. Convert the following propositions:

(1.) Every man is a biped.
(2.) No triangle has one side equal to the sum of the other two.
(3.) Some books are dictionaries.
(4.) "Every consciousness of relation is not cognition."
(5.) Vegetables only are deciduous.
(6.) A stitch in time saves nine.
(7.) Perfect happiness is impossible.
(8.) Few are acquainted with themselves.
(9.) No one is free who does not control himself.
(10.) Good orators are not always good statesmen.
(11.) Some inorganic substances do not contain carbon.
(12.) All men are not born equal.
(13.) Only the brave deserve the fair.
(14.) No one is a hero to his valet.
(15.) He jests at scars who never felt a wound.
(16.) Uneasy lies the head that wears a crown.
(17.) Better late than never.
(18.) A certain man had a fig tree.
(19.) Familiarity breeds contempt.
(20.) Every mistake is not culpable.
(21.) I shall not all die (non omnis moriar).
(22.) Not many of the metals are brittle.
(23.) Great is Diana of the Ephesians.
(24.) Talents are often misused.
(25.) Romulus and Remus were twins.
(26.) Some books are to be read only in part.
(27.) Nothing is praiseworthy but virtue.
(28.) Two blacks will not make a white.
(29.) Not one of the Greeks at Thermopylæ escaped.
(30.) No one is always happy.

(31.) Metals are all good conductors of heat.
(32.) There is none good but one.
(33.) All that glitters is not gold.
(34.) He can't be wrong whose life is in the right.

15. State the relation between the following propositions as indicated by the figures in parentheses at the end of each proposition:

(1.) Good men are wise.
(2.) Unwise men are not good (1.).
(3.) Some wise men are good (1.).
(4.) No good men are unwise (1.) (2.).
(5.) Some unwise men are not good (2.) (4.).
(6.) Some good men are wise (1.) (3.).
(7.) No good men are wise (1.) (4.) (6.) (3.).
(8.) Some good men are not wise (1.) (3.) (6.) (7.).
(9.) No unwise men are good (1.) (2.) (4.) (5.) (8.).
(10.) No wise men are good (1.) 2.) (6.) (7.) (8.).

16. What is the logical relation, if any, between the two following propositions: "A false balance is an abomination to the Lord, but a just weight is his delight."

17. What can be inferred by Obversion, Conversion, and Contraversion from the following proposition: "The angles at the base of an isosceles triangle are equal."

18. State the Contraverse and Obverse of propositions (1.) (5.) (6.) (7.) (9.) (15.) (19.) (23.) (31.) and (33.) under question 14.

19. Can we logically infer that because heat expands bodies, therefore cold contracts them?

20. State the relation between the following three propositions: "The voluntary muscles are all striped, and the unstriped are all involuntary, but a few of the involuntary muscles are striped."

CHAPTER XI.

(1.) What is meant by a middle term? Explain what is meant by major and minor terms, and show how each of the three terms can be distinguished in a syllogism.

(2.) State the rules of the syllogism.

(3.) Define what is meant by a formal fallacy, and state what three kinds of them occur.

(4.) How can we symbolize by diagrams the several kinds of formal fallacy.

(5.) In the following syllogisms and reasonings point out the major, middle, and minor terms, with the corresponding premises and conclusion:

(a) All men are fallible.
All kings are men.
∴ All kings are fallible.

(b) Platinum is a metal.
All metals are heavy.
∴ Platinum is heavy.

(c) Cattle are ruminants.
Horses are not ruminants.
∴ Cattle are not horses.

(d) Iron is a metal.
Metals are substances.
∴ Some substances are iron.

(e) He is wise, because he knows what his interest is and whoever knows his interest is wise.

CHAPTER XII.

1. Define what is meant by the Mood of a syllogism. Illustrate and show how many combinations are possible with the four propositions A, E, I, O.
2. Mark those moods which are invalid and state the reason.
3. What kind of a conclusion can be drawn from the following premises, AA, EA, IA, AE, OA, EI.
4. Define what is meant by the Figure of a syllogism, and illustrate each Figure.
5. What is meant by a weakened conclusion?
6. What peculiar value attaches to each of the first three Figures of syllogism?
7. Show in what figures the following premises give valid conclusions, AA, AE, AI, EA, OA, EI, AO.
8. Show that O cannot stand as a premise in the first Figure, as major premise in the second Figure, and as minor premise in the third Figure.
9. What fallacies would be committed by having A as a conclusion in any Figure but the first?
10. Why can we prove only negative propositions in the second Figure?
11. What fallacy is committed if the minor premise of the first Figure be negative?
12. If one premise be O, what must the other be?
13. What are the Moods and Figures of the following syllogisms? Name the valid and the invalid Moods.

(a) Some M's are P's
No S's are M's.
∴ Some P's are not S's.

(b) All P's are M's.
No M's are S's.
∴ No P's are S's.

(c) All S's are M's.
No P's are M's.
∴ Some S's are not P's.

(d) No M's are P's.
All M's are S's.
∴ Some S's are not P's.

(e) All feathered animals are vertebrates.
   No reptiles are feathered animals.
   ∴ Some reptiles are not vertebrates.
(f) All vices are reprehensible.
   Emulation is not reprehensible.
   ∴ Emulation is not a vice.
(g) All men are rational beings.
   All Caucasians are rational beings
   ∴ All Caucasians are men.
(h) All vices are reprehensible.
   Emulation is not a vice.
   ∴ Emulation is not reprehensible.
(i) Only citizens are voters.
   A, B, C are voters.
   ∴ A, B, C are citizens.

14. Deduce conclusions, stating Moods and Figures, from the following premises :

(a) All planets are heavenly bodies.
   No planets are self-luminous.
(b) All Europeans are Caucasians.
   All Caucasians are white.
(c) All lions are carnivorous animals.
   No carnivorous animals are devoid of claws.
(d) Some animals are quadrupeds.
   All quadrupeds are vertebrates.
(e) Oak-trees are not evergreen.
   Pine-trees are evergreen.

15. Invent examples showing true conclusions with false premises.

CHAPTER XIII.

1. Explain the mnemonic lines which represent the valid Moods and Figures, and the reduction of the last three Figures to the first.

2. Construct syllogisms in Camenes, Cesare, Ferison, Fesapo, Camestres, and Datisi, and reduce them to the corresponding Moods in the first Figure.

3. What is the difference between Direct and Indirect Reduction.

4. Why cannot Baroko and Bokardo be reduced directly? Show how they may be reduced indirectly.

5. Apply indirect reduction to Cesare and Camenes.

CHAPTER XIV.

1. Define Prosyllogism, Episyllogism, Enthymeme, Epicheirema, and Sorites. Illustrate each.

2. What is the difference between the regressive and the progressive Sorites?

3. Upon what is the distinction between the three orders of Enthymeme based?

4. Produce an example of syllogism in which there are two Prosyllogisms.

5. Complete the following syllogisms:

(a) Europeans are Caucasians because they are white.

(b) Since he was directed to deliver the message, and did not, I am at liberty to do as I please.

(c) We cannot know what is false because knowledge cannot be deceptive.

(d) A is B, because C is B
E is A, because G is A
∴ E is B.

(e) A manor cannot begin at this day, because a court-baron cannot now be founded.

Chapter XV.

1. What is hypothetical reasoning? Illustrate. State the signs of it and define what is meant by the antecedent and the consequent.

2. What are the valid and the invalid forms of hypothetical reasoning? What is meant by *modus ponens* and *modus tollens*?

3. How can hypothetical syllogisms be reduced to the categorical?

4. What moods and figures of the categorical syllogism do the *modus ponens* and *modus tollens* of the hypothetical syllogism belong?

5. What formal fallacies are committed by the invalid forms of hypothetical syllogism? Prove by reduction to the categorical.

6. Examine the following instances of hypothetical reasoning:

(a) Rain has fallen, if the ground is wet; but the ground is not wet; therefore rain has not fallen.

(b) If rain has fallen, the ground is wet; but rain has not fallen; therefore the ground is not wet.

(c) The ground is wet if rain has fallen; the ground is wet; therefore rain has fallen.

(d) If the ground is wet, rain has fallen; but rain has fallen; therefore the ground is wet.

(e) If a man cannot make progress toward perfection, he must be a brute; but no man is a brute; therefore every man is capable of such progress.

(f) If two and two may make five in some other planet, Mill's opinion about the matter is correct; but they do not make five in any place, and hence Mill is wrong.

Chapter XVI.

1. Define and illustrate Disjunctive Reasoning.
2. Upon what does *incomplete disjunction* depend?
3. Classify the forms of disjunctive syllogism, and show how they may be reduced to either the hypothetical or the categorical form.
4. To what fallacy is disjunctive reasoning incident?
5. Examine the following cases of reasoning:

(a) Criminals are either good or bad.
They are bad.
∴ They are not good.

(b) The weather will be either clear or warm.
It will not be warm.
∴ Therefore it will be clear.

(c) Aristotle was either very talented or very industrious.
He was very industrious.
∴ He was not very talented.

(d) If the government enacts such a law it must either adopt socialism or go into bankruptcy. But it will not enact such a law, and hence there is no danger of either socialism or bankruptcy.

(e) If pain is severe it will be brief, and if it last long it will be slight; it is either severe or it lasts long, and therefore will be either brief or slight.

(f) If capital punishment involves cruelty to its victims it ought to be abolished in favor of some other penalty; if it does no good for society it should also be abolished. But it either involves cruelty to its victims or does no good to society, and hence it ought to be abolished.

Chapter XVII.

1. Define the term "fallacy" and explain what is meant by *formal* and *material* fallacies.
2. Explain what is meant by the fallacies of Amphibology and of Accent.
3. Give outline form of the classification of fallacies.
4. Illustrate the formal fallacies and also those of Amphibology and Accent.

Chapter XVIII.

1. What are the grounds upon which a twofold division of material fallacies may rest?
2. Explain what is meant by the following terms: *petitio principii, non*

sequitur, non causa pro causa, or *post hoc, ergo propter hoc, circulus in probando, assumptio non probata,* and *ignoratio elenchi.*

3. Explain the fallacies of Quantity and Quality, or those of Composition and Division and of Accident.

CHAPTER XIX.

1. What is meant by the quantification of the predicate?
2. What effect upon the number of propositions to be considered by Logic is produced by quantifying the predicate explicitly?
3. How does the quantification of the predicate effect the process of conversion?
4. What additional rule must be added to the rules of the syllogism if we accept the doctrine of the quantification of the predicate?
5. What is peculiar about definitions and exclusive propositions in relation to this doctrine?

CHAPTER XX.

1. What is the nature of mathematical propositions?
2. What effect do they produce upon the Figures of the syllogism?
3. Write out the list of valid Moods when propositions are mathematical, and show why each one is valid that is not valid in ordinary reasoning.
4. How does mathematical reasoning simplify the symbolic representation of the syllogism?
5. What characteristic of conceptions appears in mathematical reasoning, and what is excluded?
6. What is meant by *Traduction,* or traductive reasoning?
7. How can you treat syllogisms which have been called irregular by Jevons and others? Invent instances and illustrate. How are they related to the principles of mathematical reasoning?

CHAPTER XXI.

1. What is the general nature of the "Laws of Thought" and what are their divisions?
2. Define and illustrate the laws of Identity, Contradiction, Excluded Middle, and Sufficient Reason.
3. Enumerate the secondary laws of thought.
4. How are the primary laws related to the principles enunciated in Formal Logic?

CHAPTER XXII.

1. What views have been taken regarding the nature of inductive reasoning?
2. What two recognized forms of induction are to be considered in the definition, and how are they to be treated?
3. What is the difference between *generalization* by enumeration and inductive inference?
4. What is the essential conception involved in the idea of induction since the time of Bacon, and how did it arise?
5. Illustrate Inductive reasoning and compare it with the deductive.
6. State the form of the inductive syllogism, illustrate and compare it with the deductive.
7. What are the reasons for regarding inductive reasoning as purely *qualitative*, and deductive as mainly, if not wholly, *quantitative?*
8. What is meant by the division of inductive inferences into *statical* and *dynamical?*
9. What are the so-called "principles" of induction? Define their meaning and relation to inductive reasoning.
10. What is meant by the principles or canons of Agreement and Difference, and their functions in inductive reasoning?

CHAPTER XXIII.

1. Explain the meaning of scientific method, also what is meant by the methods of *discovery* and of *instruction*.
2. Define and illustrate Deductive Method. State its divisions and explain their relation to the method.
3. Explain what is meant by the Inductive Method as compared with inductive reasoning.
4. Define and illustrate the primary processes of ascertaining new knowledge.
5. What are the rules to be observed in forming hypotheses?
6. Explain the process of verification, stating its relation to inductive inference.
7. What are the two or more kinds of verification? Explain those which seem also to be conditions or factors of the inductive inference, namely, Observation and Experiment.
8. What are the Inductive Methods, and how do they serve as verifying processes.
9. Show how deduction may enter as a process of verification.
10. Define and illustrate what are called the fallacies of induction. Why distinguish between errors of observation and errors of inference?

PRACTICAL EXERCISES

DEDUCTIVE

The student is expected to examine the following arguments; to state the mood and figure of the syllogism where necessary; to complete imperfect syllogisms; to indicate the instances of valid and invalid reasoning; and if invalid, to state whether the fallacy is formal or material, and what the particular fallacy is. He should also be prepared to make any resolution of the propositions and syllogisms which the rules of Logic would enable him to do.

1. None but animals are quadrupeds.
   Horses are quadrupeds.
   Therefore horses are animals.
2. Personal deformity is an affliction of nature.
   Disgrace is not an affliction of nature.
   Therefore personal deformity is not a disgrace.
3. All roses are beautiful.
   Lilies are not roses.
   Therefore lilies are not beautiful.
4. All paper is useful; and all that is useful is a source of comfort to men; therefore all paper is a source of comfort to men.
5. Some statesmen are also authors; for such are Burke, Macaulay, Gladstone, Lord Russell, etc.
6. Some philosophers are logicians.
   No logicians are ignorant of the works of Aristotle.
   Therefore some philosophers are not ignorant of the works of Aristotle.
7. No persons destitute of imagination are true poets.
   Some persons destitute of imagination are good logicians.
   Therefore some true poets are not good logicians.
8. This explosion must have been occasioned by gunpowder; for nothing else would have possessed sufficient force.
9. If Cæsar was a tyrant, he deserved to die.
   Cæsar was not a tyrant.
   Therefore he did not deserve to die.
10. Good is the object of moral approbation. The highest good is, therefore, the ultimate object of such approbation, the end of action.

11. Every one desires his own good.
Justice and temperance are every one's good.
Therefore every one desires to be just and temperate.
12. "But it is doubtful yet whether Cæsar will come forth to-day or not. For he is superstitious grown of late."
13. Every man should be moderate ; for excess will cause disease.
14. All Parisians are Frenchmen.
No Chinese are Parisians.
Therefore some Chinese are not Frenchmen.
15. Some men are not virtuous.
All Americans are men.
Some Americans are not virtuous.
16. Blessed are the merciful ; for they shall obtain mercy.
17. As almost all the organs of the body have a known use, the spleen must have some use.
18. Some of the inhabitants of the globe are more civilized than others.
No savages are more civilized than other races.
Some savages are not inhabitants of the globe.
19. Cogito, ergo sum (I think, therefore I am).
20. He must be a Mohammedan, for all Mohammedans hold these opinions.
21. He must be a Christian, for only Christians hold these opinions.
22. Logic is either a science or an art.
It is a science.
Therefore it is not an art.
23. No idle person can be a successful writer of history ; therefore, Hume, Macaulay, Hallam, and Grote must have been industrious.
24. Who spareth the rod hateth his child ; the parent who loveth his child therefore spareth not the rod.
25. The coronation took place either at Paris, Berlin, or Vienna ; it did not occur at Paris or Berlin, and consequently must have occurred at Vienna.
26. Every moral man obeys the law ; every citizen does not do so, and therefore is not moral.
27. Rational beings are accountable for their conduct ; brutes, not being rational, are therefore free from responsibility.
28. All valid syllogisms have three terms.
This syllogism has three terms.
This syllogism is therefore valid.
29. All syllogisms are valid that have three terms.
This syllogism has three terms.
Therefore this syllogism is valid.
30. Comets are heavy matter ; for otherwise they would not obey the law of gravitation.

31. A charitable man has no merit in relieving distress, because he merely does what is pleasing to himself.
32. None but savages were in America when it was discovered.
The Hottentots were savages.
Therefore they were in America when it was discovered.
33. None but despots possess absolute power.
The Czar of Russia is a despot.
Therefore he possesses absolute power.
34. Bacon was a great philosopher and statesman, and as he was also a lawyer we may infer that any lawyer may be a great philosopher and statesman.
35. Mathematical studies undoubtedly improve the reasoning powers; but as Logic is not a mathematical study we may conclude that it does not improve our reasoning powers.
36. If a man cannot obey the law he must be either a mere machine or a demon; but no man is either of these, and hence he must be able to obey the law.
37. Whatever tends to withdraw the mind from pursuits of a low nature deserves to be promoted; classical learning does this, since it gives us a taste for intellectual enjoyments; therefore it deserves to be promoted.
38. Alexander the Great was the son of King Philip, and therefore King Philip was the father of Alexander the Great.
39. He that withholdeth corn, the people shall curse him. But blessing shall be upon the head of him that selleth it.—*Proverbs of Solomon.*
40. If virtue is involuntary, vice is involuntary.
Vice is voluntary.
Therefore virtue is voluntary.
41. All civilized people are inhabitants of the earth. Few Indians are civilized, and therefore few Indians are inhabitants of the earth.
42. To improve is to change, and to be perfect is to have changed often. What hope can we entertain of those who oppose change?

---

43. The Germans are a nation. Bismarck, Stein, Kant, and Hegel were Germans, and hence must have been a nation.
44. The right should be enforced by law. Hence as the exercise of the suffrage is a right, it should be enforced by law.
45. Napoleon was not a great emperor; for though he would have been great had he succeeded in retaining his power, he did not do so.
46. Nothing is better than wisdom; dry bread is better than nothing; therefore dry bread is better than wisdom.
47. Knowledge is of no use to any one in preventing him from committing crimes; for we hear every day of frauds and forgeries which

never would have been committed had the person not learned to read and write.

48. The end of punishment is either the protection of society or the reformation of the individual. Capital punishment ought, therefore, to be abolished, because it neither prevents crimes of violence, nor protects society, nor does it reform the individual.

49. Wealth is value ; value is purchasing power ; purchasing power is the product of labor, and the product of labor is property ; therefore wealth is property.

50. Every rule has exceptions ; this is a rule, and therefore has exceptions ; therefore there are some rules that have no exceptions.

51. All who think this man innocent think he should not be punished ; you think he should not be punished ; therefore you think him innocent.

52. All who think this man innocent think he should not be punished ; you think he should be punished ; therefore you do not think him innocent.

53. Haste makes waste, and waste makes want. A man, therefore, never loses by delay.

54. All equilateral triangles are equiangular, and therefore all equiangular triangles are equilateral.

55. Only the virtuous are truly noble ; some who are called noble are not virtuous ; therefore some who are called noble are not truly noble.

56. For those who are bent on cultivating their minds by diligent study the incitement of academic honors is unnecessary ; and it is ineffectual for the idle and such as are indifferent to mental improvement ; therefore the incitement of academic honors is either unnecessary or ineffectual.

57. Logic as it was cultivated by the schoolmen proved a fruitless study ; therefore Logic, as it is cultivated at the present day must be a fruitless study likewise.

58. Repentance is a good quality ; wicked men abound in repentance, and therefore abound in what is good.

59. Warm countries alone produce wine. Spain is a warm country, and therefore produces wine.

60. It is an intensely cold climate that is sufficient to freeze mercury ; the climate of Siberia is sufficient to freeze it, and hence must be intensely cold.

61. No designing person ought to be trusted ; engravers are by profession designers ; therefore they ought not to be trusted.

62. I will not do this act because it is unjust ; I know it is unjust because my conscience tells me so, and my conscience tells me so because the act is wrong.

63. Is a stone a body ? Yes. Then is not an animal a body ? Yes. Are you an animal ? I think so. Ergo, you are a stone, being a body.— *Lucian.*

64. If ye were Abraham's children ye would do the works of Abraham. —*John* viii. 39.

65. He that is of God heareth God's words ; ye therefore hear them not, because you are not of God.—*John* viii. 47.

66. His imbecility of character might have been inferred from his proneness to favorites ; for all weak princes have this failing.

67. He is brave who conquers his passions , he who resists temptation conquers his passions ; so that he who resists temptation is brave.

68. Suicide is not always to be condemned ; for it is but voluntary death, and this has been gladly embraced by many of the greatest heroes of antiquity.

69 All that glitters is not gold ; tinsel glitters and is therefore not gold.

70. Meat and drink are the necessaries of life. The revenues of the king were spent on meat and drink, and were therefore spent on the necessaries of life.

71. He who calls you a man speaks truly ; he who calls you a fool calls you a man ; therefore he who calls you a fool speaks truly.

72. Theft is a crime , theft was encouraged by the laws of Sparta ; therefore the laws of Sparta encouraged crime.

73. Since all gold is a metal, the most rare of all masses of gold must be the most rare of all the metals.

74. Nothing but the express train carries the mail, and as the last train was an express it must have carried the mail.

75. Protective laws should be abolished, for they are injurious if they produce scarcity, and they are useless if they do not.

76. The Quaker asserts that if men were true Christians and acted upon their religious principles there would be no need of armies ; hence he draws the conclusion that a military force is useless, and being useless is pernicious.

77. Detention of property implies at least possession ; for detention is natural possession.

78. "Profit" is interpreted or defined to be "advantage ;" to take profit, then, is to take advantage , it is wrong to take advantage of one's neighbor ; therefore it is wrong to take profit.

79. Peel's remission of taxes was beneficial ; the taxes remitted by Peel were indirect, and therefore the remission of indirect taxes is beneficial.

80. Some poisons are vegetable , no poisons are useful drugs, and therefore some useful drugs are not vegetable.

81. Whosoever intentionally kills another should suffer death ; a soldier therefore who kills his enemy should suffer death.

82. Few towns in the country have 500,000 inhabitants, and since all such towns ought to have three representatives in Congress, it is evident that few towns ought to have three representatives.

83. If Bacon's opinion be right it is improper to stock a new colony with

criminals from prison; but this course we must allow to be proper if the method of colonizing New South Wales be a wise one. If this be wise, therefore Bacon's opinion is not right.

84. The people of the country are suffering from famine, and as A, B, C are people of the country, they must be suffering from famine.

85. You are not what I am; I am a man; therefore you are not a man.

86. Gold and silver are wealth; and therefore the diminution of the gold and silver of a country by exportation is the diminution of the wealth of the country.

87. The holder of some shares in a lottery is sure to gain a prize, and as I am the holder of some shares in a lottery I am sure to gain a prize.

88. A monopoly of the sugar-refining business is beneficial to sugar refiners; and of the corn trade to corn growers; and of the silk manufacture to the silk-weavers; of labor to the laborers. Now all these classes of men make up the whole community. Therefore a system of restrictions upon competition is beneficial to the community.

89. Over-credulous persons should never be believed; and as the ancient historians were in many instances over-credulous they ought never to be believed.

90. That is unfortunate; you insolently assert that you are a Darwinian, while the truth is that you are a poet.

91. Every incident in the narrative is probable, and hence the narrative may be believed since it is probable.

92. If a substance is solid it possesses elasticity, and so also it does if it be liquid or gaseous; but all substances are either solid, liquid, or gaseous; therefore all substances possess elasticity.

93. Who is most hungry eats most; who eats least is most hungry; therefore who eats least eats most.

94. If the Elixir of life is of any value those who take it will improve in health; now my friend who has been taking it has improved in health, and therefore the Elixir is of value as a curative agent.

95. The policy of protection was immediately followed by a great increase in the prosperity and wealth of the country, and hence we may infer that the result was due to its connection with the enactment of the protective law. In reply, however, we are told that before the passage of the law the loss by fire in Chicago in one year was $200,000,000, but was only $3,000,000 for the year after its passage, so great was the effect of this act.

96. What produces intoxication should be prohibited; the use of intoxicating liquors causes intoxication; therefore the use of spirituous liquors should be prohibited.

97. When we hear that all the righteous people are happy, it is hard to avoid exclaiming, What! are all the unhappy persons we see thought to be unrighteous?

93. Italy is a Catholic country and abounds in beggars; France is also a Catholic country, and therefore abounds in beggars.

99. The Latin word "virtus" originally meant "manliness;" hence the virtue of manliness or courage is the highest virtue and type of all other virtues.

100. If it be fated that you recover from your present disease, you will recover, whether you call in a doctor or not; again, if it be fated that you do not recover from your present disease, you will not recover, whether you call in a doctor or not. But one or the other of these contradictories is fated, and therefore it can be of no service to call in a doctor.

101. This person may reasonably be supposed to have committed the theft, for he can give no satisfactory account of himself on the night of the alleged offence; moreover he is a person of bad character, and, being poor, is naturally liable to a temptation to steal.

102. All the trees in the park make a thick shade; this oak-tree is one of them and therefore makes a thick shade.

103. All visible bodies shine by their own or by reflected light. The moon does not shine by its own; therefore it shines by reflected light; but the sun shines by its own light; therefore it cannot shine by reflected light.

104. The two propositions, "Aristotle is living," and "Aristotle is dead," are both intelligible propositions; they are both of them true or both of them false, because all intelligible propositions must be either true or false.

105. How can anyone maintain that pain is always an evil who admits that remorse involves pain, and yet may sometimes be a real good?

106. I am charged with absenteeism from my post, and on that ground I am accused of ignorance in regard to the proper duties of my office. But my accuser himself, who was my predecessor in the same office, was not in the country, of which he was the ruler, longer than five days.

107. Every law is either useless or it occasions hurt to some person; now a law that is useless ought to be abolished; and so ought every law that occasions hurt; therefore every law ought to be abolished.

108. What fallacies are implied or charged against Mr. Spencer in the following criticism?

"Mr. Spencer's distinction between objects and relations is far from satisfactory; and even if it were a true distinction, I do not see that any adequate classification of knowledge could be based upon it, because there is no science within the circle of knowledge that does not deal both with objects and relations."—*Knight: Essays in Philosophy.*

109. Does a grain of millet, when dropped on the floor, make sound? No. Does a bushel of millet make sound under the same circumstances? Yes. Is there not a determinate proportion between the bushel and the grain? There is. There must therefore be the same proportion between

the sonorousness of the two. If one grain be not sonorous, neither can ten thousand grains be so.

110. What you say is that virtue is the power of attaining good ? Yes. And you would say that goods are such as health and wealth, and the possession of gold and silver, and having office and honor in the state—these are what you call goods ? Yes, all these. Then, according to Meno, who is the hereditary friend of the great king, virtue is the power of getting silver and gold.—*Plato's Dialogues: Meno.*

111. Injustice is more profitable than justice because those who do unjust acts gain more than the just.

112. As for saying that without God man cannot have moral sentiments, or, in other words, cannot distinguish between vice and virtue, it is as if one said that without the idea of God man would not feel the necessity of eating and drinking.—*John Morley.*

113. I am offered a sum of money to assist this person in gaining the office he desires; to assist a person is to do him good, and no rule of morality forbids the doing of good; therefore no rule of morality forbids me to receive the sum of money for assisting this person.

114. Ruminant animals are those which have cloven feet, and they usually have horns; the extinct animal which left this foot-print had a cloven foot; therefore it was a ruminant animal and had horns. Again, as no beasts of prey are ruminant animals, it cannot have been a beast of prey.

115. Without order there is no living in public society, because the want thereof is the mother of confusion, whereupon division of necessity followeth, and out of division destruction.—*Hooker: Ecclesiastical Polity.*

116. The man who does any kind of work in a careless, bungling, or superficial way is not acting as a reasonable being; for the first demand of reason, as the truthful faculty in the world of action, is to realize its idea completely and thoroughly, and this no hasty and superficial handiwork will pretend to do.

117. Happiness signifies a gratified state of all the faculties. The gratification of a faculty is produced by its exercise. To be agreable that exercise must be proportionate to the power of the faculty; if it is insufficient discontent arises and its excess produces weariness. Hence to have complete felicity is to have all the faculties exerted in the ratio of their several developments.

118. We must either gratify our vicious propensities or resist them; the former course will involve us in sin and misery; the latter requires self-denial. Therefore we must either fall into sin and misery, or practise self-denial.

119. Every moral aim requires the rational means of attaining it; these means are the establishment of laws; and as happiness is the moral aim of

man it follows that the attainment of it requires the establishment of laws.

120. He that can swim needs not despair to fly ; for to swim is to fly in a grosser fluid, and to fly is to swim in a subtler fluid.

121. What fallacy was the humorist afraid of who said that he would not admit that two and two make four until he knew what use was to be made of the assertion.

122. The Good is a state of consciousness ; for the Good is a possible object of knowledge ; but all objects of knowledge are states of consciousness. Hence the Good is a state of consciousness.

123. The Good is pleasure ; for the Good results from the due performance of function ; but the Good is a state of consciousness ; therefore the Good is the state of consciousness which results from the due performance of function.

124. Riches are for spending, and spending for honor and good actions; therefore extraordinary expense must be limited by the worth of the occasion.

125. The several species of brutes being created to prey upon one another proves that the human species were intended to prey upon them.

126. If any objection can be urged to justify a change of established laws, no laws could be reasonably maintained ; but some laws can be reasonably maintained ; therefore no objection that can be urged will justify a change of established laws.

127. You are inconsistent with yourself, for you told me yesterday that there was a presumption of this man's guilt, and now when I say that I may presume his guilt, you contradict me.

128. The more correct the logic, the more certainly the conclusion will be wrong if the premises are false ; therefore where the premises are wholly uncertain the best logician is the least safe guide.

129. If our rulers could be trusted always to look to the best interests of their subjects monarchy would be the best form of government ; but they cannot be trusted ; therefore monarchy is not the best form of government.

130. He who bears arms at the command of the magistrate does what is lawful for a Christian ; the Swiss in the French service, and the British in the American service, bore arms at the command of the magistrate ; therefore they did what was lawful for a Christian.

131. A man that hath no virtue in himself envieth virtue in others; for men's minds will either feed upon their own good or upon others evil, and who wanteth the one will prey upon the other.

132. The object of war is durable peace ; therefore soldiers are the best peacemakers.

133. Confidence in promises is essential to human intercourse and commerce ; for without it the greatest part of our conduct would proceed upon

chance. But there could be no confidence in promises if men were not obliged to perform them ; the obligation, therefore, to perform promises is essential to the same ends and in the same degree.

134. If the majority of those who use public-houses are prepared to close them legislation is unnecessary ; but if they are not prepared for such a measure, then to force it on them by outside pressure is both dangerous and unjust.

135. He who believes himself to be always in the right in his opinion lays claim to infallibility ; you always believe yourself to be in the right in your opinion ; therefore you lay claim to infallibility.

136. If we never find skins except as integuments of animals we may safely conclude that animals cannot exist without skins. If color cannot exist by itself, it follows that neither can anything that is colored exist without color. So, if language without thought is unreal, thought without language must also be so.

137. If the light is not refracted near the surface of the moon there cannot be any twilight ; but if the moon has no atmosphere light is not refracted near its surface ; therefore if the moon has no atmosphere there cannot be any twilight.

138. No soldiers should be brought into the field who are not well qualified to perform their duty ; none but veterans are well qualified to perform their part ; therefore none but veterans should be brought into the field.

139. The *minimum visibile* is the least magnitude which can be seen ; no part of it alone is visible, and yet all parts of it must affect the mind in order that it may be visible ; therefore every part of it must affect the mind without being visible.

140. Improbable events happen almost every day, but what happens almost every day is a very probable event ; therefore improbable events are very probable events.

141. What fallacies are implied against an opponent in the following statement : " Each of its links is in fact unsound. And even though no flaw were visible in them, still the conclusion is demonstrably false."

142. "Now that which does not make a man worse, how can it make a man's life worse ? But neither through ignorance, nor having the knowledge but not the power to guard against or correct these things, is it possible that the nature of the universe has overlooked them ; nor is it possible that it has made so great a mistake, either through want of power or want of skill, that good and evil should happen indiscriminately to the good and the bad. But death certainly, and life, honor and dishonor, pain and pleasure—all these things happen equally to good men and bad, being things which make us neither better nor worse. Therefore they are neither good nor evil."—*Marcus Aurelius.*

143. Since there is no harm or evil to the elements themselves in their

continual changes into one another, a man should have no apprehension about the dissolution of all elements. For it is according to nature, and nothing is evil that is according to nature.—*Marcus Aurelius.*

144. Form a syllogism and show under what conditions any one of the following three fallacies may be found—*non sequitur*, *petitio principii*, and *equivocation*.

145. What fallacy is charged to the defenders of Charles the Second by Macaulay in the following statements: "We charge him with having broken his coronation oath, and we are told that he kept his marriage vow! We accuse him of having given up his people to the merciless inflictions of the most hot-headed and hard-hearted of prelates, and the defence is that he took his little son on his knee and kissed him! We censure him for having violated the articles of the Petition of Right, after having for good and valuable consideration promised to observe them, and we are informed that he was accustomed to hear prayers at six o'clock in the morning."

146. "Don't you think the possession of gold is good? Yes, said Ctesippus, and the more the better. And to have money everywhere and always is a good? Certainly, a great good, he said. And you admit that gold is a good? I have admitted that, he replied. And ought not a man then to have gold everywhere and always, and as much as possible in himself, and may not be be deemed the happiest of men who has three talents of gold in his stomach, and a talent in his head, and a stater of gold in either eye."—*Plato's Dialogues : Euthydemus.*

147. "If we are to test the truth of materialism by its outcome for well-being, we can hold it only by showing that the supreme end of man is to develop a body, and that materialism is especially useful in promoting the interests of the animal nature. The normal brain is that which takes care of itself, and the test of truth is self-preservation. Moral aims and scientific truth, so far as they have no physical value, must be voted not merely worthless, but delusion ; for the test of truth is physical preservation. Hence the inhabitant of the sty would be the prince of materialistic philosophers; he is not troubled by delusion and he preserves himself."

148. If sin *by itself* confers the right and imposes the duty of punishment, there must be the right to inflict either a definite punishment or an infinite amount. If the latter, it is obvious that the state will always have the right to inflict any quantity of punishment it pleases upon any of its citizens at any time, since all have sinned and incurred thereby an unlimited liability to punishment. If, on the other hand, wrong-doing confers a right to inflict a merely limited amount of punishment, it will not be possible to determine the amount outside of utilitarian considerations, since moral guilt cannot be measured in terms of physical pain. But it is apparent that the right to inflict an infinite punishment with-

out distinction of the crimes in regard to consequences is absurd, as also the infliction of a definite amount without regard to its utility, and hence sin by itself and independently of the advantage to society is not punishable.

149. Our tariff is found fault with because it does not make men independent and virtuous, besides giving them the opportunity to become prosperous. It is said to be responsible for the over-production which has characterized some branches of manufacture. The same evil occurs, and more frequently, in Lancashire under Free Trade.

150. The usefulness of government has been established by a long experience in the enactment and enforcing of laws against such acts as injure social order, and if anything be needed to establish the benefits of despotic governments it can be found in the power and practice exercised by them to punish crimes against life and property. Therefore we may infer their usefulness as forms of government.

151. If a debater affirm a proposition in the major premise which is true only in an abstract sense; that is, of the genus or conferentia, and the minor premise expresses what is true of unessential properties, what fallacies would be implied, first, by indicating this difference; second, by disputing the universality of the major premise, if the argument was in the first Figure; and third, by disputing the truth of either premise.

152. "Five years ago a first-class pair of nickel-plated steel skates, with the necessary clamps to fasten them to the boot or shoe, cost $15. To-day precisely the same article, and with an equal finish and completeness, can be obtained for $4. Three years ago a second grade of nickel-plated steel skates cost $4. The same article can be produced to-day for $1.50. The decline of seventy per cent. in five years, and of sixty per cent. in three years, shows just how protection cheapens prices."—*Milwaukee Evening Wisconsin*.

153. If the earth were of equal density throughout, it would be about $2\frac{1}{4}$ times as dense as water; but it is about $5\frac{1}{4}$ times as dense; therefore the earth must be of unequal density.

154. "'By open discrimination, or by secret rates, drawbacks, and rebates, a few railway managers may subject to their will every business in which transportation is a large element of cost, as absolutely as any Oriental despot ever controlled the property of his subjects. No civilized community has ever known a body of rulers with such power to distribute at pleasure, among its mercantile classes, prosperity or adversity, wealth or ruin. That this is no abstract or remote danger to society is plain to any man who will look at the condition of trade and of mercantile morals in the United States to-day.' How vivid! But how absurd! how untrue! Our commercial morals are equal to the highest in the world."— *Kirkman: Railway Rates and Government Control.*

155. "Not only the effects are good, but the agent sees beforehand

that they will be so. This may make the action indeed (done from antipathy) a perfectly right action ; but it does not make antipathy a right ground of action. For the same sentiment of antipathy, if implicitly deferred to, may be, and very frequently is, productive of the very worst effects. Antipathy, therefore, can never be a right ground of action."— *Bentham : Principles of Morals and Legislation.*

156. "Mr. Gladstone, however, commits himself to the principle that 'all protection is morally bad.' If this has been his belief ever since he became an advocate of free trade, his conscience must have received many and severe wounds, as session after session, while Chancellor of the Exchequer, he carried through Parliament a bounty—may I not say a direct protection?—of £180,000 to a line of steamers running between England and the United States—a protection that began six years before free-trade was proclaimed, and was continued nearly twenty years after."—*Mr. Blaine, in the North American Review for January,* 1890.

DEDUCTIVE AND INDUCTIVE

Examine the following arguments, stating whether they are deductive or inductive ; if deductive, show whether they are valid or invalid, and why ; if inductive, show what Method of Induction is involved.

1. Two of the wealthiest men of the West are said to have been messenger-boys. It pays to go slowly, after all.

2. Geometry contemplates figures. Figure is the termination of magnitude ; but extension in the abstract has no definite determinate magnitude. Whence it follows clearly that it can have no figure, and consequently is not the object of Geometry, whose object is commonly said to be abstract extension.

3. The newly discovered painting must be a Rubens ; for the conception, the drawing, the tone and the tints are precisely those seen in the authentic works of that master.

4. In nine counties, in which the population is from 100 to 150 per square mile, the births to 100 marriages are 396 ; in sixteen counties, with a population of 150 to 200 per square mile, the births are 390 to 100 marriages. Therefore the number of births per marriage is inversely related to the density of population, and contradicts Malthus's theory of the law of population.

5. "Cramming" for examination is detrimental rather than otherwise ; for I have noticed that no matter what the subject is, I invariably write a poor paper when I "cram," and a good one when I do not.

6. The great famine in Ireland began in 1845, and increased until it reached a climax in 1848. During this time agrarian crime increased

very rapidly until in 1848 it was more than three times as great as in 1845. After this time it decreased with the return of better crops, until in 1851 it was only fifty per cent. more than it was in 1845. It is evident from this that a close relation of cause and effect exists between famine and agrarian crime.

7. Pitt did not bribe the Irish parliament in 1800, when he so lavishly bestowed peerages on its members. For he bestowed honors on only forty of his followers, while in 1779, Lord North bestowed thirty peerages in one day, and in 1832, Lord Grey got the king's consent to the creation of a hundred.

8. "Suppose we have a southward velocity amounting, let us say, to 3 feet per second, and simultaneously an eastward velocity amounting to 4 feet per second, then we know by kinematics how to construct the single velocity which is the resultant of these two. All we have to do is to draw a line of length 3 southward and from its extremity a line of length 4 to the eastward, and then complete the triangle. In a geometrical sense, therefore, a velocity of 3 southward and a velocity of 4 eastward will be equivalent to a velocity which, if you calculate what the third side of that triangle will be, is represented by 5 on that scale."— *Recent Advances in Physical Science.*

9. On May 27, 1875, a remarkable shower of small pieces of hay occurred at Monkstown, near Dublin. They appeared floating slowly down from a great height. A similar shower occurred a few days earlier in Denbighshire. From this and many similar facts we may conclude that 'the distribution of organisms of the same species over continents and islands separated by the ocean has been effected by the agency of natural forces.

10. The influence of heat in changing the level of the ground upon which the Temple of Jupiter Serapis stands might be inferred from several circumstances. In the first place, there are numerous hot springs in the vicinity, and when we reflect on the dates of the principal oscillations of level this conclusion is made much more probable. Thus before the Christian era, when Vesuvius was regarded as a spent volcano, the ground upon which the temple stood was several feet above water. But after the eruption of Vesuvius in 79 B.C. the temple was sinking. Subsequently Vesuvius became dormant and the foundations of the temple began rising. Again Vesuvius became active, and has remained so ever since. During this time the temple has been subsiding again, so far as we know its history.

11. "I have two pendulums with very massive bobs suspended from them, and have carefully made these two pendulums as nearly as possible the same. Both pendulums are now at rest, but suppose I set one to vibrate, leaving the other at rest, you will notice, if you watch the second for a short time that it begins to vibrate in its turn, and as time goes on

it swings through larger and larger arcs of vibration till at last the first pendulum is brought to rest. Now this is quite obviously a case of transference of energy from one pendulum to the other, effected, you will see, through the wooden structure."—*Recent Advances in Physical Science*.

12. Why should any but professional moralists trouble themselves with the solution of moral difficulties ? For, as we resort to a physician in case of any physical disease, so, in case of any moral doubt or any moral disorganization, it seems natural that we should rely on the judgment of some man specially skilled in the treatment of such subjects.

13. Take a bottle of soda-water, slightly warmer than a given temperature registered by the thermopile; and mark the deflection it causes. Then cut the string which holds it, and the cork will be driven out by the elastic force of the carbonic acid gas. The gas performs its work, and in so doing it consumes heat, and the deflection of the thermopile shows that the bottle is cooler than before, heat having been lost in the process.

14. The occurrence of the Aurora Borealis under different meteorological conditions is invariably accompanied by magnetic disturbances and by the appearance of sun spots, and hence we infer that a causal connection exists between them and the sun spots.

15. It has been found that linnets when shut up and educated with singing larks—the skylark, woodlark, or titlark—will adhere entirely to the songs of these larks instead of the natural song of the linnets. We may infer, therefore, that birds learn to sing by imitation, and that their songs are no more innate than language is in man.

16. An enemy has a keener perception than a friend ; for, as Plato says, the "lover is blind as respects the loved one," and hatred is both curious and gossipy. Hiero was twitted by one of his enemies for the foulness of his breath ; so he went home and said to his wife : "How is this ? You never told me of it." But she, being pure and innocent, replied : "I thought all men's breath was like that." Thus perceptible and material things, and things that are plain to everybody, are sooner learned from enemies than from friends.—*Plutarch's Morals*.

17. A man cannot really be injured by his brethren, for no act of theirs can make him bad, and he must not be angry with them nor hate them ; for we are made for co-operation, like feet, like hands, like eyelids, like the rows of the upper and lower teeth. To act against one another, then, is contrary to nature ; and it is acting against one another to be vexed and to turn away.—*Thoughts of Marcus Aurelius*.

18. As an evidence of the remote antiquity of highly civilized man we have the following facts : On one of the remote islands of the Pacific— Easter Island—two thousand miles from South America, two thousand miles from the Marquesas, and more than one thousand miles from the Gambier Islands, are found hundreds of gigantic stone images, now

mostly in ruins. They are often forty feet high, while some seem to have been much larger, the crowns on their heads, cut out of a red stone, being sometimes ten feet in diameter, while even the head and neck of one is said to have been twenty feet high. The island containing these remarkable works has an area of about thirty square miles, and as the smallest image is about eight feet high, weighing four tons, and as the largest must weigh over a hundred tons or much more, their existence implies a large population, abundance of food, and an established government which so small an island could not supply.

19. We observe very frequently that very poor handwriting characterizes the manuscripts of able men, while the best handwriting is as frequent with those who do little mental work when compared with those whose penmanship is poor. We may, therefore, infer that poor penmanship is caused by the influence of severe mental occupation.

20. It has been shown by observation that overdriven cattle, if killed before recovery from their fatigue, become rigid and putrefy in a surprisingly short time. A similar fact has been observed in the case of animals hunted to death, cocks killed during or shortly after a fight, and soldiers slain in battle. The contrary is remarked when the muscular exercise has not been great or excessive. Hence we may infer that cadaveric rigidity depends upon a more or less unirritable condition of the muscles immediately before death.

INDEX

ABSOLUTE terms, 47–49
Abstract terms, 36–44, 65; relation to fallacy of accident, 235
Abstraction, 26, 40
Accent, fallacy of, 220, 221
Accidens, 82
Accident, 82–86; fallacies of, 231–240
Accidentia, 88
Acquisition, 341
Affirmative propositions, 109
Agreement, principle of, 331, 332; method of, 354
All, logical signification of, 111
Alternative, 109, 212
Ambiguity of terms, 50–67; of propositions, 115–121
Ambiguous middle, 226, 239
Ampliative propositions, 112
Analytic propositions, 112
Antecedent, 108, 204
Antithesis, 169
Any, logical signification of, 111
Argumentum ad rem, 250; ad judicium, 250; ad populum, 251; ad hominem, 251; ad ignorantiam, 251; ad verecundiam, 251
Aristotle, 248, 264, 267
Attributive terms, 37

BACON, 298
Bain, 32
Barbara, Celarent, etc., 190
Begging the question, 240
Benjamin Franklin, 306
Bentham, Jeremy, 243
Bentham, George, 202
Berkeley, 250
Brewster, Sir David, 332

CATEGOREMATIC terms, 31
Categorical propositions, 107; reasoning, 204
Categories, 82
Cause, fallacy of false, 255
Circulus in definiendo, 104
Circulus in probando, 241, 244
Classification, 343; of fallacies, 219–227
Cognition, 22
Collective terms, 34, 35
Comparison, 23
Complimentary propositions, 116
Complex syllogisms, 197
Composition, fallacy of, 228
Comprehension. *See* Intension
Concepts and conceptions defined, 20; formation of, 21–27; kinds of, 25–26, 31–49; denomination of, 26; intension and extension of, 68–81; analysis of, 94; infinitated, 165
Conception, 5, 16, 19–27, 28
Conclusion, 20, 171
Conclusion, weakened, 184
Concomitant variations, method of, 357
Concrete terms, 36–44; relation to fallacy of accident, 235
Conditional propositions, 107
Conferentia, 87, 88, 91, 102, 126, 127, 234, 235, 278, 323
Confusion of fallacies, 256–261
Connotation of terms, 79
Connotative, meaning of, 79–81
Consequent, 108, 204; fallacy of false, 253
Contradiction, law of, 291
Contradictory, 142
Contraposition. *See* Contraversion, note on, 168

Contraries, 142, 144
Contraversion, 161-166
Contribution, 168
Converse, 156
Converse fallacy of accident, 231
Conversion, 155-160; simple, 156, 264; per accidens or limited, 156; rules for, 156; by negation, 159; table of, 160
Convertend, 156
Co-ordinate species, 96

DEDUCTION, 270, 300, 301, 327, 358
Deductive reasoning, 15, 279, 295, 327, 328
Deductive method, 337
Definite propositions, 110; quantity, 267
Definition, 100-104; kinds of, 100; logical, 101; rules for, 103; in method, 337
Definition and division, 82-104
De Morgan, 233, 238, 243, 247, 252, 255, 262
Denomination of concepts, 26
Denotation and connotation of terms, 78
Denotative, meaning of, 79-81
Descriptive definition, 100
Destructive dilemma, 217; hypothetical reasoning, 205
Desynonymization of terms, 60
Difference, principle of, 331, 333; method of, 355
Differentia, 82, 86, 90
Differential or specific accident, fallacy of, 231
Differentiation of terms, 60
Dilemma, 217; constructive, 217; destructive, 217
Dilemmatic propositions, 108
Direct reduction, 190
Disjunction, incomplete, 214
Disjunctive propositions, 107; syllogisms, 212-218
Distribution of terms, 138; rules for, 140
Distributive terms, 34, 35

Division, definition and, 82-104; logical, 94; rules for, 95; dichotomous, 97; trichotomous, 97
Divisionis, fundamentum, 95
Dr. Johnson, 250
Duplex propositions, 116
Doctrine, elements of logical, 16-30

ENTHYMEME, 198
Epicheirema, 200
Episyllogism, 198
Equivocal terms, 51
Equivocation, fallacies of, 225, 228-240
Essence, 86
Essentia, 85, 88
Essential properties, 84; propositions, 112
Ethics, 14
Etymological definition, 100
Euler, 130
Exclamatory propositions, 107
Exceptive propositions, 120
Excluded middle, law of, 145
Exclusive propositions, 118; quantity of, 119
Experiment, 342, 352
Explicative propositions, 112
Extension of concepts, 68-81; relation between intension and, 72; law of, and intension, 73
Extension, judgments of, 123; symbolized, 130, 131

FALLACIA consequentis, 253
Fallacies, classification of, 219-227; hermeneutic, 220; formal or logical, 223; material, 225, 228-261; of amphibology, 220; of accent, 221; of four terms, 223; of illicit middle, 224; of illicit major, 224; of illicit minor, 224; of particular premises, 225; of negative premises, 225; of equivocation, 226, 228; of composition and division, 228; of accident, 231; of presumption, 240; of petitio principii, 240; of non sequitur, 253;

of false cause, 255; confusion of, 256-261; inductive, 363
Fallacy, definition of, 219
False cause, fallacy of, 255
False consequent, 253
False propositions, 144
Figures of the syllogism, 183
Form and matter, 9, 16
Formal fallacies, 223
Formal logic, 1, 12, 15
Fowler, 79, 81, 318, 321, 324, 364
Fundamentum divisionis, 95

GENERAL terms, 34-36; mathematical generals, 25, 75, 236; logical generals, 25, 101, 236; propositions, 110
Generalization of terms, 57; of observations, 296
Genus, 82, 86, 88; mathematical, 93; logical, 93; contrasted with differentia, 91, 92; summum, 90
Grammatical division of propositions, 107

HAMILTON, 1, 4, 73, 95, 154, 262, 266, 267, 270, 271, 275, 285, 291, 338
Hypothesis, 343
Hypothetical propositions, 107; reasoning, 204-211

IDENTITY, law of, 145
Ignoratio elenchi, 149, 153, 241-253
Illicit process of major term, 174, 177, 224; of middle term, 174, 177, 224; of minor term, 174, 178, 224
Immediate inference, 154-170
Imperfect induction, 295
Incomplete disjunction, 214; syllogisms, 197
Indefinite propositions, 110; quantity, 267
Indirect reduction, 193
Individual terms, 20, 32
Induction, nature of, 295-313; perfect, 295, 296, 300-302, 320, 327; principles of, 328; fallacies of, 363-366
Inductive logic, 15, 295; method, 298, 340, 353; inference, 313, 321-323; syllogism, form of, 306

Inference, nature of, 154; immediate, 155; mediate, 154, 171-189; by privative conception, 160; by contribution, 168; deductive, 15, 279, 295, 327, 328; inductive, 313, 321-325.
Infima species, 90
Infinitated conceptions, 46, 165.
Intension of terms or concepts, 68-81; relation between, and extension, 72; law of, and extension, 73
Introspection, 342
Inversion, 156
Inverted propositions, 115
Irrelevant conclusions, 245

JEVONS, 37, 50, 63, 64, 90, 96, 103, 159, 162, 193, 213, 220, 232, 239, 244, 263, 264 note, 265 note, 280, 312, 348 356
Johnson, Dr., 250
Joint method of agreement and difference, 358
Judgment, 5, 16, 28-29; intensive and extensive, 123

Κατηγορεύμενον, 106
Keynes, 13, 32, 37, 38, 79, 81, 187, 193, 213, 270

LANGUAGE, 27
Law, conception of, 6-9; of thought, 6-9; of intension and extension, 73; of contradiction, 145, 291; of excluded middle, 145, 292; of identity, 145, 291; of sufficient reason, 292, 329; of universal causation, 318, 329; of the uniformity of nature, 318, 329, 331
Laws of thought, 6-9, 290-294
Limitation, conversion by, 156, 157
Logic, definition of, 1; relation to the other sciences, 12; divisions of, 15; inductive and deductive, 15
Logical concepts, 26; 276, 277
Logical definition, 101; generals, 75, 101, 236
Logical doctrine, formal elements of, 18

MAJOR term, 171; illicit process of, 174, 224
Mal-observation, 364
"Margarita Philosophica," 233
Matter, form and, 9
Material fallacies, 225, 228-261
Mathematical concepts, 25, 26, 276, 277
Mathematical generals, 75, 236; reasoning, 275-289
Mediate reasoning, 171-189
Method, scientific, 300; inductive, 298, 340, 353; deductive, 337; of agreement, 297, 354; of difference, 297, 355; of concomitant variations, 357; of joint agreement and difference, 358; of residues, 358
Middle term, 171, 174-180, 224
Mill, J. S., 37, 41, 70, 79, 80, 81, 203, 247, 250, 317, 355
Minor term, 171; illicit process of, 174-180, 224
Mnemonic lines, 190
Miscellaneous propositions, 114
Modal propositions, 114
Modus ponendo tollens, 216
Modus ponens, 205
Modus tollendo ponens, 216
Modus tollens, 205
Moods of the syllogism, 181-189

NEGATION, conversion by, 159
Negative propositions, 109; distribution of, 140
Negative terms, 44; signs of, 45
Nego-positive terms, 44-46
Newton, 304, 350, 360
Non causa pro causa, 255
Non-observation, 363
Non sequitur, 253, 256
Notion, 21

OBSERVATION, 342, 351; errors of, 363
Obversion, 160
Opposition, 141-153; application of the principles of, 144; rules for, 144; square of, 144
Or, signification of, 109, 212

PARTICULAR propositions, 109, 110; signs of, 112
Partition, 99
Partitive propositions, 116
Per accidens, conversio, 156
Percepts, 22
Petitio argumenti, 241
Petitio principii, fallacy of, 240-245
Plurative propositions, 110
Porphyry, tree of, 97
Positive terms, 44
Post hoc, ergo propter hoc fallacy, 255
Predicables, 82-94
Predicate, 106; distribution of, 138; relation between subject and, 122-140. *See also* Quantification of predicate
Preindesignate terms, 267
Premises, 29, 171
Primary laws of thought, 290
Privative conception, 44
Probation, 338
Proof. *See* Probation
Proper names, 31-33
Property, 82; kinds of, 82-85; table, 88; propositions, 105-121; defined, 17; divisions of, 107, 121
Prosyllogism, 198
Psychology, 13
Pure propositions, 114, 277, 310-312

QUALITATIVE reasoning, 277, 310-312
Quality of propositions, 109
Quantification of the predicate, 262-274
Quantity of terms. *See* Extension; of propositions, 109
Quanto-qualitative reasoning, 277
Quaternio terminorum, 223

REASON, law of sufficient, 329
Reasoning, 5, 16, 29-30, 171-180; forms of, 197-203; hypothetical, 204-211; disjunctive, 212-218; mathematical and other, 275-289; quantitative, 277; qualitative, 277
Reduction of the moods and figures, 190-196; of hypothetical syllogisms to categorical, 209

Reisch, 233
Relative terms, 47-49
Residues, method of, 358
Rules of the syllogism, 173

SCIENCE AND ART, 1-2
Sciences, mathematical and metaphysical, 25
Scientific method, 300-366
Secondary laws of thought, 293
Simple accident, fallacy of, 231
Simple conversion, 158
Singular propositions, 110
Singular terms, 31-33
Socrates, 296
Sorites, 201
Specialization, 59-63
Species, 82, 88; contrasted with genus, 91-92; and fallacy of accident, 231, 235
Specific accident, fallacy of, 231
Subaltern propositions, 142
Subalternans, 143
Subalternate, 143
Subcontrary propositions, 143
Subject and predicate, 106; distribution of, 138; relation between, 122-140
Subordinate species, 96
Sufficient reason, law of, 329
Summum genus, 90
Superordinate conceptions, 96

Syllogism, 18, 171-180; forms of, 197-203; hypothetical, 204-211; disjunctive, 212-218; inductive, 306
Syncategorematic terms, 31
Synthesis of percepts, 20
Synthetic propositions, 112

TAUTOLOGOUS propositions, 114
Terms, 31-40; ambiguity of, 50-67; intension and extension of, 68-81; denotation and connotation of, 79; distribution of, 138
Terminorum, quaternio, 223
Theory, 344
Thompson, 1, 266
Thought, definition of, 3-6; nature of its laws, 6-9; laws of, 290-294
Traduction, 279, 310, 312
Tree of porphyry, 97
Truistic propositions, 114

UEBERWEG, 1, 134
Universal propositions, 109, 110; signs of, 111; distribution of, 140
Univocal terms, 51

ὑποκείμενον, 106
Verification, 341, 349

WATTS, DR., 1
Whately, 1, 233, 245, 247
Weakened conclusion, 184

THE END.

www.ingramcontent.com/pod-product-compliance
Lightning Source LLC
Chambersburg PA
CBHW050847300426
44111CB00010B/1157